ONE SMART WORLD
ARE YOU READY FOR IT?

ONE SMART WORLD
ARE YOU READY FOR IT?

EMIN HASIC

COPYRIGHT © 2023 EMIN HASIC

No portion of this book may be reproduced in any form without written permission from the publisher or author, except as permitted by European Union and International copyright law.

ONE SMART WORLD

Notice of Copyright

This publication is designed to provide accurate and authoritative information in regard to the subject matter covered. It is sold with the understanding that neither the author nor the publisher is engaged in rendering legal, investment, accounting or other professional services. While the publisher and author have used their best efforts in preparing this book, they make no representations or warranties with respect to the accuracy or completeness of the contents of this book and specifically disclaim any implied warranties of merchantability or fitness for a particular purpose. No warranty may be created or extended by sales representatives or written sales materials. The advice and strategies contained herein may not be suitable for your situation. You should consult with a legal or industry professional when appropriate. Neither the publisher nor the author shall be liable for any loss of profit or any other commercial damages, including but not limited to special, incidental, consequential, personal, or other damages.

Authored by: Emin Hasic
Cover Design by: Billie Rose (Rosemin Studio)
Editing by: METADPO
Printed by: MIJNBESTSELLER Netherlands

Second Updated Print Edition, 2023
ISBN 978-0-6457157-2-9
First Original Edition Published under Private Release 2016

FORWARD

As an architect and urbanist, I am deeply passionate about creating sustainable and livable cities that can meet the needs of current and future generations. That's why I am excited to introduce "One Smart World" by Emin Hasic, a comprehensive guide to building smarter and more resilient cities that truly serve the public good.

What I appreciate most about this book is the author's ability to balance practical advice with visionary thinking. They show us how to leverage technology and connectivity to create urban spaces that are more efficient, sustainable, and livable, while also addressing the ethical and social implications of these developments. Through real-world examples and case studies, the author demonstrates how Smart Cities can improve quality of life, foster innovation, and support economic growth.

"One Smart World" is not just a celebration of technological progress - it is also a call to action. The book challenges us to think beyond the traditional boundaries of urban planning and design, and to explore new approaches that are more responsive to the needs of our communities. The author encourages us to engage in collaborative decision-making, embrace diversity, work towards social equity and inclusion.

As an architect and urbanist, I find this book to be an invaluable resource. It provides practical guidance on how to design and plan for Smart Cities, while also offering a compelling vision for the future of urbanization. I appreciate the author's commitment to sustainability, resilience, and social responsibility, and their ability to inspire us to dream bigger and work towards a better future.

"One Smart World" is a must-read. It will challenge you to think differently, to act boldly, and to make a positive impact on the world.

Athanasios G. Venetis (Architect and Urbanist)
www.venetis.it

ABOUT THE AUTHOR

Emin Hasic is a respected expert in the field of data privacy with over three decades of experience and a wealth of knowledge to his name. He has dedicated himself to continuous self-education, immersing himself in the latest developments and insights on the subject.

Emin's extensive experience in data privacy and cybersecurity make him uniquely qualified to address the challenges and opportunities presented by Smart Cities. His understanding of the complex issues surrounding data privacy and cybersecurity, combined with his insights into the dynamics of urban life, enable him to provide innovative and effective solutions for Smart Cities.

Emin's deep knowledge of the subject is reflected in his reading of seminal works such as Smart Cities by Antoine Picon; The Responsive City by Stephen Goldsmith and Susan Crawford; The Future of the Professions by Richard Susskind and Daniel Susskind; Cities for People by Jan Gehl; The Death and Life of Great American Cities by Jane Jacobs; Walkable City by Jeff Speck; The Urban Apparatus by Reinhold Martin and Data and the City by Rob Kitchin, to name a few.

Emin's unique perspective on the challenges and opportunities presented by Smart Cities has enabled him to develop innovative and effective solutions to address the needs of urban communities.

Emin's expertise in data privacy enables him to provide invaluable insights into the implementation of Smart City technology and infrastructure, ensuring the safety and security of urban residents.

Emin is a valuable asset to the Smart City community, providing the expertise and knowledge necessary to navigate the complexities of modern urban life.

ACKNOWLEDGEMENTS

With great humility, Emin wishes to express his deepest gratitude to the numerous pioneers, philosophers, and visionaries who have paved the way for the exploration and advancement of Smart Cities. Their unrelenting passion and devotion to creating more intelligent, sustainable, and livable urban spaces have served as a constant source of inspiration and enlightenment throughout his journey.

With a deep sense of reverence and gratitude, Emin acknowledges the invaluable contributions and inspirations of the great minds who have paved the way for the development of Smart Cities. Without their vision and wisdom, the task of writing this book would have been daunting and insurmountable. In particular, Emin would like to give special mention to William J. Mitchell, a professor of architecture and media arts and sciences at MIT, who played a pivotal role in nurturing Emin's interest in Smart Cities during their private mentorship from 2004 to 2010. Unfortunately, Mitchell passed away on the 11th of June 2010 due to cancer. His influential writings on the intersection of technology and the built environment are widely renowned, and his impact on Emin's passion for this field is immeasurable. Emin humbly recognizes that he stands on the shoulders of giants who have laid the foundation for a new era of urban development that promises to transform the way we live, work, and interact with one another.

Emin expresses sincere appreciation to Mr. Antonis Bezas, the former Greek Deputy Minister of Economy and Finance, for his dedicated volunteerism and invaluable contributions in providing extensive insights on the Municipality of Igoumenitsa and its neighbouring towns. It was through the personal vision of Mr. Bezas that Emin was inspired to undertake a project focused on the unique hypothetical case of Igoumenitsa (Chapter XX, Pages 559 to 596), which serves as a compelling exemplar of how a small urban region can be transformed into a smart city and re-ignite its rebirth in placing a stop to the brain-drain and exodus of its youth.

Emin expresses gratitude to Athanasios Venetis for providing him with a solid understanding of the architecture and urbanization landscape of the region and his personal vision, and to Thomas Liolios for offering him an in-depth understanding of the history, arts, and cultural landscape of the region. Finally, Emin thanks Ilias Lambrousis for showing him the entire landscape of the region and its potential, and Thomas Dimas for introducing him to countless locals, enabling him to gather their views, visions and thoughts.

Emin owes a debt of gratitude that can never be fully repaid to his family, friends, and soulmate Billie Rose. Their unwavering support and encouragement have been the driving force behind his pursuit of knowledge and the realization of his dreams. Their love and unwavering belief in Emin have been his anchor and his fuel, providing him with the courage and determination to face the challenges that lie ahead.

Lastly, Emin expresses his sincerest appreciation to his readers for their interest in Smart Cities and their willingness to explore new ideas and solutions for the future of our urban world. Emin hopes that this book will serve as a catalyst for change, inspiring readers to become part of the Smart City movement and contribute to building a better world for all.

Let us embrace the challenges and opportunities that lie before us with open hearts and open minds. Together, let us create a brighter, smarter, and more sustainable future for generations to come.

May we be blessed in our endeavors.

WELCOME

Welcome to **ONE SMART WORLD**, a journey into the future of Smart Cities. As the author of this book, I am thrilled to share with you the possibilities and opportunities that a smarter world offers.

The world is rapidly evolving, and technology is at the forefront of this evolution. Smart Cities are the next big thing in this evolution, and the potential benefits they offer are vast. The vision of a Smart City is not only about technology, but it's also about creating a better life for its citizens.

This book is written with a vision to provide insights into Smart Cities – their design, development, and how they can benefit everyone. It is an inclusive guide, written for everyone from the everyday citizen to students, educators, decision makers, investors, influencers and stakeholders.

One of the most exciting aspects of Smart Cities is the opportunities they offer for innovation and entrepreneurship. By fostering a culture of innovation and collaboration, Smart Cities can be a hub for start-ups and businesses to thrive. As a result, cities can create new jobs, new products, and services that benefit everyone.

I believe that Smart Cities are the way forward, and they offer a new world of opportunities. By embracing technology and innovation, we can create cities that are smarter, more sustainable, and more equitable. I hope that this book inspires you to learn more about Smart Cities and how they can benefit our world.

Thank you for joining me on this journey into the future of Smart Cities.

Let's build a smarter world together!

Thank you for your purchase and support.

Emin Hasic (Data Privacy Executive & Author).

CONTENTS

FORWARD 4

ABOUT THE AUTHOR 5

ACKNOWLEDGEMENTS 6

WELCOME... 8

| i | INTRODUCTION.. 20

    DEFINING SMART CITIES 20

    BENEFITS OF SMART CITIES 21

    CHALLENGES AND OPPORTUNITIES... 21

| ii | HISTORY OF SMART CITIES 23

    SONGDO – SOUTH KOREA.. 23

    SANTANDER – SPAIN 24

    AMSTERDAM – NETHERLANDS 24

    RECENT YEARS 24

    THE STORY 26

| iii | HISTORICAL TIMELINE 28

    FROM 2600 BCE UNTIL 2023 CE 28

| iv | PHILOSOPHERS & VISIONARIES.. 46

    ARISTOTLE 4TH CENTURY BCE 46

    PLATO 4TH CENTURY BCE 47

    EPICURUS 3RD CENTURY BCE 47

    DIOGENES 4TH CENTURY BCE 47

    HERACLITUS 5TH CENTURY BCE 48

    PYTHAGORAS 6TH CENTURY BCE 48

    DEMOCRITUS 5TH CENTURY BCE 48

    ZENO OF CITIUM 3RD CENTURY BCE 49

    THALES OF MILETUS 6TH CENTURY BCE 49

THEOPHRASTUS 4TH CENTURY BCE ... 49

XENOPHON 4TH CENTURY BCE ... 50

CLEANTHES 3RD CENTURY BCE ... 50

PROTAGORAS 5TH CENTURY BCE ... 50

ANAXIMANDER 6TH CENTURY BCE ... 50

EMPEDOCLES 5TH CENTURY BCE ... 51

ANAXAGORAS 5TH CENTURY BCE ... 51

JANE JACOBS 1916 to 2006 ... 51

LEWIS MUMFORD 1895 to 1990 ... 52

HENRI LEFEBVRE 1901 to 1991 ... 52

FRIEDRICH ENGELS 1820 - 1895 ... 52

MANUEL CASTELLS 1942 ... 53

MARTIN HEIDEGGER 1889 - 1976 ... 53

DAVID HARVEY 1935 ... 53

RICHARD SENNETT 1943 ... 54

JANE BENNETT 1957 ... 54

MICHEL FOUCAULT 1926 - 1984 ... 54

REM KOOLHAAS 1944 ... 55

SASKIA SASSEN 1949 ... 55

MANUEL DE LANDA 1952 ... 55

WILLIAM J. MITCHELL 1944 - 2010 ... 56

DON IHDE 1934 ... 56

ANTHONY TOWNSEND 1971 ... 56

ADAM GREENFIELD 1968 ... 57

| v | EMBRACING SMART CITIES ... 59
DESIRE IS THE NEW TREND ... 59

| vi | FAILING TO EMBRACE SMART CITIES ... 62
THE BARRIERS ... 62

THE NEGATIVE IMPACT 63

START SMALL 63

| vii | PITFALLS 66

50 PITFALLS TO LOOK OUT FOR WHEN SETTING UP A SMART CITY
PROGRAM 66

| viii | BIG DATA 72

BIG DATA IN SMART CITIES 72

| ix | PROJECTS BY COUNTRY 76

AUSTRALIA 76

AUSTRIA 77

BELARUS 78

BELGIUM 79

BRAZIL 80

BULGARIA 81

CANADA 82

COSTA RICA 84

CROATIA 85

DENMARK 87

DUBAI 88

EGYPT 89

FINLAND 90

FRANCE 91

GERMANY 92

GREECE 94

HONG KONG 95

HUNGARY 96

ICELAND 97

INDIA 98

INDONESIA 99

IRELAND	100
ISRAEL	101
ITALY	102
KUWAIT	103
LITHUANIA	104
LUXEMBOURG	105
MACAO	106
MALAYSIA	107
MEXICO	108
MONACO	109
MOROCCO	110
NETHERLANDS	111
NEW ZEALAND	112
NORWAY	113
OMAN	114
PHILIPPINES	115
POLAND	116
QATAR	117
ROMANIA	118
RUSSIA	118
SAUDI ARABIA	119
SERBIA	121
SINGAPORE	122
SLOVAKIA	122
SLOVENIA	124
SOUTH AFRICA	125
SOUTH KOREA	126
SPAIN	126

SWEDEN 128

SWITZERLAND 129

THAILAND 130

TURKEY 131

UNITED ARAB EMIRATES 131

UNITED KINGDOM 132

UNITED STATES OF AMERICA 134

VIETNAM 135

| x | PROGRAMS BY INDUSTRY 137
SMART CITY 5G NETWORK (SC5GN) 137

SMART CITY AGRICULTURE (SCAG) 142

SMART CITY AIR QUALITY MONITORING (SCAQM) 146

SMART CITY ANALYTICS (SCA) 150

SMART CITY BIKE-SHARING (SCBS) 155

SMART CITY BUILDING MANAGEMENT (SCBM) 159

SMART CITY CITIZEN ENGAGEMENT (SCCE) 163

SMART CITY CITIZEN PARTICIPATION (SCCP) 167

SMART CITY CLIMATE RESILIENCE (SCCR) 172

SMART CITY COMMUNITY DEVELOPMENT (SCCD) 177

SMART CITY CONNECTED AND AUTONOMOUS VEHICLES (SCCAV) 181

SMART CITY CYBER-SECURITY (SCCS) 185

SMART CITY DATA MANAGEMENT (SCDM) 189

SMART CITY DIGITAL INCLUSION (SCDI) 193

SMART CITY DISASTER MANAGEMENT (SCDIM) 198

SMART CITY EDUCATION (SCE) 202

SMART CITY E-GOVERNANCE (SCEG) 207

SMART CITY EMERGENCY RESPONSE (SCER) 211

SMART CITY EMERGENCY SERVICES (SCES) 216

SMART CITY ENERGY DISTRIBUTION (SMED) ... 220
SMART CITY ENERGY MANAGEMENT (SCEM) ... 223
SMART CITY ENERGY STORAGE (SCENS) ... 227
SMART CITY ENVIRONMENT (SCENV) ... 231
SMART CITY ENVIRONMENTAL MONITORING (SCEVM) ... 235
SMART CITY FINANCING (SCF) ... 239
SMART CITY FLOOD MANAGEMENT (SCFM) ... 243
SMART CITY GOVERNANCE (SCG) ... 247
SMART CITY GREEN ENERGY (SCGE) ... 251
SMART CITY GREEN INFRASTRUCTURE (SCGI) ... 255
SMART CITY GRIDS (SCGR) ... 259
SMART CITY HEALTHCARE (SCHC) ... 262
SMART CITY INFRASTRUCTURE MANAGEMENT (SCIM) ... 266
SMART CITY INNOVATION (SCIN) ... 271
SMART CITY INTELLIGENT TRANSPORTATION SYSTEMS (SCITS) ... 274
SMART CITY IoT (SCIoT) ... 278
SMART CITY IRRIGATION (SCIR) ... 281
SMART CITY LIGHTING (SCL) ... 285
SMART CITY MAINTENANCE (SCM) ... 289
SMART CITY MOBILITY (SCMO) ... 293
SMART CITY PARKING MANAGEMENT (SCPM) ... 297
SMART CITY PLATFORM (SCP) ... 301
SMART CITY PUBLIC ART (SCPA) ... 305
SMART CITY PUBLIC HEALTH (SCPH) ... 309
SMART CITY PUBLIC SAFETY AND EMERGENCY MANAGEMENT (SCPSEM) ... 313
SMART CITY PUBLIC SERVICES (SCPS) ... 317
SMART CITY PUBLIC SPACES (SCPSP) ... 321

SMART CITY PUBLIC TRANSPORTATION (SCPT) 325
SMART CITY RENEWABLE ENERGY (SCRE) 329
SMART CITY RESILIENCE (SCR) 333
SMART CITY RETAIL (SCRT) 338
SMART CITY SECURITY (SCS) 342
SMART CITY SERVICES (SCSR) 346
SMART CITY STANDARDIZATION (SCSZ) 350
SMART CITY STORMWATER MANAGEMENT (SCSWM) 354
SMART CITY STREET FURNITURE (SCSF) 358
SMART CITY STREETS (SCST) 363
SMART CITY SUSTAINABILITY (SCSU) 366
SMART CITY SUSTAINABLE ENERGY (SCSE) 370
SMART CITY SUSTAINABLE TRANSPORTATION (SCSTR) 375
SMART CITY TOURISM (SCTO) 379
SMART CITY TOURIST ACCESSIBILITY (SCTA) 383
SMART CITY TOURIST ACCOMMODATION MANAGEMENT (SCTAM) 387
SMART CITY TOURIST ACTIVITY MANAGEMENT (SCTAC) 391
SMART CITY TOURIST ATTRACTIONS MANAGEMENT (SCTOA) 396
SMART CITY TOURIST CROWD MANAGEMENT (SCTCM) 399
SMART CITY TOURIST DATA ANALYSIS (SCTDA) 403
SMART CITY TOURIST DESTINATION MARKETING (SCTDM) 407
SMART CITY TOURIST EMERGENCY SERVICES (SCTES) 411
SMART CITY TOURIST EVENT MANAGEMENT (SCTEM) 415
SMART CITY TOURIST EXPERIENCE MANAGEMENT (SCTXM) 419
SMART CITY TOURIST FEEDBACK AND REVIEWS (SCTFR) 422
SMART CITY TOURIST INFORMATION SERVICES (SCTIS) 426
SMART CITY TOURIST LANGUAGE ASSISTANCE (SCTLA) 430
SMART CITY TOURIST NAVIGATION (SCTN) 433

SMART CITY TOURIST PARKING MANAGEMENT (SCTPM) 437

SMART CITY TOURIST PERSONALIZATION (SCTP) 440

SMART CITY TOURIST SAFETY AND SECURITY (SCTSS) 444

SMART CITY TOURIST SOCIAL MEDIA INTEGRATION (SCTSM) 448

SMART CITY TOURIST SUSTAINABLE TOURISM (SCTST) 452

SMART CITY TOURIST TRANSPORT MANAGEMENT (SCTTM) 456

SMART CITY TOURIST WAYFINDING (SCTW) 459

SMART CITY TRANSPORT MANAGEMENT (SCTM) 463

SMART CITY URBAN AGRICULTURE (SCUA) 467

SMART CITY URBAN MOBILITY (SCUM) 471

SMART CITY URBAN PLANNING AND DESIGN (SCUPD) 474

SMART CITY WASTE MANAGEMENT (SCWM) 478

SMART CITY WATER QUALITY MONITORING (SCWQM) 482

SMART CITY WORKFORCE DEVELOPMENT (SCWD) 485

| xi | **PROGRAM COST** **490**
ESTIMATED PROGRAM COST PER CAPITA 490

| xii | **KEY DEPARTMENTS** **520**
SMART CITY KEY DEPARTMENTS 520

| xiii | **VENDOR LANDSCAPE** **525**
TRENDS 525

| xiv | **INVESTOR OPPORTUNITIES** **528**
INVESTOR OPTIONS 528

| xv | **CAREER OPPORTUNITIES** **532**
SMART CITY CAREERS 532

| xvi | **DATA PRIVACY** **538**
HUMAN ERROR AND MISUSE IN SMART CITIES 538
MITIGATING RISKS OF HUMAN ERROR AND MISUSE 539
CONCLUSION 540

| xvii | **CYBER SECURITY** **542**

SMART CITIES CYBERSECURITY RISK...542

TYPES OF DATA COLLECTED IN SMART CITIES 543

CHALLENGES OF SECURING DATA IN SMART CITIES 544

BEST PRACTICES FOR SECURING DATA IN SMART CITIES 546

ROLE OF STAKEHOLDERS 547

| xviii | THE DARK SIDE 550
SURVEILLANCE AND PRIVACY CONCERNS 550

BIAS AND DISCRIMINATION IN DECISION-MAKING...551

CYBER-SECURITY RISKS..551

LACK OF TRANSPARENCY AND ACCOUNTABILITY 551

NEED FOR CITIZEN EMPOWERMENT AND PARTICIPATION 551

SOCIAL AND ETHICAL IMPLICATIONS OF SMART CITIES...............................552

ENVIRONMENTAL IMPACT OF SMART CITIES..552

ECONOMIC DISRUPTION AND JOB LOSSES..552

INSIDER THREATS AND EMPLOYEE SABOTAGE..553

DARK WEB AND ILLICIT ACTIVITIES IN SMART CITIES 553

THE RISK OF AI DOMINANCE IN SMART CITIES...554

IDENTITY LOSS AND TAKEOVER IN SMART CITIES..554

| xix | SMART CITIES 2030 556
SMART CITY GROWTH...556

| xx | GUIDE IN SETTING UP A SMART CITY 559
BRANDING & MARKETING 560

THE BUSINESS PLAN 561

THE THINK TANK... 564

BUILDING A SMART CITY..569

SEEKING THE INVESTMENT 585

THE BUY-IN 587

CITY COUNCIL REVENUE STREAMS 589

HELLENIC CITY COUNCIL NATIONAL LAWS 593
CONCLUSION 595

| xxi | KEY TAKEAWAYS 597
KEY TAKEAWAYS 597

EPILOGUE **600**
BONUS OFFER **601**
VALUABLE EXTERNAL LINKS **602**
RESOURCES **605**
ACRONYMS **608**
SMART CITY ALGORITHMS **609**
SELF ASSESSMENT **610**
SMART CITY CONTACTS **611**
INDEX **612**

ONE SMART WORLD
ARE YOU READY FOR IT?

ONE SMART WORLD
ARE YOU READY FOR IT?

| i |
INTRODUCTION

As our world becomes more urbanized, there is an increasing need for cities that are sustainable, efficient, and technologically advanced. Smart cities are the solution to this problem, offering innovative solutions that can help us create more livable, resilient, and prosperous urban environments. By leveraging the power of data, connectivity, and digital technologies, these cities are able to improve the quality of life of their citizens, optimize resource use, and promote economic growth.

Moving towards a future that is increasingly shaped by technology, smart cities represent an exciting opportunity to reimagine our urban landscapes and build a more sustainable and equitable world. This book aims to explore the concept of smart cities, their potential benefits, and the challenges that need to be overcome to make them a reality.

DEFINING SMART CITIES

Smart cities are urban and rural areas that leverage cutting-edge technologies and data analytics to improve the quality of life for their residents. These cities integrate various types of infrastructure, including transportation, energy, and communication systems, to create a more sustainable and efficient living environment.

Smart cities rely on an interconnected network of devices, sensors, and data centers to collect and analyze real-time data on everything from traffic patterns to air quality. This information is then used to optimize systems, improve public services, and enhance overall quality of life.

Smart cities represent the future of living, where advanced technologies are leveraged to create a more connected, efficient, livable and safe environment.

BENEFITS OF SMART CITIES

Smart cities offer a multitude of benefits, ranging from reducing traffic congestion and air pollution to improving public safety and healthcare. In this chapter, we will examine the various ways in which smart cities can enhance the quality of life of their citizens and promote economic growth. We will also look at case studies of cities that have successfully implemented smart city solutions and the impact they have had on their communities.

CHALLENGES AND OPPORTUNITIES

While smart cities hold great promise, they also face significant challenges, from data privacy concerns to the digital divide. In this chapter, we will explore the obstacles that must be overcome to make smart cities a reality, as well as the opportunities they present for creating more sustainable and equitable urban environments. We will also discuss the role of government, businesses, and citizens in building smart cities and ensuring that they serve the needs of all members of the community.

Summary: We introduce the concept of smart cities, highlighting their potential to create more sustainable, efficient, and technologically advanced urban environments. We define the term "smart city" and explore its core components, including advanced technologies, data, and connectivity.

We also examine the potential benefits of smart cities, including improvements to quality of life, resource efficiency, and economic growth. Finally, we discuss the challenges and opportunities associated with smart cities, including data privacy and security concerns, the digital divide, the opportunities, and the need for citizen-centric design.

Conclusion: *The concept of smart cities represents a promising solution to the challenges faced by urban environments in the 21st century. By leveraging advanced technologies and data-driven solutions, smart cities have the potential to significantly improve the quality of life of citizens, enhance resource efficiency, and promote economic growth.*

However, to achieve this vision, there are significant challenges that must be addressed. These include ensuring data privacy and security, addressing the digital divide, and ensuring that smart city solutions are designed with the needs of all citizens in mind.

Despite these challenges, the potential benefits of smart cities are clear, and as technology continues to advance, the opportunity to create more sustainable and equitable urban environments has never been greater.

Quote: *" Imagination is more important than knowledge. The concept of smart cities represents a fusion of imagination and knowledge, a vision for the future where technology and data-driven solutions are harnessed to create more sustainable, efficient, and equitable urban environments." Emin Hasic*

ONE SMART WORLD
ARE YOU READY FOR IT?

| ii |
HISTORY OF SMART CITIES

Let me take you on a journey through the fascinating history of modern day smart cities, starting from its birth and up to the present day, where the concept of smart cities can be traced back to the late 1990s and early 2000s, when the term "smart city" was first introduced and used to describe cities that leverage advanced technology and data analysis to enhance the quality of life for its citizens. It was a vision of a future where cities would use technology not just to improve their own efficiency and sustainability, but to create better lives for the people who lived in them.

SONGDO – SOUTH KOREA

One of the earliest examples of a smart city was Songdo in South Korea, which was planned and built from the ground up as a fully integrated smart city. The city, which opened in 2009, was designed to be a hub for international business, and it features state-of-the-art technology, including ubiquitous broadband connectivity and a range of sensors and monitoring systems. This made Songdo one of the first cities in the world to have a truly integrated and interconnected infrastructure, where everything from transportation to waste management was connected and optimized using data and technology.

SANTANDER – SPAIN

Another early example of a smart city was Santander in Spain, which became one of the first cities in the world to implement a smart city program in 2006. The program, which was a collaboration between the city government and local universities, aimed to improve the efficiency of city services and reduce the environmental impact of the city. This was achieved through the use of innovative technologies such as intelligent lighting systems, smart energy grids, and real-time traffic management. These technologies allowed the city to collect and analyze vast amounts of data in real-time, which was then used to make informed decisions about the allocation of resources and the optimization of city services.

AMSTERDAM – NETHERLANDS

In 2008, Amsterdam in the Netherlands launched the Amsterdam Smart City initiative, which aimed to use technology to improve the quality of life for the city's residents. The initiative was one of the first large-scale smart city projects in Europe, and it has since become a model for other cities looking to implement similar programs. Amsterdam's smart city initiative focused on several key areas, including energy efficiency, waste management, and transportation. One of the most notable aspects of the initiative was the city's use of smart energy grids, which allowed it to monitor and optimize energy use in real-time, reducing waste and increasing efficiency.

RECENT YEARS

In recent years, the concept of smart cities has gained widespread popularity, and many cities around the world have begun to implement smart city programs of their own. Some of the most notable smart cities in the world today include Singapore, Barcelona, and Helsinki. These cities have implemented a range of innovative technologies, including advanced transportation systems, smart energy grids, and innovative waste management systems. They have also focused on using technology to improve the delivery of city services, such as health care and education, and to create more inclusive and equitable communities.

Smart cities have been praised for their potential to improve the quality of life for residents and to make cities more efficient and sustainable. For example, the use of advanced transportation systems has made it easier for people to get around cities, reducing traffic congestion and increasing the speed and efficiency of transportation. The use of smart energy grids has made it possible to monitor and optimize energy use, reducing waste and helping to address the global challenge of climate change. And the use of innovative waste management systems has made it possible to reduce waste and increase recycling, creating a cleaner and more sustainable city environment.

However, the development of smart cities has also been criticized for its potential to exacerbate social and economic inequalities and to further centralize control over city services. For example, there are concerns that the use of technology to monitor and manage city services could lead to a loss of privacy for citizens, and that decisions about the allocation of resources and services may be made based on data-driven algorithms that are not transparent or accountable. There are also concerns that the focus on technology-driven solutions may distract from more fundamental issues of urban planning and governance, such as affordable housing and inclusive economic development.

Despite these challenges, the trend towards smart cities is likely to continue as cities around the world seek to address the complex and interrelated challenges of urbanization, including economic growth, social cohesion, and environmental sustainability. The development of smart cities is also likely to continue to be shaped by advances in technology, such as the Internet of Things (IoT) and artificial intelligence (AI), which are enabling new and more sophisticated forms of data collection, analysis, and decision-making.

As cities around the world continue to evolve and become more connected and technologically advanced, the concept of smart cities will continue to evolve and mature. While the definition of a smart city may change over time, the fundamental goal of using technology to improve the quality of life for citizens is likely to remain unchanged.

THE STORY

The history of smart cities is a story of innovation and progress, as cities around the world have sought to leverage technology to enhance the quality of life for their residents and to create more sustainable and efficient urban environments. While the development of smart cities has been accompanied by challenges and controversies, it is clear that this trend will continue to shape the future of cities in the years to come. As trusted and knowledgeable experts in the field, it is our responsibility to ensure that the development of smart cities is guided by principles of transparency, accountability, and inclusiveness, so that the benefits of technology can be shared by all citizens.

Summary: Smart cities emerged in the late 1990s and early 2000s, with cities like Songdo in South Korea, Santander in Spain, and Amsterdam in the Netherlands leading the way. In recent years, many cities worldwide have implemented smart city programs, using technology to improve the quality of life for their residents and make cities more efficient and sustainable. While smart cities offer many benefits, they also face challenges such as potential exacerbation of social and economic inequalities and centralization of control over city services. Despite these challenges, smart cities will continue to evolve and shape the future of cities, guided by principles of transparency, accountability, and inclusiveness.

Conclusion: The development of smart cities is an ongoing process, and its success will depend on how well it addresses the complex challenges of urbanization while also ensuring transparency, accountability, and inclusiveness. While technology plays a critical role in creating smart cities, it is also important to remember that it is just one tool in the larger effort to create more livable, sustainable, and equitable cities. As we continue to innovate and improve our cities, it is essential to prioritize the needs and perspectives of all citizens, ensuring that technology serves everyone's best interests. Ultimately, the success of smart cities depends on our ability to strike a balance between technological innovation and social responsibility, creating a better future for all.

Quote: " Technology is a wonderful servant but a dangerous master. The development of smart cities is a testament to the potential of technology to improve our lives, but we must be vigilant in ensuring that it serves the best interests of all citizens, and not just a privileged few." Emin Hasic

ONE SMART WORLD
ARE YOU READY FOR IT?

| iii |
HISTORICAL TIMELINE

The idea of smart cities, where technology is used to improve the quality of life for its inhabitants, has a long and rich history dating back to 2600 BCE. Throughout the centuries, cities around the world have evolved to incorporate new technologies, from the invention of the wheel to the implementation of modern-day smart city solutions. As we look back through history, we can see how innovation has played a critical role in shaping the way we live in urban environments today. From ancient cities like Mohenjo-Daro to modern-day metropolises like Tokyo, the development of smart cities has been a continuous process that has fundamentally changed the way we interact with our surroundings.

FROM 2600 BCE UNTIL 2023 CE

2600 BCE: MOHENJO-DARO is an ancient city in present-day Pakistan that was built around 2600 BCE. It is considered one of the earliest examples of urban planning and design, with a grid-like street pattern, a sophisticated water management system, and public buildings such as granaries and baths. The design principles used in Mohenjo-daro have influenced the development of many smart city initiatives that prioritize efficient infrastructure and public services.

4TH CENTURY BCE: PLATO was a Greek philosopher who wrote extensively about politics, ethics, and the nature of reality. He believed that cities should be organized according to a strict social hierarchy, with each member of society playing a specific role based on their innate abilities. Plato's ideas have influenced the development of many smart city initiatives that prioritize efficient governance and the allocation of resources based on need and merit.

4TH CENTURY BCE: ARISTOTLE was a Greek philosopher who wrote extensively about politics, ethics, and the natural world. He believed that cities should be organized into a hierarchy of administrative units, with each unit responsible for a specific set of functions such as defense, education, and public works. Aristotle's ideas have influenced the development of many smart city initiatives that prioritize efficient governance and public services.

5TH CENTURY BCE: HIPPODAMUS OF MILETUS was an ancient Greek architect and urban planner who is credited with developing the grid-like street pattern that is now common in many modern cities. He argued that cities should be organized into a rational and symmetrical layout, with public spaces and buildings located at regular intervals. Hippodamus' ideas have influenced the development of many smart city initiatives that prioritize efficient land use and public space design.

5TH CENTURY BCE: KAUTILYA was an Indian philosopher and statesman who is credited with writing the Arthashastra, a treatise on politics, economics, and governance. He believed that cities should be designed with a focus on security and defense, with walls, gates, and watchtowers to protect against external threats. Kautilya's ideas have influenced the development of many smart city initiatives that prioritize public safety and security.

5TH CENTURY BCE: SUN TZU was a Chinese general and military strategist who wrote The Art of War, a treatise on warfare and strategy. He believed that cities should be designed with a focus on defensive fortifications and strategic positioning, to prevent attacks from enemy forces. Sun Tzu's ideas have influenced the development of many smart city initiatives that prioritize public safety and security.

6TH CENTURY BCE: CONFUCIUS was a Chinese philosopher who wrote extensively about social order and governance. He believed that cities should be organized into well-defined districts, with each district having its own administration and set of laws. Confucius' ideas have influenced the development of many smart city initiatives that prioritize efficient governance and public services.

6TH CENTURY BCE: CYRUS THE GREAT was the founder of the Persian Empire, and he is credited with creating one of the first large-scale urban centers in the ancient world, the city of Pasargadae. Cyrus' approach to urban planning emphasized the use of public spaces and amenities, such as gardens, parks, and fountains, to promote social interaction and community building. His ideas have influenced the development of many smart city initiatives that prioritize public space design and community engagement.

6th CENTURY BCE: SIDDHARTHA GAUTAMA (BUDDHA) was a spiritual leader who taught about the importance of community and compassion. He believed that cities should be designed to promote social harmony and reduce suffering. Buddha's ideas have influenced the development of many smart city initiatives that prioritize community engagement, social inclusion, and the promotion of mental and physical wellbeing.

1ST CENTURY CE: VITRUVIUS was a Roman architect and engineer who wrote the influential work "De architectura" which included ideas about city planning. He believed that cities should be designed with a focus on functionality and security, with wide, straight streets and public spaces that could be easily monitored by authorities. Vitruvius' ideas have influenced the development of many smart city initiatives that prioritize safety and security.

3RD CENTURY CE: DIOCLETIAN was a Roman emperor who initiated significant urban planning projects throughout the empire. He was responsible for the construction of several new cities, including the city of Diocletianopolis (modern-day Hisarya, Bulgaria), which was designed with a focus on functionality and defense. Diocletian's ideas have influenced the development of many smart city initiatives that prioritize defense and protection.

4TH CENTURY CE: AUGUSTINE OF HIPPO was a Christian theologian and philosopher who wrote extensively about social justice and morality. He believed that cities should be organized according to principles of justice and fairness, with equal access to resources and opportunities for all residents. Augustine's ideas have influenced the development of many smart city initiatives that prioritize social inclusion and the reduction of inequalities.

6TH CENTURY CE: PROCOPIUS was a Byzantine historian who wrote extensively about the construction of cities and public works in the Eastern Roman Empire. He believed that cities should be designed with a focus on practicality and functionality, with an emphasis on infrastructure such as roads, aqueducts, and public buildings. Procopius' ideas have influenced the development of many smart city initiatives that prioritize infrastructure and connectivity.

8TH CENTURY CE: AL-FARABI was an Islamic philosopher and political theorist who wrote extensively about governance and the organization of society. He believed that cities should be designed to promote social harmony and cultural diversity, with different communities living and working together in a spirit of cooperation. Al-Farabi's ideas have influenced the development of many smart city initiatives that prioritize community engagement and intercultural dialogue.

8TH CENTURY CE: CHARLEMAGNE was a Frankish king who initiated significant urban planning projects throughout his empire. He believed that cities should be designed with a focus on cultural and intellectual enrichment, with public spaces and institutions dedicated to education, research, and the arts. Charlemagne's ideas have influenced the development of many smart city initiatives that prioritize innovation, creativity, and cultural enrichment.

9TH CENTURY CE: ABU YUSUF was an Islamic jurist and scholar who wrote extensively about Islamic law and governance. He believed that cities should be organized according to principles of fairness and efficiency, with equal access to resources and opportunities for all residents. Abu Yusuf's ideas have influenced the development of many smart city initiatives that prioritize the efficient use of resources and the promotion of social equity.

9TH CENTURY CE: AL-KINDI was an Islamic philosopher, mathematician, and scientist who wrote extensively about the relationship between nature and technology. He believed that cities should be designed with a focus on environmental sustainability, with buildings and infrastructure constructed in harmony with the natural world. Al-Kindi's ideas have influenced the development of many smart city initiatives that prioritize environmental sustainability and the use of renewable energy sources.

1853: GEORGE PULLMAN builds a factory town for his workers outside of Chicago. The town is designed with parks, schools, and other amenities to improve the quality of life for workers.

1857: HENRY DAVID THOREAU was an American author, philosopher, and naturalist who wrote extensively about the need for humans to live in harmony with nature. His book "Walden" described his experiences living in a small cabin in the woods, and argued that urban life was too detached from the natural world. Thoreau's ideas have influenced the development of many smart city initiatives that prioritize sustainability and environmental protection.

1898: EBENEZER HOWARD publishes "Tomorrow, A Peaceful Path to Real Reform," where he proposes the idea of garden cities. These cities would be self-contained communities surrounded by green belts, where people could live, work, and play. His vision was to create self-contained, planned communities that combined the benefits of city and country living. Howard's ideas about urban design and community planning have influenced many smart city initiatives that prioritize human well-being and social cohesion.

1900S: THE RISE OF ELECTRIC POWER and telecommunication infrastructure leads to the development of early smart city technologies, such as streetlights and telegraphs.

1927: THE CITY OF FRANKFURT, GERMANY introduces a traffic management system that uses sensors and algorithms to optimize traffic flow.

1945: LEWIS MUMFORD was an American historian, sociologist, and philosopher who wrote extensively about the role of technology in society. He argued that technology should be used to promote human flourishing and social justice, rather than purely for economic growth. Mumford's ideas have influenced the development of many smart city initiatives that prioritize the well-being of residents over economic goals.

1960S: THE DEVELOPMENT OF COMPUTERS and the internet paves the way for the creation of digital smart city technologies, such as electronic toll collection and online public services.

1961: JANE JACOBS was an American-Canadian journalist, author, and activist who wrote a book called "The Death and Life of Great American Cities." In the book, Jacobs criticized the urban renewal projects of the time and argued that cities should be designed to prioritize the needs of their residents, rather than the needs of developers or planners. Jacobs' ideas have influenced the development of many smart city initiatives that prioritize citizen engagement and community empowerment.

1969: IVAN ILLICH was an Austrian philosopher and critic of industrial society. He argued that the modern world had become too dependent on technology and that this dependence was damaging to human well-being. Illich's ideas have influenced the development of many smart city initiatives that seek to balance the benefits of technology with the need for human connection and social interaction.

1970: PAOLO SOLERI was an Italian architect and urban designer who developed the concept of "arcology," or the integration of architecture and ecology. His vision was to create dense, compact, and self-sufficient cities that minimized the environmental impact of urban living. Soleri's ideas have influenced the development of many smart city initiatives that prioritize sustainability, resource efficiency, and environmental protection.

1972: DONELLA MEADOWS was an American environmental scientist who wrote a book called "The Limits to Growth," which argued that unchecked economic growth would ultimately lead to environmental and social collapse. Meadows' ideas have influenced the development of many smart city initiatives that prioritize sustainability and resource efficiency over unlimited growth.

1977: CHRISTOPHER ALEXANDER was a British architect and urban designer who developed the concept of "pattern language," or a set of design principles that could be used to create more humane and livable cities. His ideas have influenced the development of many smart city initiatives that prioritize citizen engagement and participatory design.

1980S: THE CONCEPT OF "INTELLIGENT TRANSPORTATION SYSTEMS" (ITS) emerges, which uses technology to improve the efficiency and safety of transportation systems.

1987: MANUEL CASTELLS is a Spanish sociologist who has written extensively about the role of information and communication technologies (ICTs) in shaping urban development. He argues that ICTs have the potential to transform urban life by enabling more efficient and sustainable forms of living. Castells' ideas have influenced the development of many smart city initiatives that prioritize the use of ICTs to improve urban services and infrastructure.

1990S: CITIES BEGIN TO USE GEOGRAPHIC INFORMATION SYSTEMS (GIS) to map and analyze urban data, leading to better urban planning and decision-making.

1994: SASKIA SASSEN is a Dutch sociologist who has written about the role of global cities in the global economy. She argues that cities are becoming increasingly important as sites of economic activity, and that this is leading to new forms of social inequality and urban fragmentation. Sassen's ideas have influenced the development of many smart city initiatives that seek to promote social inclusion and reduce inequality.

2000S: THE CONCEPT OF "SMART CITIES" gains popularity, with cities around the world investing in technology and data-driven solutions to improve efficiency, sustainability, and quality of life.

2006: RICHARD FLORIDA is an American urban studies theorist who has written extensively about the role of creativity and innovation in shaping urban development. He argues that cities that foster creativity and innovation are more likely to attract talented people and businesses, leading to economic growth and prosperity. Florida's ideas have influenced the development of many smart city initiatives that seek to promote innovation and entrepreneurship.

2008: IBM introduces the Smarter Cities Challenge, a program that provides funding and support to cities to develop and implement smart city solutions.

2010: JEREMY RIFKIN is an American economist and futurist who has written about the potential for a "Third Industrial Revolution" based on renewable energy, the Internet of Things, and smart cities. He argues that this revolution could transform the way we live, work, and interact with one another, leading to a more sustainable and equitable future. Rifkin's ideas have influenced the development of many smart city initiatives that seek to create more sustainable and livable urban environments.

2010: THE EUROPEAN COMMISSION launches its "Smart Cities" initiative to encourage the development of smart city projects in Europe, with a focus on energy efficiency, sustainable transportation, and innovation.

2010s: THE RISE OF THE INTERNET OF THINGS (IoT) enables the integration of sensors and devices into urban infrastructure, allowing for real-time monitoring and control of urban systems.

2011: THE INDIAN GOVERNMENT launches its "100 Smart Cities Mission" to develop 100 smart cities across the country, with a focus on sustainable development and improving quality of life.

2012: THE SMART CITIES AND COMMUNITIES ACT is introduced in the US Congress, which would provide funding and support for smart city projects in the United States.

2013: THE CITY OF BARCELONA, SPAIN launches its "smart city" initiative, which includes a range of smart city solutions such as smart parking, smart lighting, and smart waste management.

2013: THE CITY OF RIO DE JANEIRO, BRAZIL launches its "Porto Maravilha" project, which transforms a neglected area of the city into a smart city neighborhood with improved mobility, public spaces, and sustainability.

2014: THE EUROPEAN UNION launches its "Smart Cities and Communities" initiative, which aims to support the development of smart city projects across Europe.

2015: THE CITY OF AMSTERDAM, NETHERLANDS launches its "Amsterdam Smart City" initiative, which includes a range of smart city projects such as smart mobility, smart energy, and smart citizens.

2015: THE CITY OF DUBAI, UAE hosts the "Smart City Expo Dubai," which brings together city officials, urban planners, and technology experts to discuss the future of smart cities.

2015: THE UNITED NATIONS launch its "Sustainable Development Goals" (SDGs), which include a specific goal (#11) focused on making cities and human settlements inclusive, safe, resilient, and sustainable. This goal has since become a driving force behind many smart city initiatives around the world.

2016: THE CITY OF BARCELONA, SPAIN launches its "Superblock" project to reclaim streets for pedestrians and cyclists and reduce air pollution and noise in the city.

2016: THE CITY OF MANCHESTER, UK launches its "CityVerve" project, which develops a smart city platform that uses sensors, data analytics, and citizen engagement to improve health, energy, and transportation.

2016: THE CITY OF STOCKHOLM, SWEDEN launches its "Smart Stockholm" initiative, which includes a range of smart city projects such as smart transportation, smart buildings, and smart waste management.

2016: THE CITY OF TORONTO, CANADA launches its "Quayside" project, which develops a smart city district that uses sensors, data analytics, and innovative design to improve urban livability and sustainability.

2016: THE SMART CITIES CHALLENGE in the United States awards $50 million to the winning city (Columbus, Ohio) to develop and implement innovative smart city solutions.

2016: THE SMART CITIES COUNCIL releases the Smart Cities Readiness Guide, a comprehensive guidebook for cities to plan, implement, and measure smart city initiatives.

2017: THE CITY OF BARCELONA, SPAIN launches its "Superblocks" project, which creates car-free zones in the city to improve air quality, reduce traffic noise, and enhance pedestrian and cycling infrastructure.

2017: THE CITY OF DUBAI launches its "Smart Dubai" initiative, which aims to make Dubai the happiest and smartest city in the world by 2021.

2017: THE CITY OF HELSINKI, FINLAND launches its "Smart Kalasatama" project, which transforms an old industrial area into a smart city district that focuses on sustainability, mobility, and citizen participation.

2017: THE CITY OF TORONTO, CANADA launches its "Quayside" project in partnership with Sidewalk Labs, a subsidiary of Alphabet Inc. (Google's parent company), to develop a smart city neighborhood that uses technology to improve sustainability, mobility, and quality of life.

2018: THE CITY OF BARCELONA, SPAIN launches its "Superblock" project, which transforms the city's streets into car-free zones and repurposes the space for parks, public spaces, and pedestrian areas.

2018: THE CITY OF DUBAI, UNITED ARAB EMIRATES launch its "Dubai 10X" initiative, which aims to use technology and innovation to transform the city into a global leader in smart city solutions.

2018: THE CITY OF HELSINKI, FINLAND launches its "Smart Kalasatama" initiative, which develops a smart city district that uses data and technology to improve energy efficiency, mobility, and citizen services.

2018: THE CITY OF KANSAS CITY, MISSOURI, USA launches its "Smart City" initiative, which installs a network of sensors, kiosks, and digital infrastructure to improve mobility, public safety, and quality of life for residents.

2018: THE CITY OF SAN DIEGO, CALIFORNIA, USA launches its "Smart City San Diego" initiative, which includes a range of smart city projects such as smart streetlights, electric vehicle charging stations, and smart water management.

2018: THE CITY OF STOCKHOLM, SWEDEN introduces its "Zero Waste" initiative to use technology and innovation to reduce waste and improve recycling in the city.

2018: THE CITY OF TORONTO, CANADA announces plans to build a "smart neighborhood" called Quayside, in partnership with Alphabet's Sidewalk Labs. The neighborhood will use sensors and data to optimize everything from traffic flow to waste management.

2019: THE CITY OF AMSTERDAM, NETHERLANDS launches its "Amsterdam Smart City" initiative, which develops a smart city ecosystem that involves citizens, businesses, and government to co-create innovative solutions for urban challenges.

2019: THE CITY OF DUBAI, UAE launches its "Dubai Paperless Strategy," which aims to use digital technology and smart city solutions to reduce paper usage and improve government services for residents.

2019: THE CITY OF HELSINKI, FINLAND launches its "1,000 Urban Challenges" initiative to invite citizens, businesses, and organizations to propose ideas and solutions to urban challenges using technology and innovation.

2019: THE CITY OF HONG KONG launches its "Smart City Blueprint" to use technology and innovation to enhance sustainability, mobility, and citizen engagement.

2019: THE CITY OF LOS ANGELES, USA launches its "Urban Movement Labs" initiative, which uses data and technology to improve mobility and transportation systems in the city.

2019: THE CITY OF NEW YORK, USA launches its "LinkNYC" project, which installs public Wi-Fi kiosks throughout the city to provide free internet access to residents and visitors.

2019: THE CITY OF SAN FRANCISCO, USA launches its "Digital Services Strategy" to use technology to improve access to city services, enhance public safety, and promote economic development.

2019: THE CITY OF SHENZHEN, CHINA is named the world's smartest city in the Smart City Index, which evaluates cities based on their use of technology and data to improve the lives of their citizens.

2019: THE CITY OF SINGAPORE is named the world's smartest city in the Smart City Index, which evaluates cities based on their use of technology and data to improve the lives of their citizens.

2019: THE CITY OF SYDNEY, AUSTRALIA launches its "Smart City" strategy, which uses technology and innovation to improve public services, economic development, and quality of life for residents.

2019: THE WORLD ECONOMIC FORUM launches its "Global Smart City Index," which evaluates cities based on their readiness for and implementation of smart city technologies.

2020: THE CITY OF AMSTERDAM, NETHERLANDS launches its "Amsterdam Smart City" project, which uses data and technology to improve sustainability, mobility, and citizen participation in urban development.

2020: THE CITY OF LOS ANGELES, USA launches its "LA Cyber Lab," which provides cybersecurity resources and information sharing for small and medium-sized businesses in the city.

2020: THE CITY OF NEW YORK, USA launches its "Citywide Internet of Things" project, which installs a network of sensors and digital infrastructure to collect data on various aspects of urban life such as air quality, traffic, and waste management.

2020: THE CITY OF SEATTLE, WASHINGTON, USA launches its "Smart Cities 2020" initiative, which includes a range of smart city projects such as smart traffic management, smart lighting, and smart public safety.

2020: THE CITY OF SHENZHEN, CHINA launches its "Smart Shenzhen" initiative to use technology and innovation to improve urban governance, sustainability, and quality of life for residents.

2020: THE CITY OF SINGAPORE launches its "Smart Nation" initiative, which aims to use digital technology and smart city solutions to enhance urban living, economic growth, and national resilience.

2020: THE COVID-19 PANDEMIC accelerates the adoption of smart city technologies and solutions, as cities use technology to monitor the spread of the virus, enforce social distancing measures, and support remote work and learning, manage the crisis and improve public health and safety.

2021: THE CITY OF BOSTON, USA launches its "Vision Zero" initiative, which aims to eliminate traffic fatalities and serious injuries by 2040 through the use of smart city technologies such as traffic sensors and automated enforcement.

2021: THE CITY OF COPENHAGEN, DENMARK launches its "Copenhagen Connecting" project, which uses technology and innovation to improve the city's transportation systems, public spaces, and sustainability.

2021: THE CITY OF HELSINKI, FINLAND launches its "Climate-neutral Helsinki 2035" initiative, which aims to make Helsinki carbon-neutral by

2035 through the use of smart city technologies and sustainable urban planning.

2021: THE CITY OF MONTREAL, CANADA launches its "Smart City Strategy," which aims to use technology and data to improve sustainability, reduce inequality, and enhance citizen engagement.

2021: THE CITY OF PARIS, FRANCE launches its "Reinventer Paris" project, which transforms underutilized urban spaces into smart city districts that incorporate green spaces, community amenities, and sustainable technologies.

2021: THE CITY OF RIYADH, SAUDI ARABIA launches its "Smart Riyadh" initiative, which aims to use technology and innovation to enhance urban living, economic growth, and environmental sustainability.

2021: THE CITY OF SEOUL, SOUTH KOREA launches its "Green New Deal" to use technology and innovation to achieve carbon neutrality and promote sustainability in the city.

2021: THE CITY OF SINGAPORE launches its "Smart Nation" initiative to use technology and innovation to enhance sustainability, mobility, and quality of life for residents.

2021: THE CITY OF SYDNEY, AUSTRALIA launches its "Smart Sydney" initiative, which aims to use smart city technologies to improve sustainability, reduce congestion, and enhance quality of life for residents.

2021: THE CITY OF TORONTO, CANADA launches its "Smart City Accelerator Program" to support startups and entrepreneurs in developing smart city solutions that address urban challenges.

2022: THE CITY OF BEIJING, CHINA hosts the "World Smart City Expo," which brings together experts and leaders from around the world to discuss the latest innovations and trends in smart city development.

2022: THE CITY OF BERLIN, GERMANY launches its "Smart City Berlin" initiative, which includes a range of smart city projects such as smart energy, smart transportation, and smart buildings.

2022: THE CITY OF COPENHAGEN, DENMARK launches its "Copenhagen Connecting" initiative, which develops a smart transportation system that integrates various modes of mobility, data, and user experience to promote sustainable and efficient transportation.

2022: THE CITY OF DUBAI, UNITED ARAB EMIRATES launches its "Dubai Autonomous Transportation Strategy" to use autonomous vehicles and smart transportation systems to improve mobility and reduce congestion in the city.

2022: THE CITY OF GUANGZHOU, CHINA launches its "Guangzhou Smart City" initiative, which uses artificial intelligence and big data to improve urban management, public services, and citizen participation.

2022: THE CITY OF LONDON, UK launches its "Smart London" initiative, which aims to use smart city technologies to improve the city's air quality, reduce traffic congestion, and enhance public safety.

2022: THE CITY OF LOS ANGELES, CALIFORNIA, USA launches its "LA Smart City" initiative, which includes a range of smart city projects such as smart transportation, smart buildings, and smart energy management.

2022: THE CITY OF MELBOURNE, AUSTRALIA launches its "Smart City" strategy, which aims to use technology and innovation to improve urban livability, economic development, and environmental sustainability.

2022: THE CITY OF MOSCOW, RUSSIA launches its "Smart City Moscow" project, which uses technology and innovation to improve urban governance, sustainability, and quality of life for residents.

2022: THE CITY OF NEW YORK, USA launches its "Smart City Challenge," which invites technology companies and startups to develop innovative smart city solutions to improve quality of life for New York City residents.

2022: THE CITY OF PARIS, FRANCE launches its "Smart Paris 2030" initiative, which uses technology and innovation to improve urban mobility, public services, and environmental sustainability.

2022: THE CITY OF RIYADH, SAUDI ARABIA launches its "Riyadh Smart City" initiative, which aims to use technology and innovation to improve mobility, sustainability, and quality of life for residents.

2022: THE CITY OF SAN FRANCISCO, CALIFORNIA, USA launches its "SF Smart City" initiative, which includes a range of smart city projects such as smart lighting, smart parking, and smart waste management.

2022: THE CITY OF SEOUL, SOUTH KOREA launches its "Seoul Smart City" initiative, which aims to use technology and innovation to enhance urban mobility, public services, and economic growth.

2022: THE CITY OF SHENZHEN, CHINA launches its "Smart Shenzhen" initiative, which aims to use advanced technologies such as 5G, artificial intelligence, and the Internet of Things to enhance urban governance, public services, and economic development.

2022: THE CITY OF STOCKHOLM, SWEDEN launches its "Smart Stockholm" initiative, which uses technology and innovation to improve urban governance, sustainability, and quality of life for residents.

2022: THE CITY OF SYDNEY, AUSTRALIA launches its "Smart Sydney" initiative to use technology and innovation to improve urban living and address social and environmental challenges.

2022: THE CITY OF TEL AVIV, ISRAEL launches its "Digital Tel Aviv" project, which aims to use technology and innovation to improve urban mobility, sustainability, and quality of life for residents.

2022: THE CITY OF TOKYO, JAPAN launches its "Tokyo Smart City" initiative, which includes a range of smart city projects such as smart transportation, smart energy, and smart waste management.

2022: THE CITY OF TOKYO, JAPAN launches its "Zero Emission Tokyo" initiative to use technology and innovation to achieve net-zero carbon emissions and promote sustainability in the city.

2022: THE CITY OF VANCOUVER, CANADA launches its "Greenest City Action Plan 2022-2030" to use technology and innovation to address climate change and promote sustainability.

2023: THE CITY OF RIO DE JANEIRO, BRAZIL launches its "Rio 2030" initiative to use technology and innovation to transform the city into a more sustainable, equitable, and livable place by the year 2030.

Summary: The concept of smart cities, defined as urban areas that utilize technology to enhance sustainability, efficiency, and quality of life for its residents, has a long and complex history that dates back to 2600 BCE. The ancient Indus Valley Civilization, for example, had advanced systems for urban planning and sanitation that were considered ahead of their time. The ancient Greeks and Romans also made significant contributions to urban planning and engineering, with the development of aqueducts and other infrastructure.

During the industrial revolution, cities grew rapidly and faced new challenges such as pollution and congestion. In the early 20th century, urban planning and architecture emerged as disciplines to address these issues. In the 1960s, the concept of the "computerized city" emerged, as researchers began exploring the potential for computer systems to manage and optimize urban infrastructure.

In the 21st century, the emergence of the Internet of Things (IoT) and other advanced technologies has transformed the concept of smart cities. Today, cities around the world are implementing a range of initiatives to improve sustainability, safety, and quality of life for their residents. These initiatives include smart transportation systems, energy-efficient buildings, and real-time data analysis to improve public services.

Conclusion: The history of smart cities is a testament to the importance of innovation and adaptation in response to the changing needs of urban areas. From ancient civilizations to the present day, humans have sought to create cities that are more efficient, sustainable, and livable. While there have been many challenges along the way, such as pollution, congestion, and resource scarcity, advancements in technology have allowed cities to overcome these obstacles and continue to evolve.

Today, the concept of smart cities is more important than ever, as the world faces complex challenges such as climate change and urbanization. By implementing innovative technologies and strategies, cities can become more resilient and sustainable, while also providing a better quality of life for their residents. As we look to the future, it is clear that the development of smart cities will continue to be a key priority for governments, businesses, and communities around the world.

Quote: "Intelligence is not limited to the human brain, but can be harnessed and integrated into the design of our cities. The evolution of smart cities throughout history is a testament to our ability to use technology to enhance the sustainability, efficiency, and quality of life for all. Let us continue to innovate and adapt in response to the changing needs of our urban areas, and build a better future for generations to come." Emin Hasic

ONE SMART WORLD
ARE YOU READY FOR IT?

| iv |
PHILOSOPHERS & VISIONARIES

As our cities become increasingly complex and interconnected, the need for smart city visionaries has never been greater. These individuals are able to draw on the insights of past visionaries and philosophers to envision and implement innovative solutions that utilize advanced technology and data-driven insights to create more efficient, sustainable, and livable urban environments. From Plato's recognition of the importance of good governance to Jane Jacobs' championing of community participation in urban planning, and Marshall McLuhan's insight into the power of communication technologies to shape social structures, these visionaries provide a rich philosophical heritage that informs the work of today's smart city leaders. Through the development of strategies that address complex urban challenges, these leaders promote the well-being of citizens and work to create more just and equitable cities. In this chapter, we will explore the contributions of some of the most influential smart city visionaries, and examine how their ideas continue to shape the way we think about urban development today.

ARISTOTLE 4TH CENTURY BCE

Like his teacher Plato, Aristotle was concerned with the organization and governance of the city-state. In his book "Politics," Aristotle discussed the role of the city in promoting human flourishing and emphasized the importance of community and social bonds. His ideas about citizenship and governance have influenced modern urban planning and the idea of the smart city as a place where people can live healthy, happy, and productive lives. Aristotle believed that the city was the highest form of community, and he wrote extensively on the subject of politics and governance. His ideas on civic virtue, the importance of a well-ordered society, and the role of the state in promoting the common good is viewed as relevant to modern-day debates about smart cities and their impact on social and political life.

PLATO 4TH CENTURY BCE

In his book "The Republic," Plato envisioned an ideal city-state that was well-organized and governed by a philosopher-king. He emphasized the importance of planning and architecture in creating a harmonious and efficient society. Plato believed that the layout of a city should be carefully designed to promote social order and facilitate communication between citizens. This idea of a well-organized and just city is viewed as a foundation for modern urban planning principles.

EPICURUS 3RD CENTURY BCE

Epicurus believed in the importance of living a simple and self-sufficient life, and his philosophy emphasized the importance of communal living and cooperation. His ideas about creating communities that are self-sufficient and sustainable is viewed as a precursor to modern-day concepts of eco-cities and green urbanism.

DIOGENES 4TH CENTURY BCE

Diogenes was a Cynic philosopher who lived a simple, ascetic lifestyle and rejected materialism. His philosophy emphasized the importance of living in harmony with nature and living a self-sufficient, frugal lifestyle. His ideas about living in a simple and sustainable way have relevance to modern concepts of eco-cities and sustainable urban planning.

HERACLITUS 5TH CENTURY BCE

Heraclitus believed that change was the only constant in life, and that everything was constantly in flux. His philosophy emphasized the importance of adapting to change and finding balance between opposing forces. This idea of finding balance and adapting to change have relevance to modern urban planning, where cities must constantly adapt to new challenges and technologies.

PYTHAGORAS 6TH CENTURY BCE

Pythagoras was a philosopher and mathematician who believed in the importance of harmony and balance. His philosophy emphasized the importance of order and symmetry, and his ideas about the mathematical relationships between objects have relevance to modern urban design principles.

DEMOCRITUS 5TH CENTURY BCE

Democritus was a philosopher who believed that everything was made up of atoms and that the universe was composed of an infinite number of these tiny particles. His philosophy emphasized the importance of understanding the natural world through observation and experimentation. His ideas about the importance of scientific inquiry and understanding the natural world have relevance to modern urban planning, where scientific research can be used to inform decisions about sustainability and environmental protection.

ZENO OF CITIUM 3RD CENTURY BCE

Zeno was the founder of the Stoic philosophy, which emphasized the importance of living in accordance with nature and the acceptance of the natural order of things. His philosophy emphasized the importance of self-control, rationality, and a sense of community. His ideas about living in harmony with nature and accepting the natural order of things have relevance to modern urban planning, where cities must be designed to work in harmony with the natural environment.

THALES OF MILETUS 6TH CENTURY BCE

Thales was one of the earliest Greek philosophers who believed that the natural world could be explained through observation and reason. His philosophy emphasized the importance of using empirical evidence to understand the world. His ideas about observation and experimentation have relevance to modern urban planning, where cities can be analysed and improved through data analysis and experimentation.

THEOPHRASTUS 4TH CENTURY BCE

Theophrastus was a philosopher who focused on the natural world and botany. His philosophy emphasized the importance of observing and understanding the natural world. His ideas about the importance of biodiversity and understanding the natural environment have relevance to modern urban planning, where cities must be designed to work in harmony with the natural environment.

XENOPHON 4ᵀᴴ CENTURY BCE

Xenophon was a historian and philosopher who wrote extensively about politics and leadership. His philosophy emphasized the importance of leadership and governance in creating successful societies. His ideas about the importance of good leadership and governance have relevance to modern urban planning, where cities must be governed effectively to ensure their sustainability and success.

CLEANTHES 3ᴿᴰ CENTURY BCE

Cleanthes was a philosopher who believed in the unity of all things and the interconnectedness of the natural world. His philosophy emphasized the importance of living in harmony with nature and understanding our place in the natural world. His ideas about living in harmony with nature have relevance to modern urban planning, where cities must be designed to work in harmony with the natural environment.

PROTAGORAS 5ᵀᴴ CENTURY BCE

Protagoras was a philosopher who believed that humans were the measure of all things. His philosophy emphasized the importance of subjective experience and individual perspective. His ideas about subjectivity have relevance to modern urban planning, where cities must be designed to meet the needs and desires of their diverse inhabitants.

ANAXIMANDER 6ᵀᴴ CENTURY BCE

Anaximander was a philosopher who believed in the concept of the infinite and the eternal. His philosophy emphasized the importance of balance and the idea that everything in the universe was connected. His ideas about balance and connectedness have relevance to modern urban planning, where cities must be designed to balance the needs of their inhabitants with the needs of the natural environment.

EMPEDOCLES 5TH CENTURY BCE

Empedocles was a philosopher who believed that the universe was made up of four elements: earth, air, fire, and water. His philosophy emphasized the importance of balance and harmony between these elements. His ideas about balance and harmony have relevance to modern urban planning, where cities must be designed to balance the needs of their inhabitants with the needs of the natural environment.

ANAXAGORAS 5TH CENTURY BCE

Anaxagoras was a philosopher who believed that the universe was made up of an infinite number of tiny particles. His philosophy emphasized the importance of understanding the fundamental building blocks of nature. His ideas about understanding the fundamental nature of things have relevance to modern urban planning, where cities can be analysed and improved through data analysis and experimentation.

JANE JACOBS 1916 to 2006

Although not a philosopher per se, Jane Jacobs was a writer and activist who challenged traditional urban planning practices. In her book "The Death and Life of Great American Cities," Jacobs argued that cities should be designed to promote social interaction and diversity. She emphasized the importance of mixed-use neighbourhoods, pedestrian-friendly streets, and local businesses. Her ideas have influenced modern urban planning and the concept of the "smart city" that prioritizes sustainability, livability, and social inclusion.

LEWIS MUMFORD 1895 to 1990

Mumford was an American historian and philosopher who wrote about the history of cities and the impact of technology on urban life. In his book "The City in History," Mumford argued that cities should be designed to promote human flourishing and creativity, rather than efficiency and productivity. He believed that the growth of cities should be guided by ethical and aesthetic principles, and that the development of technology should be balanced with a concern for social and environmental sustainability.

HENRI LEFEBVRE 1901 to 1991

Lefebvre was a French philosopher and sociologist who wrote extensively on the urban environment. His book "The Production of Space" argued that the built environment is not just a physical space, but a social and cultural product that reflects the values and priorities of the society that creates it. Lefebvre's ideas about the social and cultural dimensions of urban space have influenced modern urban planning and the concept of the smart city as a place that reflects the needs and desires of its citizens.

FRIEDRICH ENGELS 1820 - 1895

Engels was a German philosopher and social theorist who, along with Karl Marx, developed the theory of communism. In his book "The Condition of the Working Class in England," Engels criticized the urban environment of the industrial revolution for its lack of planning, sanitation, and social welfare. His ideas about the need for social and economic equality have influenced modern urban planning and the concept of the smart city as a place that promotes social inclusion and economic opportunity.

MANUEL CASTELLS 1942

Castells is a Spanish sociologist who has written extensively on the impact of technology on society. In his book "The Rise of the Network Society," Castells argued that the digital revolution is transforming the way we live and work, and that cities are at the forefront of this transformation. His ideas about the role of information technology in shaping urban life have influenced modern urban planning and the concept of the smart city as a place that is connected, efficient, and innovative.

MARTIN HEIDEGGER 1889 - 1976

Heidegger was a German philosopher who wrote extensively on the concept of technology and its impact on human life. In his essay "The Question Concerning Technology," he argued that technology is not just a means to an end, but a way of understanding and being in the world. Heidegger's ideas about the relationship between technology and human existence have influenced our understanding of the role of technology in shaping the urban environment and the concept of the smart city.

DAVID HARVEY 1935

Harvey is a British geographer who has written extensively on the political economy of urbanization. In his book "The Condition of Postmodernity," he argued that the global economy and the rise of neoliberalism have transformed the way we think about and inhabit cities. Harvey's ideas about the social and economic dimensions of urbanization have influenced modern urban planning and the concept of the smart city as a place that is connected, sustainable, and equitable.

RICHARD SENNETT 1943

Sennett is an American sociologist and urbanist who has written extensively on the social and cultural dimensions of urban life. In his book "The Uses of Disorder," he argued that cities should not be designed to eliminate disorder and diversity, but rather to embrace them. Sennett's ideas about the importance of social and cultural diversity in the urban environment have influenced modern urban planning and the concept of the smart city as a place that fosters creativity, innovation, and collaboration.

JANE BENNETT 1957

Bennett is an American political theorist who has written extensively on the relationship between materiality and agency. In her book "Vibrant Matter," she argued that non-human objects and materials have their own agency and vitality, and that they play an important role in shaping human life and society. Bennett's ideas about the agency of non-human objects and materials have influenced our understanding of the relationship between technology and the urban environment, and the concept of the smart city as a place that is responsive and adaptive to the needs of its citizens.

MICHEL FOUCAULT 1926 - 1984

Foucault was a French philosopher and social theorist who wrote extensively on power and the ways in which it is exercised in society. In his book "Discipline and Punish," he examined the development of disciplinary institutions such as prisons and schools, and how they shape individual behaviour and social norms. His ideas about the relationship between power, knowledge, and the built environment have influenced modern urban planning and the concept of the smart city as a place that is transparent and accountable.

REM KOOLHAAS 1944

Koolhaas is a Dutch architect and urbanist who has designed buildings and urban plans around the world. In his book "Delirious New York," he examined the cultural and social dimensions of urbanization in New York City, and how the city has shaped and been shaped by popular culture. Koolhaas's ideas about the cultural and social dimensions of urban life have influenced modern urban planning and the concept of the smart city as a place that fosters creativity, innovation, and cultural exchange.

SASKIA SASSEN 1949

Sassen is a Dutch-American sociologist and urbanist who has written extensively on the globalization of urbanization. In her book "The Global City," she examined the ways in which cities have become key nodes in the global economy, and how they are linked to each other through flows of capital, people, and information. Sassen's ideas about the global dimensions of urbanization have influenced modern urban planning and the concept of the smart city as a place that is connected, dynamic, and globally oriented.

MANUEL DE LANDA 1952

De Landa is a Mexican-American philosopher and urbanist who has written extensively on the relationship between technology and society. In his book "A Thousand Years of Nonlinear History," he examined the ways in which technologies and social structures have evolved over time, and how they are intertwined with each other. De Landa's ideas about the complex and nonlinear relationship between technology and society have influenced our understanding of the role of technology in shaping the urban environment and the concept of the smart city.

WILLIAM J. MITCHELL 1944 - 2010

Mitchell was a professor of architecture and media arts and sciences at MIT who wrote extensively on the relationship between technology and the built environment. In his book "City of Bits," he explored the impact of digital technology on urban life, and how it is transforming the ways in which we live, work, and communicate. Mitchell's ideas about the role of digital technology in shaping the urban environment have influenced modern urban planning and the concept of the smart city as a place that is responsive, sustainable, and digitally connected.

DON IHDE 1934

Ihde is an American philosopher of science and technology who has written extensively on the relationship between humans and technology. In his book "Technology and the Lifeworld," he argued that technology is not a neutral tool, but rather an extension of human experience that shapes our perception of the world. Ihde's ideas about the relationship between humans and technology have influenced our understanding of the role of technology in shaping the urban environment and the concept of the smart city as a place that is designed with human needs and experiences in mind.

ANTHONY TOWNSEND 1971

Townsend is an American urbanist and futurist who has written extensively on the future of cities and the impact of technology on urban life. In his book "Smart Cities: Big Data, Civic Hackers, and the Quest for a New Utopia," he examined the ways in which data and technology are transforming the urban environment, and how cities can use these tools to become more sustainable, efficient, and responsive. Townsend's ideas about the transformative potential of data and technology in the urban environment have influenced modern urban planning and the concept of the smart city as a place that is data-driven, open, and participatory.

ADAM GREENFIELD 1968

 Greenfield is an American writer and urbanist who has written extensively on the intersection of technology and urban life. In his book "Against the Smart City," he critiqued the uncritical embrace of technology in urban planning, arguing that it often reinforces existing power structures and inequalities. Greenfield's ideas about the need for critical reflection and public engagement in the development of smart cities have influenced our understanding of the ethical and social dimensions of the smart city concept.

Summary: Smart city visionaries draw on the insights of past visionaries and philosophers to implement innovative solutions that utilize advanced technology and data-driven insights to create more efficient, sustainable, and livable urban environments. This chapter explores the contributions of some of the most influential smart city visionaries, including Plato, Jane Jacobs, Lewis Mumford, Aristotle, Henri Lefebvre, Friedrich Engels, Manuel Castells, Martin Heidegger, David Harvey, Richard Sennett, and Jane Bennett. Their ideas continue to shape the way we think about urban development today.

Conclusion: The contributions of current and past visionaries and philosophers have significantly influenced modern urban planning and the concept of the smart city. Today's smart city leaders promote the well-being of citizens and work to create more just and equitable cities. By developing strategies that address complex urban challenges, smart city visionaries utilize advanced technology and data-driven insights to create more efficient, sustainable, and livable urban environments. As cities become increasingly complex and interconnected, the need for smart city visionaries has never been greater.

Furthermore, while Greece is commonly known as the birthplace of democracy, the country's contribution to modern society extends far beyond the realm of politics. Through the works of ancient Greek philosophers, we can see that they were well ahead of their time in conceptualizing the idea of a smart city. Their visions for urban planning and design focused on creating livable, efficient, and sustainable communities, which are the same goals that modern smart cities strive to achieve. With this in mind, it's clear that Greece's influence on contemporary urban planning and design cannot be understated, and the country deserves recognition as the birthplace of smart cities.

Quote: *" As our cities grow increasingly complex and interconnected, we require visionaries who can draw on the insights of past philosophers and utilize advanced technology to create more efficient, sustainable, and livable urban environments. The true sign of intelligence is not knowledge but imagination." Emin Hasic*

| V |
EMBRACING SMART CITIES

Smart cities have become an increasingly popular concept in recent years as cities and towns around the world seek to leverage technology to improve the lives of their citizens. With the rise of urbanization and the challenges that come with it, the development of smart cities offers an opportunity to create more sustainable and efficient urban environments. However, the path to creating a truly smart city is not without its challenges and controversies. To make the most of the opportunities offered by smart cities, it is important for cities to approach this journey with a clear understanding of what it truly means to embrace this approach.

DESIRE IS THE NEW TREND

As we are witnessing, the trend towards smart cities is being driven by the desire to improve the quality of life for citizens and to create more sustainable and efficient urban environments. Despite the challenges and controversies that accompany the development of smart cities, it is clear that this trend is here to stay and that cities and towns around the world will continue to embrace this approach.

So, what does it mean to truly embrace smart cities and to make the most of the opportunities they offer? There are a number of key considerations that cities should keep in mind as they embark on this journey.

First, cities must prioritize the needs and priorities of their residents. The development of smart cities must be driven by a deep understanding of what citizens want and need, and it must be guided by their needs and aspirations. This requires meaningful engagement with residents and the creation of opportunities for citizens to participate in decision-making processes.

Second, cities must ensure that the development of smart cities is guided by principles of transparency, accountability, and inclusiveness. The collection and use of data must be done in a way that is transparent and that protects the privacy of citizens. The decisions about the allocation of resources and services must be made in a way that is accountable and that is based on a clear understanding of the needs of the community.

Third, cities must embrace a holistic and integrated approach to urban planning and governance. The development of smart cities must be integrated with other urban planning and governance initiatives, such as affordable housing and inclusive economic development, in order to create a more holistic and integrated approach to city-building.

Fourth, cities must embrace new technologies and innovative approaches in order to stay ahead of the curve and to continue to improve the quality of life for citizens. This requires a commitment to ongoing learning and development, as well as a willingness to experiment and take calculated risks.

Finally, cities must work together to share best practices and to collaborate on the development of smart cities. This requires the creation of networks and partnerships between cities, as well as the sharing of information and expertise.

Embracing smart cities is about much more than simply incorporating technology into urban environments. It requires a deep commitment to understanding the needs and priorities of citizens, to transparency and accountability, to a holistic and integrated approach to urban planning and governance, to embracing new technologies and innovative approaches, and to collaboration and partnership.

Summary: Embracing smart cities requires a deep commitment to understanding the needs and priorities of citizens, transparency, accountability, a holistic and integrated approach to urban planning and governance, embracing new technologies and innovative approaches, and collaboration and partnership. The development of smart cities must be guided by principles of transparency, accountability, and inclusiveness, with the collection and use of data done in a way that is transparent and that protects the privacy of citizens. The decisions about the allocation of resources and services must be made in a way that is accountable and based on a clear understanding of the needs of the community. Additionally, it requires the creation of networks and partnerships between cities, as well as the sharing of information and expertise.

Conclusion: As trusted and knowledgeable experts in the field of smart cities, it is our responsibility to guide cities on this journey and to ensure that they are able to fully embrace the opportunities and benefits of smart cities. By doing so, we can help to create a future in which technology is used to improve the quality of life for all citizens, and in which cities are more sustainable, efficient, and inclusive. Ultimately, embracing smart cities is about more than just incorporating technology into urban environments, it's about creating a better future for all.

Quote: "Embracing smart cities is not just about incorporating technology into urban environments. It requires a deep understanding of the needs and aspirations of citizens, transparency, accountability, a holistic approach to urban planning and governance, embracing new technologies, and collaboration. Only then can we create a sustainable and efficient future for all." Emin Hasic

ONE SMART WORLD
ARE YOU READY FOR IT?

| vi |
FAILING TO EMBRACE SMART CITIES

The concept of smart cities has been around for several years, and it has been increasingly gaining traction. Smart cities are urban areas that utilize advanced technology and data to improve the quality of life for citizens, enhance the environment, and promote economic growth. Despite the numerous benefits that smart cities offer, some cities have been reluctant to embrace the concept, and this failure to adopt smart city technology has hindered progress.

THE BARRIERS

One of the main reasons why some cities have been reluctant to embrace smart city technology is the cost of implementation. Implementing smart city technology requires a significant investment in infrastructure, hardware, and software. This can be a significant hurdle for many cities, particularly those with limited budgets or other financial constraints.

Another barrier to embracing smart city technology is a lack of technical expertise. Implementing smart city technology requires specialized technical skills that may not be readily available in all cities. Many cities lack the expertise necessary to deploy and manage smart city infrastructure, which can be a significant barrier to adoption.

In addition, there is often a lack of political will to implement smart city technology. Some policymakers may not fully understand the benefits of smart city technology or may not see it as a priority. This can lead to a lack of investment in smart city infrastructure and a failure to fully embrace the concept of smart cities.

THE NEGATIVE IMPACT

The failure to embrace the concept of smart cities can have a negative impact on citizens and the economy. Smart city technology has the potential to improve the quality of life for citizens, enhance the environment, and promote economic growth. Without the benefits of smart city technology, cities may struggle to attract new businesses, retain residents, and compete in the global economy.

One of the most significant benefits of smart city technology is the ability to improve public safety. Smart city technology can help cities to identify and respond to potential safety threats, such as natural disasters or terrorist attacks. It can also help to reduce crime and improve emergency response times, which can save lives and protect citizens.

Smart city technology can also improve transportation systems, reducing traffic congestion and improving mobility. This can help to reduce the time and cost of commuting, which can improve quality of life and promote economic growth. In addition, smart city technology can help to reduce energy consumption and greenhouse gas emissions, promoting sustainability and environmental stewardship.

START SMALL

Despite the barriers to adopting smart city technology, there are several steps that cities can take to overcome these challenges and embrace the concept of smart cities.

One approach is to start small, implementing pilot projects to test the feasibility and potential benefits of smart city technology. This can help to build momentum and demonstrate the benefits of smart city technology to policymakers and citizens.

Cities can also leverage public-private partnerships to help finance and implement smart city technology. By working with private companies and investors, cities can tap into additional resources and expertise that may not be available through traditional government channels.

Cities can invest in developing the necessary technical expertise to deploy and manage smart city technology. This may involve investing in training programs or partnering with universities and research institutions to develop the necessary skills.

It is important for cities to recognize the potential benefits of smart city technology and prioritize its adoption. Failure to do so can result in falling behind other cities and regions that have embraced smart city technology, reducing their ability to attract new businesses, retain residents, and compete in the global economy.

Summary: *Smart cities are urban areas that use advanced technology and data to improve the quality of life for citizens, enhance the environment, and promote economic growth. However, some cities are reluctant to adopt smart city technology due to barriers such as the cost of implementation, lack of technical expertise, and a lack of political will. The failure to embrace smart city technology can have a negative impact on citizens and the economy, including reduced ability to attract businesses and retain residents. Smart city technology can improve public safety, transportation systems, energy consumption, and greenhouse gas emissions. Cities can overcome these barriers by starting small with pilot projects, leveraging public-private partnerships, and investing in the necessary technical expertise.*

Conclusion: *The concept of smart cities has the potential to greatly benefit citizens and the economy. However, the barriers to adopting smart city technology, such as cost, lack of technical expertise, and a lack* *of political will, can hinder progress. Cities must recognize the potential benefits of smart city technology and prioritize its adoption to avoid falling behind other cities and regions. By starting small, leveraging public-private partnerships, and investing in technical expertise, cities can overcome these barriers and fully embrace the concept of smart cities. This will improve the quality of life for citizens, enhance the environment, and promote economic growth.*

Quote: *" Any intelligent fool can make things bigger and more complex, but it takes a touch of genius and a lot of courage to move in the opposite direction. Embracing the concept of smart cities may present challenges, but the potential benefits to improve the quality of life for citizens, enhance the environment, and promote economic growth make it worth pursuing. Let us not be afraid to start small and invest in the necessary technical expertise, for it is through these efforts that we can create a better world for ourselves and future generations." Emin Hasic*

ONE SMART WORLD
ARE YOU READY FOR IT?

| vii |
PITFALLS

Smart cities, use technology and data to improve urban services and enhance the quality of life for their residents, have become an increasingly popular concept in recent years. The promise of smart cities is to use technology to create more efficient, sustainable, and livable urban environments. However, as with any new concept, there are potential pitfalls that must be addressed in order to fully realize the benefits of smart cities. We explore 50 potential pitfalls of smart cities.

50 PITFALLS TO LOOK OUT FOR WHEN SETTING UP A SMART CITY PROGRAM

1. **Inadequate stakeholder engagement and community participation**. This can result in the smart city program not meeting the needs and preferences of the community, leading to a lack of support or even opposition.
2. **Insufficient attention to data privacy and security**. Without proper safeguards in place, sensitive information could be vulnerable to cyberattacks and misuse, eroding public trust.
3. **Failure to address the digital divide and ensure equitable access to technology**. This can result in disparities in access to services and opportunities, exacerbating existing inequalities.
4. **Lack of interoperability and data standardization**. This can make it difficult to integrate data and systems across different departments

and agencies, reducing the effectiveness and efficiency of the smart city program.
5. **Poorly planned and implemented infrastructure and technology.** This can result in suboptimal performance, high costs, and low user adoption rates.
6. **Over-reliance on technology without considering the human element.** This can result in technology solutions that do not meet the needs of users or take into account the broader social and environmental context.
7. **Lack of a clear business model and financial sustainability plan.** This can result in a failure to attract private investment or secure public funding, leading to financial challenges and program cuts.
8. **Ignoring potential negative social and environmental impacts.** Smart city programs may have unintended consequences that can harm the environment or negatively impact the community if not properly addressed.
9. **Failure to address legal and regulatory barriers.** Smart city programs may be subject to various laws and regulations that can limit their effectiveness or pose legal risks if not properly addressed.
10. **Ineffective governance and lack of leadership.** Poor governance can lead to a lack of accountability, confusion, and inefficiencies, hindering the success of the smart city program.
11. **Lack of transparency and public trust.** Without transparency, the public may lack trust in the program, which can harm its effectiveness and sustainability.
12. **Lack of scalability and adaptability.** The smart city program should be designed with scalability and adaptability in mind, to ensure that it can evolve with changing technologies and community needs.
13. **Poor data management and analysis.** Effective data management and analysis are critical to the success of a smart city program, allowing for insights and informed decision-making.
14. **Insufficient focus on the user experience.** Without considering the user experience, smart city solutions may not be intuitive or user-friendly, reducing their effectiveness and adoption rates.
15. **Lack of interdepartmental collaboration.** The success of a smart city program depends on collaboration across departments and agencies, to ensure a comprehensive and coordinated approach.
16. **Failure to consider the ethical implications of smart city technologies.** Smart city programs must consider ethical implications such as data privacy, bias, and autonomy to ensure that they are not violating individual rights.

17. **Inadequate planning for emergencies and disasters**. Smart city programs should have contingency plans in place to respond to emergencies and disasters effectively.
18. **Poor community outreach and education**. Without adequate community outreach and education, the public may not understand the benefits of the smart city program, leading to lack of support.
19. **Lack of focus on sustainability and resilience**. Smart city programs should prioritize sustainability and resilience, to ensure that they can be maintained and adapted over time.
20. **Failure to account for cultural differences**. Smart city programs must be sensitive to cultural differences, to ensure that they are inclusive and effective across diverse communities.
21. **Insufficient attention to the environmental impact of technology**. Smart city programs should consider the environmental impact of technology solutions, to ensure that they are not contributing to environmental degradation.
22. **Inadequate measurement and evaluation**. A lack of measurement and evaluation can make it difficult to assess the effectiveness and impact of the smart city program.
23. **Failure to consider the needs of vulnerable populations**. Smart city programs must be designed to meet the needs of vulnerable populations, such as the elderly, disabled, and low-income communities.
24. **Lack of a comprehensive data governance framework.** The smart city program should have a comprehensive data governance framework in place to ensure the responsible collection, storage, and use of data.
25. **Inability to demonstrate the value and benefits of the smart city program**. The smart city program should be able to demonstrate the value and benefits it provides to the community, to ensure continued support and funding.
26. **Failure to involve the private sector.** The smart city program should engage and collaborate with private sector partners, such as technology companies and infrastructure providers, to ensure a comprehensive and effective approach.
27. **Inadequate consideration of the regulatory landscape.** Smart city programs may face a complex regulatory landscape that can impact their design and implementation, requiring careful consideration and planning.

28. **Lack of focus on inclusivity and diversity.** Smart city programs should prioritize inclusivity and diversity, to ensure that they are accessible and beneficial to all members of the community.
29. **Failure to address potential job displacement.** Smart city programs may have the unintended consequence of displacing jobs or changing the nature of work, requiring careful consideration and planning to mitigate negative impacts.
30. **Lack of community ownership and engagement.** Smart city programs should engage and involve the community in the design, implementation, and evaluation of the program, to ensure that it is responsive to their needs and priorities.
31. **Insufficient attention to cybersecurity.** Smart city programs are vulnerable to cyber-attacks, making it crucial to prioritize and invest in robust cybersecurity measures.
32. **Inadequate consideration of privacy concerns.** Smart city programs collect and use vast amounts of data, raising concerns about individual privacy and requiring thoughtful consideration and management of data privacy risks.
33. **Overreliance on technology.** Smart city programs should avoid overreliance on technology, and instead ensure that technological solutions are integrated with human expertise, to ensure that programs are effective and responsive to community needs.
34. **Insufficient long-term planning.** Smart city programs require long-term planning and investment, and should be designed with a clear understanding of future needs and challenges, to ensure that they remain effective and sustainable.
35. **Inadequate focus on environmental sustainability.** Smart city programs should be designed with a focus on environmental sustainability, to ensure that they contribute to long-term environmental goals and reduce the carbon footprint of the community.
36. **Lack of interoperability.** Smart city programs often rely on a range of disparate technologies and systems, making it important to ensure that they are interoperable and can communicate with each other effectively.
37. **Insufficient consideration of physical infrastructure.** Smart city programs should be designed with a clear understanding of the physical infrastructure required to support them, such as fiberoptic cables and wireless networks.
38. **Inadequate consideration of the social infrastructure.** Smart city programs should be designed with a clear understanding of the

social infrastructure required to support them, such as community engagement and education.
39. **Overreliance on centralized control.** Smart city programs should avoid overreliance on centralized control and instead prioritize decentralization and collaboration, to ensure that programs are responsive and adaptive to community needs.
40. **Insufficient attention to public safety.** Smart city programs should be designed with a focus on public safety, to ensure that they do not compromise the safety of community members or the overall security of the city.
41. **Lack of scalability.** Smart city programs should be designed with a clear understanding of scalability, to ensure that they can be expanded or replicated as needed, and to avoid the costs and complexities associated with reinventing the wheel.
42. **Inadequate monitoring and evaluation.** Smart city programs should be subject to ongoing monitoring and evaluation, to ensure that they are effective and responsive to community needs, and to identify opportunities for improvement.
43. **Insufficient investment in skills development.** Smart city programs often require new skills and expertise, making it important to invest in skills development and training for community members, workers, and other stakeholders.
44. **Failure to balance short-term and long-term needs.** Smart city programs should be designed with a clear understanding of both short-term and long-term needs, to ensure that they are effective and sustainable over time.
45. **Lack of flexibility and adaptability.** Smart city programs should be designed to be flexible and adaptable, to ensure that they can respond to changing community needs and priorities, and to avoid becoming obsolete.
46. **Insufficient consideration of cultural factors.** Smart city programs should be designed with a clear understanding of cultural factors, to ensure that they are compatible with community values, traditions, and practices.
47. **Failure to address ethical concerns.** Smart city programs may raise ethical concerns related to data privacy, equity, and access, making it important to ensure that they are designed and implemented with clear ethical principles.
48. **Inadequate collaboration and coordination between stakeholders.** Smart city programs involve multiple stakeholders, including government agencies, community members, and private sector

partners, making it important to prioritize collaboration and coordination to avoid duplication of effort and conflicts of interest.
49. **Failure to leverage open data.** Smart city programs should leverage open data, to ensure that data is available to the community and to foster innovation and collaboration among stakeholders.
50. **Lack of prioritization of citizen engagement.** Smart city programs should prioritize citizen engagement, to ensure that the community is involved in decision-making and that the program is designed to meet their needs and priorities.

Summary: *Smart cities are a concept that seeks to leverage technology and data to create more efficient, sustainable, and livable urban environments. However, there are potential pitfalls associated with the implementation of smart city technologies. In this article, we have explored 50 of these potential pitfalls, ranging from issues related to data privacy and security, to concerns about equity and accessibility. By identifying these challenges, we hope to promote a more thoughtful and deliberate approach to the development of smart cities, one that takes into account the potential risks as well as the potential benefits.*

Conclusion: *Smart cities have the potential to revolutionize urban life, but realizing this potential requires careful consideration of the potential pitfalls associated with these technologies. In this article, we have explored 50 potential challenges associated with the implementation of smart city technologies. While some of these challenges may be difficult to overcome, many can be mitigated through careful planning, collaboration, and community engagement. By addressing these challenges head-on, we can create smart cities that are not only efficient and sustainable, but also equitable and accessible, and that improve the quality of life for all residents.*

Quote: *"Technological progress is not synonymous with human progress. While the promise of smart cities is to improve the efficiency and sustainability of urban environments, we must not forget to consider the potential pitfalls that may arise. Only through careful consideration and collaboration can we create truly smart cities that enhance the quality of life for all residents." Emin Hasic*

| viii |
BIG DATA

Smart cities are a relatively new concept in urban planning and development that has gained significant attention in recent years. The idea is to use digital technologies to create more efficient and sustainable cities, with the ultimate goal of improving the quality of life for citizens. The term "smart city" refers to a city that is connected to the internet of things, which enables various devices and systems to communicate with each other and collect data. This data can then be analyzed to provide insights into the functioning of the city and to identify areas for improvement.

As the world becomes more digitally connected, cities are expected to adopt and implement new technologies to keep up with the changing times. Big data is one of the most critical components of a smart city, as it allows for the collection, analysis, and interpretation of vast amounts of data. This, in turn, can be used to make informed decisions about city planning, transportation, public safety, healthcare, and energy consumption, among other things.

BIG DATA IN SMART CITIES

In the past few years, the amount of data being generated has grown exponentially, and it is expected to continue increasing at an unprecedented rate. This is due to the widespread use of digital devices such as smartphones, laptops, and tablets, which are now an integral part of daily life for millions of people worldwide. In addition, the internet of things (IoT) has led to the proliferation of internet-connected devices, ranging from smart homes to smart cars and even smart cities. These devices collect and transmit data, creating a massive amount of data that can be used for various purposes, including urban planning and development.

In smart cities, big data is collected from various sources, including sensors, cameras, and other internet-connected devices. For example, traffic sensors can provide real-time information about traffic flow and congestion, while weather sensors can provide information about weather patterns and conditions. This data can be used to optimize the functioning of the city, such as improving traffic flow, managing energy consumption, and reducing waste.

Another source of data is social media. By analyzing social media data, cities can gain valuable insights into the needs and preferences of their citizens. For example, by monitoring social media activity during a public event, city officials can better understand the type of services and infrastructure needed to make the event more successful. Social media data can also be used to track the spread of diseases, allowing officials to take appropriate measures to prevent further outbreaks.

Once the data is collected, it must be processed and analyzed. This involves using various analytical tools and techniques to extract meaningful insights from the data. For example, machine learning algorithms can be used to identify patterns and trends in the data, while predictive modeling can be used to make predictions about future events or trends. These insights can then be used to make informed decisions about city planning and development.

The growth in digital devices and the internet of things has led to a massive increase in the amount of data being generated. In smart cities, big data is collected from various sources, including sensors, cameras, and social media, and is then processed and analyzed to provide insights into the functioning of the city. This data can be used to improve the quality of life for citizens by optimizing traffic flow, managing energy consumption, and reducing waste, among other things.

The use of big data in smart cities spans across different sectors and industries, with various applications in transportation, healthcare, energy, and public safety.

In transportation, data collected from sensors, cameras, and other sources can be used to optimize traffic flow and reduce congestion. Real-time data can also be used to improve public transportation systems, such as predicting bus or train arrival times, optimizing routes, and scheduling maintenance.

In healthcare, big data can be used to track the spread of diseases, monitor public health trends, and predict future outbreaks. By analyzing data from various sources, including social media and public health records, cities can identify potential hotspots for diseases and take proactive measures to prevent the spread of infections. Additionally, big data can be used to develop personalized healthcare plans, improve patient outcomes, and reduce healthcare costs.

In energy, big data can be used to optimize the use of resources and reduce waste. By collecting and analyzing data from smart grids and other sources, cities can identify patterns in energy usage, detect inefficiencies, and adjust energy distribution accordingly. This can lead to significant energy savings, reduced carbon emissions, and a more sustainable future.

In public safety, big data can be used to monitor crime and improve emergency response times. By analyzing data from surveillance cameras, social media, and other sources, cities can identify potential threats, predict crime hotspots, and respond to emergencies more effectively. However, the use of big data in public safety raises concerns about privacy and data security. Citizens' personal data must be protected from unauthorized access or misuse, and the use of facial recognition technology, for example, should be closely monitored and regulated.

Summary: Big data is critical in the development of smart cities, providing insights into various aspects of urban planning and development. It's used in transportation, healthcare, energy, and public safety. However, privacy concerns and data security are significant challenges facing the use of big data. Citizens' personal data must be protected, and data ownership and access must be addressed.

Conclusion: The use of big data in smart cities is essential to improve the lives of citizens. It provides valuable insights into various aspects of urban planning and development and helps to create more efficient and sustainable cities. However, it's crucial to address the challenges facing the use of big data, such as privacy and security concerns. Smart cities must prioritize the protection of citizens' personal data and ensure that they have access to the data generated.

Quote: "Data is not just a collection of numbers and figures, but rather a key to unlocking the mysteries of the universe. In the realm of smart cities, big data holds the potential to unravel the complexities of urban life and bring order to chaos." Emin Hasic

ONE SMART WORLD
ARE YOU READY FOR IT?

| ix |
PROJECTS BY COUNTRY

As cities around the world continue to grow in population, the need for efficient and sustainable urban development has become increasingly pressing. Smart City projects offer innovative solutions to improve the quality of life for urban citizens while also reducing the environmental impact of urbanization. While Smart City projects can take many different forms, they generally involve the integration of advanced technologies, data analytics, and citizen engagement to improve the delivery of public services and infrastructure. Following is a short overview of Smart City projects from around the world, with a focus on the initiatives that have been launched in different countries.

AUSTRALIA

1. **Smart Gold Coast:** An initiative to make the Gold Coast a smart city by integrating technology to improve sustainability, safety, and livability.
2. **Smart Adelaide:** A project aimed at using data and technology to enhance economic growth, social and environmental sustainability, and quality of life in Adelaide.
3. **Digital Twin Victoria:** A project that creates a virtual replica of Victoria's infrastructure, allowing real-time monitoring, simulations and predictive analysis of events and trends.
4. **Darwin Smart City:** A project that involves integrating technology into Darwin's urban infrastructure, such as smart streetlights and waste management systems.

5. **Perth Smart City:** A project aimed at making Perth a more connected and sustainable city through the use of technology, data and innovation.
6. **Smart Hobart:** A project that uses technology to enhance Hobart's livability, sustainability, and economic growth.
7. **Sydney Water Smart:** A project that uses smart technology to optimize Sydney's water supply and improve the efficiency of its water system.
8. **Sunshine Coast Smart City:** A project aimed at enhancing the Sunshine Coast's livability, sustainability, and economic growth through the use of technology.
9. **City of Newcastle Smart City Strategy:** A project that focuses on creating a smart city that enhances the quality of life for residents, visitors and businesses.
10. **Smart Western Sydney:** A project aimed at creating a more connected and sustainable Western Sydney through the use of technology, data and innovation.

AUSTRIA

1. **Intelligent Traffic Management System, Vienna:** Implementing an intelligent traffic management system in the capital city of Vienna to improve traffic flow, reduce congestion and improve road safety.
2. **Smart Energy Management System, Graz:** Implementing a smart energy management system in the city of Graz to optimize energy usage and reduce costs.
3. **Smart Lighting, Linz:** Implementing smart lighting systems in the city of Linz to reduce energy consumption and improve public safety.
4. **E-Governance, Nationwide:** Implementing e-governance systems to provide citizens with online access to government services and information.
5. **Waste Management, Nationwide:** Implementing smart waste management systems to improve waste collection, recycling and disposal.
6. **Smart Grid, Nationwide:** Developing a smart grid system to improve energy distribution and reliability.
7. **Smart Health, Nationwide:** Implementing smart health systems to improve access to healthcare and reduce healthcare costs.

8. **Smart Agriculture, Nationwide:** Implementing smart agriculture systems to improve crop yields, reduce waste and increase sustainability.
9. **Intelligent Water Management, Salzburg:** Implementing an intelligent water management system in the city of Salzburg to optimize water usage and reduce waste.
10. **Smart Parks, Nationwide:** Developing smart park systems to improve park management, reduce waste and increase sustainability.

BELARUS

1. **Minsk Smart City:** This project aims to create a smart city in Minsk, the capital of Belarus, by implementing innovative solutions for transportation, energy, and waste management. The project aims to improve the quality of life of the citizens and create a more sustainable city.
2. **Smart Transportation for Belarus:** This project aims to improve the transportation system in Belarus, by implementing smart solutions for public transportation and reducing the use of private cars. The project aims to create a more sustainable and efficient transportation system.
3. **Green Energy for Belarus:** This project aims to increase the use of renewable energy sources in Belarus, such as solar and wind energy, in order to reduce the country's dependence on fossil fuels and improve energy efficiency.
4. **Smart Water Management for Belarus:** This project aims to improve the water management system in Belarus, by implementing smart solutions for water distribution, treatment, and reuse. The project aims to create a more sustainable and efficient water management system.
5. **Sustainable Tourism for Belarus:** This project aims to improve the tourism industry in Belarus, by implementing sustainable solutions for transportation, energy, and waste management. The project aims to enhance the quality of tourism services and create a more sustainable and livable environment for the citizens and visitors.
6. **Intelligent Building Management System:** This project aims to implement smart solutions for heating, cooling, and lighting in buildings across Belarus, in order to reduce energy consumption and improve comfort levels.

7. **Smart Grid for Belarus:** This project aims to upgrade the electrical grid in Belarus, by implementing smart solutions for energy distribution and management, to improve reliability and efficiency.
8. **E-Government for Belarus:** This project aims to improve the delivery of government services to the citizens of Belarus, by implementing electronic services and portals, to make it easier for citizens to access government services and information.
9. **Intelligent Traffic Management System:** This project aims to improve the traffic management system in Belarus, by implementing smart solutions for traffic control, monitoring, and prediction, to reduce congestion and improve traffic flow.
10. **Environmental Monitoring for Belarus:** This project aims to monitor and analyze the environmental impact of various activities in Belarus, by implementing smart solutions for data collection, analysis, and reporting, to ensure a healthy and sustainable environment.

BELGIUM

1. **Brussels Smart City:** Brussels Smart City is a program aimed at making Brussels a more sustainable and livable city through the use of technology. The initiative focuses on areas such as mobility, energy, and environmental sustainability.
2. **Antwerp Smart Zone:** Antwerp Smart Zone is a project aimed at creating a smart, sustainable and innovative zone in the city of Antwerp. The initiative focuses on areas such as mobility, energy, and urban innovation.
3. **Ghent Smart City:** Ghent Smart City is a program aimed at using technology to create a more sustainable, livable and inclusive city. The initiative focuses on areas such as mobility, energy, and social innovation.
4. **Liège Smart City:** Liège Smart City is a project aimed at using technology to improve the quality of life for citizens and promote economic growth in the city of Liège. The initiative focuses on areas such as mobility, energy, and urban innovation.
5. **Ostend Smart City:** Ostend Smart City is a program aimed at creating a sustainable and innovative city that is resilient to climate change. The initiative focuses on areas such as mobility, energy, and environmental sustainability.
6. **Smart Flanders:** Smart Flanders is a program aimed at creating a more innovative, sustainable and prosperous Flanders region. The

initiative focuses on areas such as mobility, energy, and social innovation.
7. **Leuven 2030:** Leuven 2030 is a program aimed at making the city of Leuven a more sustainable and livable place by 2030. The initiative focuses on areas such as mobility, energy, and environmental sustainability.
8. **Mechelen Smart City:** Mechelen Smart City is a project aimed at using technology to create a more sustainable and innovative city in the province of Antwerp. The initiative focuses on areas such as mobility, energy, and urban innovation.
9. **Charleroi Smart City:** Charleroi Smart City is a program aimed at using technology to create a more sustainable and inclusive city in the province of Hainaut. The initiative focuses on areas such as mobility, energy, and social innovation.
10. **Hasselt Smart City:** Hasselt Smart City is a program aimed at using technology to create a more sustainable, livable, and innovative city in the province of Limburg. The initiative focuses on areas such as mobility, energy, and environmental sustainability.

BRAZIL

1. **Porto Maravilha:** Located in Rio de Janeiro, this project aims to revitalize the city's port area, using technology to improve the quality of life for residents and visitors. It includes smart parking, public Wi-Fi, and an intelligent lighting system.
2. **Smart City Laguna:** Located in the city of São Gonçalo do Amarante, this project is focused on sustainable living, using renewable energy and innovative transportation solutions. It features a smart grid, electric transportation, and a waste management system.
3. **Curitiba:** This city in southern Brazil is known for its innovative public transportation system, which includes a dedicated bus lane and real-time information displays. It has also implemented smart parking, smart lighting, and a digital library system.
4. **Santos:** This coastal city has implemented a smart parking system, allowing residents to find available parking spots in real-time using a mobile app. It also features a smart bus system and an intelligent lighting system.
5. **Rio de Janeiro Operations Center:** This is a command center that uses technology to monitor the city's infrastructure and respond to

emergencies in real-time. It collects data from various sensors and cameras around the city, allowing for quick and effective responses.
6. **São Paulo Smart City:** This project aims to improve transportation, public safety, and environmental sustainability in São Paulo, Brazil's largest city. It includes smart traffic management, a bike-sharing system, and an intelligent lighting system.
7. **Belo Horizonte Smart City:** This project focuses on sustainability, using renewable energy and promoting green initiatives. It features a smart grid, a waste management system, and a water conservation program.
8. **Brasília:** The capital city of Brazil has implemented a smart traffic management system, using real-time data to improve traffic flow and reduce congestion. It also features a smart lighting system and a mobile app that provides information on city services.
9. **Digital Port:** Located in the city of Vitória, this project aims to create a digital hub for innovation and entrepreneurship. It includes a smart city lab, a co-working space, and a startup accelerator.
10. **Parque Cidade:** This smart city project in Brasília focuses on sustainability and social inclusion. It includes a smart grid, a waste management system, and a mobile app that provides information on city services. It also features a community center and a public park.

BULGARIA

1. **Smart City Plovdiv:** This project aims to create a smart city in Plovdiv, the second-largest city in Bulgaria, by implementing solutions for transportation, energy, and waste management. The project aims to create a more livable and sustainable city for its residents.
2. **Smart Sofia:** This project aims to create a smart city in Sofia, the capital of Bulgaria, by implementing innovative solutions for transportation, energy, and waste management. The project aims to improve the quality of life of the citizens and create a more sustainable city.
3. **Digital Transformation for Bulgaria:** This project aims to improve the digital infrastructure in Bulgaria, by implementing smart solutions for communication, transportation, and energy. The project aims to create a more connected and digital country, and to enhance the quality of life of its citizens.
4. **Green Energy for Bulgaria:** This project aims to increase the use of renewable energy sources in Bulgaria, such as solar and wind energy,

in order to reduce the country's dependence on fossil fuels and improve energy efficiency.

5. **Sustainable Transportation for Bulgaria:** This project aims to improve the transportation system in Bulgaria, by implementing smart solutions for public transportation and reducing the use of private cars. The project aims to create a more sustainable and efficient transportation system.
6. **Smart Waste Management for Bulgaria:** This project aims to improve the waste management system in Bulgaria, by implementing smart solutions for waste collection and recycling. The project aims to create a cleaner and more sustainable environment for the citizens.
7. **Smart Lighting for Bulgaria:** This project aims to improve the lighting system in Bulgaria, by implementing smart and energy-efficient lighting solutions. The project aims to reduce energy consumption and improve the quality of life of the citizens.
8. **Sustainable Urban Planning for Bulgaria:** This project aims to improve the urban planning of Bulgaria, by implementing sustainable solutions for land use, transportation, and waste management. The project aims to create a more livable and sustainable city for its residents.
9. **Smart Water Management for Bulgaria:** This project aims to improve the water management system in Bulgaria, by implementing smart solutions for water distribution, treatment, and reuse. The project aims to create a more sustainable and efficient water management system.
10. **Smart Healthcare for Bulgaria:** This project aims to improve the healthcare system in Bulgaria, by implementing smart solutions for patient care and medical treatment. The project aims to enhance the quality of healthcare services and improve the health of the citizens.

CANADA

1. **Waterfront Toronto's Quayside Project:** Waterfront Toronto is working with Google's sister company Sidewalk Labs to develop a smart city district in Quayside, Toronto. The project aims to incorporate innovative technologies and sustainable infrastructure to create a connected, data-driven and human-centered urban environment.

2. **Smart Traffic Management in Edmonton:** Edmonton has implemented a smart traffic management system that uses real-time data and intelligent transportation systems to optimize traffic flow, reduce congestion and improve safety. The system uses connected vehicle technology and AI algorithms to predict traffic patterns and adjust traffic signals accordingly.
3. **Montreal's Underground City:** Montreal's Underground City is an extensive network of subterranean pedestrian walkways and shopping centers that spans over 30 kilometers. The city is developing smart technologies that will allow the Underground City to be used for community events, provide emergency services, and enhance visitor experiences.
4. **Brampton's Innovation District:** Brampton's Innovation District is a collaboration between the city, local businesses, and academic institutions to create a hub for innovation, technology, and entrepreneurship. The district includes the creation of a smart transportation network, with electric and autonomous vehicles, as well as smart infrastructure such as energy-efficient buildings.
5. **Calgary's Smart Park:** Calgary's Smart Park is a parking management system that uses smart sensors to detect vehicle presence and adjust pricing based on demand. The system allows drivers to find available parking spots and make payments via mobile app, reducing congestion and improving traffic flow.
6. **Smart Energy in Vancouver:** Vancouver is developing a smart energy system that incorporates renewable energy sources and integrates energy storage, demand response, and real-time data analytics. The system aims to reduce energy consumption, lower greenhouse gas emissions and provide reliable and affordable energy to residents and businesses.
7. **Halifax's SmartCity Initiative:** Halifax's SmartCity Initiative is a multi-year project that involves the development of innovative smart city technologies to enhance public services, promote economic growth, and improve quality of life for residents. The initiative includes the implementation of smart transportation solutions, the use of connected devices for public safety, and the creation of an open data platform.
8. **Ottawa's Smart Farming:** Ottawa is developing a smart farming system that uses sensors and data analytics to optimize crop production, reduce waste, and enhance environmental sustainability. The system allows farmers to monitor soil conditions, irrigation, and

weather patterns to make data-driven decisions about planting and harvesting.
9. **The Connected North in Canada's Northern Communities:** The Connected North is a project that uses smart technologies to connect remote communities in Canada's north to essential services, education, and cultural resources. The project includes the deployment of videoconferencing, remote learning, and telemedicine services to improve access to healthcare and education.
10. **Niagara Falls' Smart City Strategy:** Niagara Falls' Smart City Strategy aims to transform the city into a connected and sustainable urban environment by implementing smart technologies in transportation, energy, and public services. The strategy includes the development of a smart transportation network, the use of smart grid technology to optimize energy use, and the implementation of an open data platform for public access to city information.

COSTA RICA

1. **Smart San Jose:** This project aims to create a smart city in San Jose, the capital of Costa Rica, by implementing innovative solutions for transportation, energy, and waste management. The project aims to improve the quality of life of the citizens and create a more sustainable city.
2. **Sustainable Transportation for Costa Rica:** This project aims to improve the transportation system in Costa Rica, by implementing smart solutions for public transportation and reducing the use of private cars. The project aims to create a more sustainable and efficient transportation system.
3. **Green Energy for Costa Rica:** This project aims to increase the use of renewable energy sources in Costa Rica, such as solar and wind energy, in order to reduce the country's dependence on fossil fuels and improve energy efficiency.
4. **Smart Water Management for Costa Rica:** This project aims to improve the water management system in Costa Rica, by implementing smart solutions for water distribution, treatment, and reuse. The project aims to create a more sustainable and efficient water management system.
5. **Sustainable Tourism for Costa Rica:** This project aims to improve the tourism industry in Costa Rica, by implementing sustainable solutions for transportation, energy, and waste management. The

project aims to enhance the quality of tourism services and create a more sustainable and livable environment for the citizens and visitors.

6. **Digital Transformation for Costa Rica:** This project aims to improve the digital infrastructure in Costa Rica, by implementing smart solutions for communication, transportation, and energy. The project aims to create a more connected and digital country, and to enhance the quality of life of its citizens.
7. **Smart Waste Management for Costa Rica:** This project aims to improve the waste management system in Costa Rica, by implementing innovative solutions for waste collection, treatment, and recycling. The project aims to create a more sustainable and efficient waste management system.
8. **Smart Buildings for Costa Rica:** This project aims to improve the energy efficiency of buildings in Costa Rica, by implementing smart solutions for heating, cooling, and lighting. The project aims to reduce the energy consumption of buildings and create a more sustainable and livable environment.
9. **Smart Healthcare for Costa Rica:** This project aims to improve the healthcare system in Costa Rica, by implementing innovative solutions for health monitoring, diagnosis, and treatment. The project aims to enhance the quality of healthcare services and create a more sustainable and livable environment for the citizens.
10. **Smart Agriculture for Costa Rica:** This project aims to improve the agriculture industry in Costa Rica, by implementing smart solutions for water management, crop monitoring, and yield optimization. The project aims to enhance the quality of agriculture services and create a more sustainable and livable environment for the citizens.

CROATIA

1. **Smart Street Lights in Split:** In 2019, the city of Split installed smart streetlights that use sensors to detect human movement and adjust the lighting accordingly. This has resulted in significant energy savings and increased safety in the city.
2. **Smart Waste Management in Zagreb:** The city of Zagreb has implemented a smart waste management system that uses sensors to monitor waste levels in containers and optimize waste collection routes. This has reduced the number of garbage trucks on the road, resulting in cost savings and a reduction in carbon emissions.

3. **Smart Parking in Osijek:** The city of Osijek has implemented a smart parking system that uses sensors to detect available parking spaces and provide real-time information to drivers via a mobile app. This has reduced traffic congestion and made it easier for drivers to find parking.
4. **Green Energy in Dubrovnik:** The city of Dubrovnik has implemented a number of green energy initiatives, including the installation of solar panels on public buildings and the use of electric vehicles for public transportation. These initiatives have reduced the city's carbon footprint and contributed to a more sustainable future.
5. **Digital City Platform in Rijeka:** The city of Rijeka has developed a digital platform that provides citizens with access to a range of city services, including public transportation, parking, and waste management. This has made it easier for citizens to access city services and has improved overall quality of life in the city.
6. **Smart City Pilot in Pula:** The city of Pula has implemented a pilot project for a smart city that includes smart street lighting, smart waste management, and a mobile app for citizens to access city services. The pilot has been successful in reducing energy consumption and increasing citizen engagement with city services.
7. **Smart Traffic Management in Zagreb:** The city of Zagreb has implemented a smart traffic management system that uses real-time data from traffic sensors to optimize traffic flow and reduce congestion. This has resulted in reduced travel times and improved air quality in the city.
8. **Smart Building Management in Split:** The city of Split has implemented a smart building management system in public buildings that uses sensors to monitor energy use and optimize building operations. This has resulted in significant energy savings and a reduction in greenhouse gas emissions.
9. **Smart Waste Separation in Varaždin:** The city of Varaždin has implemented a smart waste separation system that uses sensors to detect the type of waste being disposed of and guide citizens on how to properly sort their waste. This has increased recycling rates and reduced the amount of waste sent to landfills.
10. **Smart Public Transportation in Rijeka:** The city of Rijeka has implemented a smart public transportation system that uses real-time data to optimize bus routes and schedules. This has resulted in reduced waiting times and increased ridership, making public transportation a more convenient and attractive option for citizens.

DENMARK

1. **Copenhagen Connecting:** This project aims to make Copenhagen a smarter and more sustainable city by integrating digital solutions across transportation, energy, and the environment.
2. **The Climate Districts project:** The Climate Districts project is an effort to create more sustainable urban areas in Denmark, with a focus on reducing energy consumption, promoting renewable energy sources, and improving public transport infrastructure.
3. **Smart Aarhus:** Smart Aarhus is a city-wide initiative aimed at transforming Aarhus into a smart city. The project includes a range of initiatives such as using data analytics to improve public services, implementing a city-wide intelligent lighting system, and creating a network of smart bike-sharing stations.
4. **EnergyLab Nordhavn:** EnergyLab Nordhavn is a project in the Nordhavn district of Copenhagen aimed at creating a sustainable, energy-efficient urban area. The project includes the implementation of renewable energy sources such as wind and solar power, energy-efficient buildings, and smart energy management systems.
5. **CityFlow:** CityFlow is a smart city project that aims to improve traffic flow and reduce congestion in Aalborg, Denmark. The project includes the use of sensors, data analytics, and real-time traffic management to optimize the flow of vehicles and reduce travel times.
6. The Smart Energy Network project: The Smart Energy Network project is an effort to create a more sustainable energy infrastructure in Denmark by integrating renewable energy sources and promoting energy efficiency. The project includes the implementation of a smart energy grid, as well as energy storage and management systems.
7. **Silkeborg Intelligent Traffic System:** The Silkeborg Intelligent Traffic System is a project that uses real-time traffic data to optimize traffic flow and reduce congestion in the city of Silkeborg. The system includes sensors, traffic management software, and data analytics to improve traffic flow and reduce travel times.
8. **The Internet of Things Aarhus:** The Internet of Things Aarhus project is aimed at creating a network of connected devices throughout the city of Aarhus. The project includes the use of sensors, data analytics, and machine learning to improve public services and enhance the overall quality of life in the city.
9. **The Smart Hospital project:** The Smart Hospital project is an initiative to create more efficient and effective healthcare services in Denmark.

The project includes the use of electronic health records, telemedicine, and advanced data analytics to improve patient care and reduce costs.
10. **The Smart Campus project:** The Smart Campus project is an effort to create more sustainable and connected university campuses in Denmark. The project includes the use of smart buildings, renewable energy sources, and data analytics to improve energy efficiency and enhance the learning experience for students.

DUBAI

1. **Smart Dubai:** Smart Dubai is the city's overarching Smart City program. It aims to transform Dubai into a smart city by using innovative technologies and data to enhance government services and the quality of life for citizens and visitors.
2. **Dubai Pulse:** Dubai Pulse is a data-sharing platform that integrates data from multiple government and private sector organizations in Dubai. The platform enables real-time data analysis and visualization to inform decision-making, policy-making, and improve public services.
3. **Dubai Electricity and Water Authority (DEWA) Smart Grid:** DEWA's Smart Grid project aims to create a reliable, efficient, and sustainable power grid by using advanced technologies and data analytics. The project includes initiatives such as smart meters, advanced sensors, and AI-based analytics to improve energy efficiency and reduce carbon emissions.
4. **Dubai Autonomous Transportation Strategy:** Dubai's Autonomous Transportation Strategy aims to transform 25% of the city's transportation to autonomous modes by 2030. The strategy includes initiatives such as the deployment of autonomous vehicles, drones, and smart transportation systems.
5. **Dubai Internet of Things (IoT) Strategy:** Dubai's IoT strategy aims to create an open and secure IoT ecosystem that supports innovation and economic growth. The strategy includes initiatives such as the deployment of IoT devices, smart sensors, and data analytics to improve public services and create new business opportunities.
6. **Dubai Blockchain Strategy:** Dubai's Blockchain Strategy aims to make the city a global leader in blockchain technology by using it to improve government services, secure transactions, and reduce fraud. The strategy includes initiatives such as the deployment of

blockchain-based smart contracts, digital identity systems, and payment solutions.
7. **Dubai Paperless Strategy:** Dubai's Paperless Strategy aims to eliminate paper transactions and make all government services digital by 2021. The strategy includes initiatives such as the implementation of electronic payment systems, digital signatures, and secure document management solutions.
8. **Dubai Air Quality Monitoring Network:** Dubai's Air Quality Monitoring Network is a city-wide network of sensors that continuously monitor air quality and provide real-time data to residents, businesses, and government agencies. The network aims to improve public health by identifying pollution hotspots and informing policy-making.
9. **Dubai Smart Lighting Project:** The Dubai Smart Lighting Project aims to reduce energy consumption and enhance public safety by deploying smart LED lighting solutions. The project includes the installation of smart streetlights, sensors, and a centralized control system.
10. **Dubai Green Mobility Initiative:** The Dubai Green Mobility Initiative aims to increase the use of electric and hybrid vehicles in the city to reduce carbon emissions and improve air quality. The initiative includes initiatives such as the deployment of electric charging stations, incentives for purchasing electric vehicles, and the implementation of smart transportation systems.

EGYPT

1. **New Administrative Capital, New Capital City:** A new city being developed as Egypt's administrative capital, with a focus on sustainability and smart city technologies.
2. **Smart City Cairo, Cairo:** A smart city project aimed at transforming Cairo into a smart city through the use of innovative technologies.
3. **Sixth of October Smart City, Sixth of October City:** A smart city project aimed at transforming the city of Sixth of October into a smart city through the use of innovative technologies.
4. **Smart Traffic Management System, Nationwide:** Implementing a smart traffic management system to improve traffic flow, reduce congestion and improve road safety.

5. **Smart Energy Management System, Nationwide:** Implementing a smart energy management system to optimize energy usage and reduce costs.
6. **Smart Buildings, Nationwide:** Implementing smart building technologies to optimize energy usage and reduce costs.
7. **Smart Water Management, Nationwide:** Implementing smart water management systems to improve water distribution and reduce waste.
8. **Smart Waste Management, Nationwide:** Implementing smart waste management systems to improve waste collection, recycling and disposal.
9. **Port Said Smart City, Port Said:** A smart city project aimed at transforming the city of Port Said into a smart city through the use of innovative technologies.
10. **Smart Agriculture, Nationwide:** Implementing smart agriculture systems to improve crop yields, reduce waste and increase sustainability.

FINLAND

1. **Helsinki Smart Region, Helsinki:** A smart city project aimed at transforming the capital city of Helsinki into a smart city through the use of innovative technologies.
2. **Smart Energy Management System, Nationwide:** Implementing a smart energy management system to optimize energy usage and reduce costs.
3. **Smart Transportation System, Nationwide:** Implementing a smart transportation system to improve traffic flow, reduce congestion and improve road safety.
4. **Smart Waste Management, Nationwide:** Implementing smart waste management systems to improve waste collection, recycling and disposal.
5. **Smart Buildings, Nationwide:** Implementing smart building technologies to optimize energy usage and reduce costs.
6. **Oulu Health and Wellbeing Ecosystem, Oulu:** A smart city project aimed at improving the health and wellbeing of citizens through the use of innovative technologies.
7. **Tampere Smart City, Tampere:** A smart city project aimed at transforming the city of Tampere into a smart city through the use of innovative technologies.

8. **Espoo Smart City, Espoo:** A smart city project aimed at transforming the city of Espoo into a smart city through the use of innovative technologies.
9. **Vantaa Smart City, Vantaa:** A smart city project aimed at transforming the city of Vantaa into a smart city through the use of innovative technologies.
10. **Smart Agriculture, Nationwide:** Implementing smart agriculture systems to improve crop yields, reduce waste and increase sustainability.

FRANCE

1. **Lyon's Smart City:** Lyon's Smart City program aims to improve public services, transportation, and sustainability in the city. The program includes initiatives such as the development of a smart transportation system, the implementation of smart streetlights, and the use of data analytics to inform policy-making.
2. **Nice's Smart City:** Nice's Smart City program focuses on creating a sustainable, connected, and innovative city that improves the quality of life for residents. The program includes initiatives such as the implementation of smart transportation systems, the deployment of smart parking solutions, and the use of data analytics to improve public services.
3. **Paris' Smart City:** Paris' Smart City program aims to create a more sustainable and connected city by using technology to improve public services and transportation. The program includes initiatives such as the implementation of smart traffic management systems, the use of sensors to monitor air quality, and the deployment of smart street furniture.
4. **Grenoble's Smart City:** Grenoble's Smart City program focuses on using technology to improve public services, sustainability, and economic growth. The program includes initiatives such as the implementation of a smart grid, the deployment of electric buses, and the development of a smart lighting system.
5. **Toulouse's Smart City:** Toulouse's Smart City program aims to create a more sustainable, connected, and innovative city by using technology to improve public services and transportation. The program includes initiatives such as the deployment of smart parking solutions, the implementation of smart transportation systems, and the use of data analytics to inform policy-making.

6. **Marseille's Smart City:** Marseille's Smart City program focuses on using technology to improve public services, transportation, and economic growth. The program includes initiatives such as the implementation of a smart transportation system, the deployment of smart waste management solutions, and the development of a smart lighting system.
7. **Nantes' Smart City:** Nantes' Smart City program aims to create a more sustainable and connected city by using technology to improve public services and transportation. The program includes initiatives such as the implementation of a smart transportation system, the deployment of smart parking solutions, and the use of data analytics to inform policy-making.
8. **Strasbourg's Smart City:** Strasbourg's Smart City program focuses on using technology to improve public services, sustainability, and economic growth. The program includes initiatives such as the deployment of smart energy management solutions, the implementation of smart transportation systems, and the development of a smart lighting system.
9. **Bordeaux's Smart City:** Bordeaux's Smart City program aims to create a more sustainable, connected, and innovative city by using technology to improve public services and transportation. The program includes initiatives such as the deployment of smart parking solutions, the implementation of smart transportation systems, and the use of data analytics to inform policy-making.
10. **Rouen's Smart City:** Rouen's Smart City program focuses on using technology to improve public services, sustainability, and economic growth. The program includes initiatives such as the deployment of smart energy management solutions, the implementation of smart transportation systems, and the development of a smart lighting system.

GERMANY

1. **Smart Green Tower:** The Smart Green Tower is a mixed-use building in Freiburg that uses smart technology to optimize energy consumption, reduce carbon emissions, and increase the quality of life for residents.
2. **Smart Grids for Renewable Energy:** Several German cities, such as Hamburg and Berlin, have implemented smart grids that use renewable energy sources such as wind and solar power. These smart

grids are designed to improve the efficiency and reliability of the energy grid, while reducing carbon emissions.

3. **Smart Mobility:** German cities like Munich and Stuttgart have implemented smart mobility solutions, such as intelligent traffic management systems, car-sharing services, and electric vehicle charging infrastructure. These solutions aim to reduce traffic congestion and air pollution, while improving transportation options for residents.

4. **Smart Waste Management:** German cities like Berlin and Hamburg have implemented smart waste management solutions that use sensors and data analytics to optimize waste collection and disposal. This reduces costs, improves efficiency, and reduces environmental impact.

5. **Smart Buildings:** German cities like Frankfurt and Berlin have implemented smart building solutions that use sensors and automation to optimize energy consumption, reduce operating costs, and increase the comfort of occupants.

6. **Smart Water Management:** German cities like Hamburg and Berlin have implemented smart water management solutions that use sensors and data analytics to optimize water usage and reduce waste. This helps to conserve water resources and reduce costs.

7. **Digital Citizen Services:** German cities like Munich and Berlin have implemented digital citizen services that use technology to improve access to public services such as healthcare, education, and social welfare.

8. **Smart Lighting:** German cities like Hamburg and Frankfurt have implemented smart lighting solutions that use LED lighting and sensors to optimize energy consumption, reduce light pollution, and increase safety.

9. **Smart Tourism:** German cities like Berlin and Munich have implemented smart tourism solutions that use technology to improve the visitor experience, such as augmented reality tours, smart maps, and digital information kiosks.

10. **Smart Security:** German cities like Munich and Berlin have implemented smart security solutions that use technology to improve public safety, such as video surveillance systems, smart emergency services, and digital crime prevention initiatives.

GREECE

1. **Smart Policing System in Athens:** The Athens Smart Policing System uses advanced analytics to predict and prevent crime. It also helps police respond more quickly and effectively to incidents by providing real-time information and communication tools.
2. **Intelligent Transport System (ITS) in Thessaloniki:** The ITS in Thessaloniki aims to improve traffic flow, reduce congestion, and enhance road safety by using real-time data and analytics to optimize traffic signals, manage parking, and provide real-time travel information to drivers.
3. **Smart Lighting System in Larissa:** The Smart Lighting System in Larissa uses sensors and advanced analytics to optimize lighting levels and reduce energy consumption. The system can also detect faults and alert maintenance teams for prompt repair.
4. **Integrated Waste Management System in Halandri:** The Integrated Waste Management System in Halandri aims to improve waste collection and disposal through the use of smart bins equipped with sensors and real-time monitoring. This helps reduce waste accumulation, minimize health risks, and promote sustainable waste management practices.
5. **Smart Parking System in Patras:** The Smart Parking System in Patras uses sensors and real-time data to optimize parking spaces, reduce congestion, and improve traffic flow. Drivers can also use a mobile app to locate available parking spaces and pay for parking.
6. **Air Quality Monitoring System in Athens:** The Air Quality Monitoring System in Athens uses sensors and real-time data to monitor air quality levels in different parts of the city. The system can also alert authorities and citizens in case of dangerous levels of pollution.
7. **Smart Water Management System in Thessaloniki:** The Smart Water Management System in Thessaloniki uses sensors and real-time data to optimize water distribution, reduce leaks, and detect faults in the water network. The system can also help conserve water by identifying areas with high consumption and recommending measures to reduce usage.
8. **Intelligent Energy Management System in Corinth:** The Intelligent Energy Management System in Corinth uses sensors and real-time data to optimize energy consumption, reduce waste, and promote renewable energy sources. The system can also help households and businesses track their energy consumption and identify ways to reduce it.

9. **Smart Healthcare System in Heraklion:** The Smart Healthcare System in Heraklion uses digital technologies to improve healthcare services, reduce waiting times, and enhance patient outcomes. The system can also provide remote healthcare services and support for patients with chronic conditions.
10. **Intelligent Public Transport System in Rhodes:** The Intelligent Public Transport System in Rhodes uses sensors and real-time data to optimize bus routes, reduce waiting times, and improve passenger experience. The system can also provide real-time travel information to passengers and help transport operators identify areas for improvement.

HONG KONG

1. **Hong Kong Smart City Blueprint:** The Hong Kong Smart City Blueprint was launched in 2017, and it aims to make Hong Kong a world-class smart city by developing and deploying smart technology solutions in various areas such as mobility, living, environment, and government.
2. **Smart Traffic Management:** The Hong Kong government has implemented a smart traffic management system that uses sensors and data analytics to optimize traffic flow, reduce congestion, and improve air quality.
3. **Smart Waste Management:** Hong Kong has implemented a smart waste management system that uses sensors and data analytics to optimize waste collection and disposal. This helps to reduce costs, improve efficiency, and reduce environmental impact.
4. **Smart Lampposts:** Hong Kong has installed smart lampposts that are equipped with sensors for monitoring weather, air quality, traffic, and more. The data collected is used to improve urban planning and provide better services to residents.
5. **Smart Public Transport:** Hong Kong has implemented a smart public transport system that uses real-time data to optimize routes, reduce wait times, and improve the overall passenger experience.
6. **Smart Water Management:** Hong Kong has implemented a smart water management system that uses sensors and data analytics to optimize water usage and reduce waste. This helps to conserve water resources and reduce costs.
7. **Smart Healthcare:** Hong Kong has implemented a smart healthcare system that uses technology to improve patient care, such as

electronic health records, telemedicine, and mobile health applications.
8. **Smart Education:** Hong Kong has implemented a smart education system that uses technology to improve learning outcomes and make education more accessible, such as e-learning platforms and digital classrooms.
9. **Smart Tourism:** Hong Kong has implemented a smart tourism system that uses technology to improve the visitor experience, such as smart maps, mobile applications, and digital information kiosks.
10. **Smart Safety and Security:** Hong Kong has implemented a smart safety and security system that uses technology to improve public safety, such as video surveillance systems, smart emergency services, and digital crime prevention initiatives.

HUNGARY

1. **Budapest Smart City:** This project aims to modernize the infrastructure of Budapest, making it a more sustainable, liveable, and attractive city. The project focuses on the development of sustainable transportation systems, efficient energy management, and the optimization of public services.
2. **Smart City Miskolc:** This initiative is focused on modernizing the city of Miskolc through the implementation of smart city technologies and the development of sustainable infrastructure. The project includes the creation of a smart grid, the optimization of energy consumption, and the development of smart transportation systems.
3. **Smart City Szeged:** The city of Szeged is using smart city technologies to improve the quality of life for its citizens. The project focuses on reducing energy consumption, improving public services, and increasing the efficiency of transportation systems.
4. **Smart City Debrecen:** This initiative aims to make Debrecen a more sustainable and livable city through the use of smart city technologies. The project includes the development of a smart grid, the optimization of energy consumption, and the creation of new public services.
5. **Smart City Pécs:** The city of Pécs is working towards becoming a smart city through the implementation of sustainable technologies and the development of efficient infrastructure. The project includes the creation of a smart grid, the optimization of energy consumption, and the development of smart transportation systems.

6. **Budapest Smart City Platform:** Budapest has developed a smart city platform that uses various data sources, such as energy consumption, air quality, and traffic management, to optimize city operations and improve the quality of life for citizens.
7. **Sustainable Energy Solutions in Szeged:** Szeged has implemented a range of smart energy solutions, including the use of renewable energy sources, energy-efficient buildings, and smart grid technologies. The goal is to reduce the city's carbon footprint and make it more energy-independent.
8. **Smart Lighting in Debrecen:** The city of Debrecen has implemented a smart lighting system that adjusts the brightness and color of streetlights based on the time of day, weather conditions, and pedestrian traffic. This helps reduce energy consumption and improve visibility and safety in the city.
9. **Smart Traffic Management in Pécs:** The city of Pécs has implemented a smart traffic management system that uses real-time traffic data and predictive analytics to optimize traffic flow, reduce congestion, and improve road safety.
10. **Eco-friendly Public Transport in Miskolc:** Miskolc has implemented a range of eco-friendly public transport solutions, including electric buses, bike-sharing schemes, and park-and-ride facilities. The goal is to reduce air pollution and make public transport more accessible and convenient for citizens.

ICELAND

1. **Reykjavik Energy:** Reykjavik Energy is a smart grid project in Reykjavik that aims to optimize energy consumption and reduce energy waste. The project involves the deployment of advanced metering infrastructure and energy management systems to monitor and control energy usage in real-time.
2. **Smart Lighting:** Reykjavik has implemented a smart lighting system to optimize energy usage and reduce maintenance costs. The system includes LED lights that are connected to a centralized control system, which can be programd to adjust lighting levels based on weather conditions and other factors.
3. **Smart Parking:** Reykjavik has implemented a smart parking system to help drivers find available parking spaces and reduce traffic congestion. The system uses sensors and cameras to monitor parking

spaces in real-time, and provides information to drivers through a mobile app.
4. **Intelligent Transport System:** Reykjavik has implemented an intelligent transport system to improve traffic flow and reduce congestion. The system uses real-time traffic data to optimize traffic signal timings and provide information to drivers to help them make informed decisions about their travel.
5. **Renewable Energy:** Iceland is known for its abundant renewable energy resources, including geothermal and hydropower. The country is a leader in the development of sustainable energy solutions, and many smart city projects in Iceland aim to further advance these efforts.

INDIA

1. **Surat Smart City:** Surat Smart City in Gujarat is aimed at improving the quality of life for citizens through better infrastructure, housing, and public transportation.
2. **Bhopal Smart City:** Bhopal Smart City in Madhya Pradesh aims to provide efficient public transportation and improved waste management, along with creating a business-friendly environment and promoting tourism.
3. **Pune Smart City:** The Pune Smart City project in Maharashtra aims to provide better public transport systems, safer roads, and reliable water and electricity supply.
4. **Chandigarh Smart City:** Chandigarh Smart City is focused on creating a sustainable and eco-friendly city with improved traffic management, waste management, and public safety.
5. **Indore Smart City:** The Indore Smart City project in Madhya Pradesh is working on initiatives to improve urban mobility, solid waste management, and energy efficiency.
6. **Jaipur Smart City:** Jaipur Smart City is focused on improving public transport and the quality of urban spaces through a range of initiatives, including the renovation of public parks and the creation of pedestrian-friendly zones.
7. **Visakhapatnam Smart City:** Visakhapatnam Smart City in Andhra Pradesh is focused on creating a sustainable city through initiatives such as the development of eco-friendly transportation, improved public safety, and better waste management.

8. **Varanasi Smart City:** The Varanasi Smart City project in Uttar Pradesh is aimed at revitalizing the city's historic core through improved infrastructure, better public transportation, and enhanced tourist facilities.
9. **Kochi Smart City:** The Kochi Smart City project in Kerala is focused on creating a sustainable and livable city through initiatives such as the creation of bike lanes, improved waste management, and the development of renewable energy sources.
10. **Nagpur Smart City:** Nagpur Smart City in Maharashtra is focused on transforming the city into a hub for business, tourism, and innovation through a range of initiatives such as the creation of smart infrastructure, improved urban mobility, and enhanced public safety.

INDONESIA

1. **Smart City Jakarta:** Jakarta is one of the largest cities in Indonesia, and the Smart City Jakarta project aims to use technology to improve the city's infrastructure and services. The project includes initiatives such as smart transportation, e-governance, and citizen services.
2. **Kota Tanpa Sampah (Zero Waste City):** This is an initiative in the city of Bandung to reduce waste and promote sustainable living. The project involves educating residents on recycling and composting, as well as introducing smart waste management systems.
3. **Smart City Surabaya:** Surabaya is the second-largest city in Indonesia, and the Smart City Surabaya project aims to improve the city's public services and infrastructure. The project includes initiatives such as smart transportation, e-governance, and citizen services.
4. **Makassar Smart City:** Makassar is a city in South Sulawesi, and the Makassar Smart City project aims to use technology to improve public services and citizen engagement. The project includes initiatives such as smart transportation, e-governance, and citizen services.
5. **Smart City Semarang:** Semarang is a city in Central Java, and the Smart City Semarang project aims to improve the city's public services and infrastructure. The project includes initiatives such as smart transportation, e-governance, and citizen services.
6. **E-Budgeting:** This is a national initiative in Indonesia to promote transparency and citizen participation in the budgeting process. The

project involves creating an online platform for citizens to provide feedback and input on the government's budget.

7. **Smart Water Management:** This is an initiative in the city of Malang to improve the management of water resources. The project involves introducing smart sensors and meters to monitor water usage and quality, as well as promoting water conservation.
8. **Smart Traffic Management:** This is an initiative in the city of Makassar to reduce traffic congestion and improve safety. The project involves introducing smart traffic lights and sensors to monitor traffic flow, as well as promoting the use of public transportation.
9. **Green City:** This is an initiative in the city of Surabaya to promote sustainable living and reduce pollution. The project involves introducing green spaces and promoting eco-friendly transportation, as well as educating residents on recycling and waste reduction.
10. **Smart City Yogyakarta:** Yogyakarta is a city in Central Java, and the Smart City Yogyakarta project aims to use technology to improve public services and citizen engagement. The project includes initiatives such as smart transportation, e-governance, and citizen services.

IRELAND

1. **Dublin Smart City, Dublin:** An initiative aimed at transforming Dublin into a smart and sustainable city through the use of innovative technologies.
2. **Cork Smart City, Cork:** A smart city initiative aimed at transforming Cork into a leading smart city through the integration of smart technologies and infrastructure.
3. **Limerick Smart City, Limerick:** An initiative aimed at transforming Limerick into a smart, sustainable and inclusive city.
4. **Smart Energy Management System, Nationwide:** Implementing a smart energy management system to optimize energy usage and reduce costs.
5. **Smart Transportation System, Nationwide:** Implementing a smart transportation system to improve traffic flow, reduce congestion and improve road safety.
6. **Smart Water Management System, Nationwide:** Implementing a smart water management system to optimize water usage and reduce waste.

7. **E-Governance, Nationwide:** Implementing e-governance systems to provide citizens with online access to government services and information.
8. **Waste Management, Nationwide:** Implementing smart waste management systems to improve waste collection, recycling and disposal.
9. **Smart Health, Nationwide:** Implementing smart health systems to improve access to healthcare and reduce healthcare costs.
10. **Smart Agriculture, Nationwide:** Implementing smart agriculture systems to improve crop yields, reduce waste and increase sustainability.

ISRAEL

1. **Smart Transportation System, Nationwide:** Implementing a smart transportation system across the country to improve traffic flow, reduce congestion and improve road safety.
2. **Smart Energy Management System, Nationwide:** Implementing a smart energy management system to optimize energy usage and reduce costs.
3. **Intelligent Water Management System, Nationwide:** Implementing an intelligent water management system to optimize water usage and reduce waste.
4. **Smart Lighting, Nationwide:** Implementing smart lighting systems to reduce energy consumption and improve public safety.
5. **E-Governance, Nationwide:** Implementing e-governance systems to provide citizens with online access to government services and information.
6. **Waste Management, Nationwide:** Implementing smart waste management systems to improve waste collection, recycling and disposal.
7. **Smart Agriculture, Nationwide:** Implementing smart agriculture systems to improve crop yields, reduce waste and increase sustainability.
8. **Smart Health, Nationwide:** Implementing smart health systems to improve access to healthcare and reduce healthcare costs.
9. **Smart Park, Nationwide:** Developing smart park systems to improve park management, reduce waste and increase sustainability.
10. **Smart Grid, Nationwide:** Developing a smart grid system to improve energy distribution and reliability.

ITALY

1. **Smart District:** The Smart District project in Milan is focused on developing sustainable and livable neighbourhoods. The project aims to promote energy efficiency, reduce greenhouse gas emissions, and improve public transportation.
2. **Smart Parking:** Milan has implemented a smart parking system that uses sensors and mobile applications to improve the parking experience for residents and visitors.
3. **Smart Lighting:** Turin has implemented a smart lighting system that uses LED lights and sensors to optimize energy usage and reduce costs. The system can also be used for other purposes, such as public safety.
4. **Smart Water Management:** The city of Bologna has implemented a smart water management system that uses sensors and data analytics to monitor water usage, reduce waste, and improve water quality.
5. **Smart Mobility:** The Smart Mobility project in Florence aims to improve public transportation by using real-time data and mobile applications to provide residents with better information about transportation options and schedules.
6. **Smart Waste Management:** The city of Pisa has implemented a smart waste management system that uses sensors and data analytics to optimize waste collection and disposal. The system can also be used to promote recycling and reduce waste.
7. **Smart Tourism:** The city of Venice has implemented a smart tourism system that uses technology to improve the visitor experience, such as digital maps, mobile applications, and virtual tours.
8. **Smart Healthcare:** The Smart Healthcare project in Rome aims to improve patient care by using technology such as electronic health records and telemedicine.
9. **Smart Energy:** The Smart Energy project in Naples is focused on developing renewable energy sources and reducing greenhouse gas emissions.
10. **Smart Agriculture:** The Smart Agriculture project in Parma is focused on using technology to improve the efficiency and sustainability of agricultural practices. This includes using sensors and data analytics to optimize irrigation, fertilizer usage, and crop management.

KUWAIT

1. **Kuwait Smart City:** This project is located in South Surra, and it aims to create a smart and sustainable city by integrating smart technologies in various fields, including energy, transportation, healthcare, and education.
2. **Al-Zour New Refinery City:** This project is located in Al-Zour, and it will be a smart and sustainable city that will house the largest oil refinery in Kuwait. The city will feature smart buildings, smart transportation systems, and sustainable energy solutions.
3. **Kuwait's Silicon Oasis:** This project is located in the Al-Rai area and is designed to be a hub for technology, innovation, and entrepreneurship. The city will feature smart buildings, smart transportation, and sustainable energy solutions.
4. **Al-Sulaymaniya Smart City:** This project is located in Al-Sulaymaniya and aims to create a smart and sustainable city that will feature smart transportation, renewable energy solutions, and smart healthcare facilities.
5. **Sabah Al-Ahmad City:** This project is located in South Sabah Al-Ahmad and aims to create a smart and sustainable city that will feature smart buildings, smart transportation, and sustainable energy solutions.
6. **Kuwait Smart Village:** This project is located in the Al-Shadadiya area and aims to create a smart and sustainable village that will feature smart homes, smart transportation, and sustainable energy solutions.
7. **Sabah Al-Salem University City:** This project is located in Sabah Al-Salem and aims to create a smart university city that will house several universities and research centers. The city will feature smart buildings, smart transportation, and sustainable energy solutions.
8. **Al-Dasma Smart City:** This project is located in Al-Dasma and aims to create a smart and sustainable city that will feature smart homes, smart transportation, and sustainable energy solutions.
9. **Al-Farwaniya Smart City:** This project is located in Al-Farwaniya and aims to create a smart and sustainable city that will feature smart homes, smart transportation, and sustainable energy solutions.
10. **Jahra Smart City:** This project is located in Jahra and aims to create a smart and sustainable city that will feature smart homes, smart transportation, and sustainable energy solutions.

LITHUANIA

1. **Smart Energy Management:** This project aims to improve the energy efficiency of public buildings, street lighting and other infrastructure in Lithuania, by implementing smart energy management systems and renewable energy solutions.
2. **Intelligent Traffic Management:** This project aims to improve the traffic management system in Lithuania, by implementing smart solutions for traffic monitoring, control, and optimization, to reduce congestion, enhance public safety, and improve air quality.
3. **Smart Healthcare for Lithuania:** This project aims to improve the health care system in Lithuania, by implementing smart solutions for telemedicine, disease prevention, and health monitoring, to provide better health care services to the citizens.
4. **Digital Education for Lithuania:** This project aims to improve the education system in Lithuania, by implementing smart solutions for student assessment, teacher development, and educational resources, to enhance the quality of education.
5. **Sustainable Development for Lithuania:** This project aims to promote sustainable development in Lithuania, by implementing smart solutions for energy, transportation, waste, and water management, to reduce the impact of human activities on the environment and improve the quality of life.
6. **Smart Lighting:** This project involves upgrading street lighting with smart lighting systems that can be remotely monitored and controlled, reducing energy consumption and costs, and improving public safety.
7. **Smart Waste Management:** This project involves implementing smart waste management solutions that can monitor waste levels and schedule waste collection based on real-time data, reducing the environmental impact of waste and improving public health and hygiene.
8. **Green Mobility:** This project aims to promote sustainable transportation in Lithuania, by implementing smart solutions for electric vehicle charging, bike-sharing, and public transportation, to reduce greenhouse gas emissions and improve air quality.
9. **Digital Inclusion:** This project aims to improve digital literacy and access to information and services for all citizens, by implementing smart solutions for e-governance, e-health, and e-commerce, to enhance the quality of life and promote social and economic development.

10. **Water Management:** This project involves implementing smart solutions for water monitoring, treatment, and distribution, to improve water quality and reduce water waste, ensuring sustainable water management for the benefit of all citizens.

LUXEMBOURG

1. **Smart City Luxembourg:** This project aims to create a smart city in Luxembourg by implementing innovative solutions for energy efficiency, transportation, and waste management. The project aims to create a more sustainable and livable city for its residents.
2. **City Lab Luxembourg:** This project is a collaboration between the government of Luxembourg and several research institutions. The aim of the project is to create an innovation center for smart city solutions, where new technologies and ideas can be tested and implemented.
3. **Smart Energy for Luxembourg:** This project aims to implement smart energy solutions in Luxembourg, such as smart grids and renewable energy sources, in order to reduce the country's dependence on fossil fuels and improve energy efficiency.
4. **Luxembourg Connected:** This project aims to enhance the connectivity and digitalization of the country, by implementing smart solutions for transportation, energy, and communication. The project aims to improve the quality of life of the citizens and create a more sustainable city.
5. **Sustainable Mobility for Luxembourg:** This project aims to improve the transportation system in Luxembourg, by implementing smart solutions for public transportation and reducing the use of private cars. The project aims to create a more sustainable and efficient transportation system.
6. **EcoCity 2030:** This project aims to create a sustainable and eco-friendly city in Luxembourg, by implementing solutions for energy efficiency, waste management, and sustainable transportation. The project aims to reduce the city's carbon footprint and create a more livable environment.
7. **Smart Building Solutions for Luxembourg:** This project aims to implement smart building solutions in Luxembourg, such as smart heating and cooling systems, energy-efficient lighting, and renewable energy sources. The project aims to reduce the energy consumption of buildings and create a more sustainable city.

8. **Sustainable Urban Planning for Luxembourg:** This project aims to improve the urban planning of Luxembourg, by implementing sustainable solutions for land use, transportation, and waste management. The project aims to create a more livable and sustainable city for its residents.
9. **Green Infrastructure for Luxembourg:** This project aims to enhance the green infrastructure in Luxembourg, by implementing solutions for parks, green roofs, and sustainable transportation. The project aims to improve the quality of life of the citizens and create a more sustainable city.
10. **Digital Transformation for Luxembourg:** This project aims to improve the digital infrastructure in Luxembourg, by implementing smart solutions for communication, transportation, and energy. The project aims to create a more connected and digital city, and to enhance the quality of life of its residents.

MACAO

1. **Smart Transportation for Macao:** This project aims to improve the transportation system in Macao, by implementing smart solutions for traffic management, public transportation, and parking, to reduce congestion and improve mobility.
2. **Green Energy for Macao:** This project aims to increase the use of renewable energy sources in Macao, such as solar and wind energy, to reduce the dependence on fossil fuels and improve energy efficiency.
3. **Intelligent Building Management System:** This project aims to implement smart solutions for heating, cooling, and lighting in buildings across Macao, in order to reduce energy consumption and improve comfort levels.
4. **E-Government for Macao:** This project aims to improve the delivery of government services to the citizens of Macao, by implementing electronic services and portals, to make it easier for citizens to access government services and information.
5. **Smart Waste Management for Macao:** This project aims to improve the waste management system in Macao, by implementing smart solutions for waste collection, treatment, and disposal, to reduce waste and improve the environment.
6. **Public Safety for Macao:** This project aims to improve public safety in Macao, by implementing smart solutions for emergency response,

crime prevention, and public surveillance, to enhance security and reduce risks.
7. **Health Management for Macao:** This project aims to improve the health care system in Macao, by implementing smart solutions for telemedicine, disease prevention, and health monitoring, to provide better health care services to the citizens.
8. **Education Management for Macao:** This project aims to improve the education system in Macao, by implementing smart solutions for online learning, educational resources, and teacher development, to enhance the quality of education.
9. **Smart Parking for Macao:** This project aims to improve the parking management system in Macao, by implementing smart solutions for parking availability, payment, and enforcement, to reduce congestion and improve the convenience of parking.
10. **Cultural and Tourism Management for Macao:** This project aims to promote the cultural and tourism industries in Macao, by implementing smart solutions for cultural preservation, tourist information, and entertainment, to attract more visitors and enhance the quality of life.

MALAYSIA

1. **Smart City, Kuala Lumpur:** An initiative aimed at transforming Kuala Lumpur into a smart city through the integration of smart technologies and infrastructure.
2. **Cyberjaya Smart City, Cyberjaya:** An initiative aimed at transforming Cyberjaya into a leading smart city through the use of innovative technologies.
3. **Smart Energy Management System, Nationwide:** Implementing a smart energy management system to optimize energy usage and reduce costs.
4. **Smart Transportation System, Nationwide:** Implementing a smart transportation system to improve traffic flow, reduce congestion and improve road safety.
5. **Smart Water Management System, Nationwide:** Implementing a smart water management system to optimize water usage and reduce waste.
6. **E-Governance, Nationwide:** Implementing e-governance systems to provide citizens with online access to government services and information.

7. **Smart Health, Nationwide:** Implementing smart health systems to improve access to healthcare and reduce healthcare costs.
8. **Smart Agriculture, Nationwide**: Implementing smart agriculture systems to improve crop yields, reduce waste and increase sustainability.
9. **Smart Waste Management, Nationwide:** Implementing smart waste management systems to improve waste collection, recycling and disposal.
10. **Smart Buildings, Nationwide:** Implementing smart building technologies to optimize energy usage and reduce costs.

MEXICO

1. **Mexico City Digital Agenda:** The Mexico City government's plan to create a smart city, including digital infrastructure, open data, and e-government services.
2. **Ciudad Creativa Digital:** A project in Guadalajara that aims to establish a technology innovation hub, with focus on software development, video games, and digital media.
3. **Guadalajara Smart City:** The government of Guadalajara has launched a project to enhance mobility, public safety, and environmental sustainability, by integrating technology and data analytics.
4. **Queretaro Smart City:** The municipality of Queretaro has partnered with private sector companies to develop smart infrastructure, digital services, and innovation programs for urban development.
5. **Aguascalientes Smart City:** The city of Aguascalientes has implemented smart solutions for water management, energy efficiency, and public transportation, among other areas.
6. **León Smart City:** The municipality of León has deployed IoT sensors, mobile apps, and other digital tools to improve traffic flow, waste management, and citizen engagement.
7. **Tijuana Innovation and Technology Park:** A public-private initiative to establish a technology cluster in Tijuana, focusing on electronics, aerospace, and biotech industries.
8. **Puebla Digital City:** The government of Puebla has launched a digital transformation program, aiming to enhance public services, e-commerce, and connectivity for citizens and businesses.

9. **Morelia Smart City:** The city of Morelia has implemented a series of projects for intelligent mobility, smart lighting, and public safety, using sensors, cameras, and data analysis.
10. **Hermosillo Smart City:** The municipality of Hermosillo has developed a plan for smart infrastructure, sustainable energy, and urban revitalization, leveraging technology and innovation.

MONACO

1. **Monaco 3.0:** Monaco 3.0 is a smart city initiative that aims to make Monaco a sustainable, connected, and innovative city by leveraging technology to improve public services, transportation, and energy efficiency.
2. **Smart Lighting:** The smart lighting project in Monaco involves the installation of energy-efficient LED streetlights that are equipped with sensors to detect movement and adjust lighting levels accordingly. This helps to reduce energy consumption and light pollution.
3. **Smart Bus:** The Smart Bus project in Monaco involves the installation of GPS devices on buses to provide real-time information to passengers about the location and arrival time of buses. This helps to improve public transportation and reduce congestion.
4. **MonacoTech:** MonacoTech is an incubator for start-ups and innovative companies that are developing technology solutions for smart cities. It provides support, mentoring, and access to resources for entrepreneurs.
5. **Digital Security:** The Digital Security project in Monaco involves the implementation of cybersecurity measures to protect against cyber threats and ensure the safety and privacy of residents' data.
6. **Smart Parking:** The Smart Parking project in Monaco involves the installation of sensors in parking spaces to detect the presence of vehicles and provide real-time information to drivers about available parking spaces. This helps to reduce traffic congestion and air pollution.
7. **Oceanographic Museum:** The Oceanographic Museum in Monaco has implemented a number of smart city technologies, including an interactive exhibition that uses virtual reality to educate visitors about marine life and environmental sustainability.
8. **Waste Management:** The Waste Management project in Monaco involves the implementation of smart waste management solutions,

including sensors on waste bins that detect when they are full and need to be emptied. This helps to improve the efficiency of waste collection and reduce environmental impact.
9. **Smart Grid:** The Smart Grid project in Monaco involves the installation of a smart energy grid that uses sensors and data analytics to optimize energy consumption and reduce costs.
10. **E-health:** The E-health project in Monaco involves the implementation of digital health solutions that leverage technology to improve healthcare services and make them more accessible to residents. This includes initiatives such as telemedicine and e-prescriptions.

MOROCCO

1. **Casablanca Finance City:** This project is located in Casablanca and aims to create a smart financial city that will feature smart buildings, smart transportation, and sustainable energy solutions. The city will be a hub for financial services and will attract international financial institutions.
2. **Tangier Tech City:** This project is located in Tangier and aims to create a smart city that will be a hub for technology and innovation. The city will feature smart buildings, smart transportation, and sustainable energy solutions.
3. **Marrakech Digital City:** This project is located in Marrakech and aims to create a smart city that will be a hub for digital innovation and technology. The city will feature smart buildings, smart transportation, and sustainable energy solutions.
4. **Rabat Smart City:** This project is located in Rabat and aims to create a smart city that will feature smart buildings, smart transportation, and sustainable energy solutions. The city will be a hub for technology, innovation, and entrepreneurship.
5. **Agadir Smart City:** This project is located in Agadir and aims to create a smart and sustainable city that will feature smart homes, smart transportation, and sustainable energy solutions. The city will be a hub for tourism and will attract international tourists.
6. **Kenitra Smart City:** This project is located in Kenitra and aims to create a smart city that will feature smart homes, smart transportation, and sustainable energy solutions. The city will be a hub for technology, innovation, and entrepreneurship.

7. **Oujda Smart City:** This project is located in Oujda and aims to create a smart and sustainable city that will feature smart homes, smart transportation, and sustainable energy solutions. The city will be a hub for tourism and will attract international tourists.
8. **Nador Smart City:** This project is located in Nador and aims to create a smart and sustainable city that will feature smart homes, smart transportation, and sustainable energy solutions. The city will be a hub for tourism and will attract international tourists.
9. **Fez Smart City:** This project is located in Fez and aims to create a smart and sustainable city that will feature smart homes, smart transportation, and sustainable energy solutions. The city will be a hub for tourism and will attract international tourists.
10. **Tétouan Smart City:** This project is located in Tétouan and aims to create a smart and sustainable city that will feature smart homes, smart transportation, and sustainable energy solutions. The city will be a hub for technology, innovation, and entrepreneurship.

NETHERLANDS

1. **Amsterdam Smart City:** Amsterdam Smart City is a collaboration between various public and private organizations aimed at promoting sustainable and innovative urban development in Amsterdam. The initiative focuses on areas such as energy, mobility, and circular economy.
2. **Rotterdam Smart City:** Rotterdam Smart City is a program aimed at improving quality of life, promoting economic growth, and reducing environmental impact in Rotterdam. The initiative focuses on areas such as mobility, energy, and digital innovation.
3. **Utrecht Smart City:** Utrecht Smart City is a platform that promotes collaboration between local government, businesses, and citizens to create a more sustainable and livable city. The initiative focuses on areas such as mobility, energy, and social innovation.
4. **Eindhoven Smart City:** Eindhoven Smart City is a program focused on promoting innovation, creativity, and entrepreneurship in the city. The initiative focuses on areas such as mobility, energy, and health.
5. **The Hague Smart City:** The Hague Smart City is a program aimed at using technology to improve the city's sustainability and quality of life. The initiative focuses on areas such as energy, mobility, and safety.
6. **Amsterdam ArenA:** Amsterdam ArenA is a stadium that has implemented a smart lighting system, which uses LED lights and

sensors to optimize energy usage and create an immersive fan experience.
7. **BikeScout:** BikeScout is a system that uses sensors and data analytics to help cyclists find the safest and fastest routes through the city.
8. **FlexPower:** FlexPower is a pilot project that aims to create a more flexible and sustainable energy grid in the city of Arnhem. The project uses smart grids and energy storage solutions to optimize energy usage and reduce carbon emissions.
9. **Circular Economy:** The Circular Economy program in the city of Almere aims to create a more sustainable and resilient city by promoting circular economy principles. The initiative focuses on areas such as waste reduction, energy efficiency, and sustainable urban design.
10. **Smart Grid:** The Smart Grid project in the city of Groningen is a pilot program that aims to create a more sustainable energy system by using smart grid technology to optimize energy usage and reduce carbon emissions. The project includes a mix of renewable energy sources, energy storage solutions, and smart meters.

NEW ZEALAND

1. **Smart Transport in Auckland:** Auckland has implemented a smart transport system that uses real-time data to optimize traffic flow, reduce congestion, and improve public transportation. This includes integrated fare systems, bus priority lanes, and real-time passenger information.
2. **Smart Lighting in Christchurch:** Christchurch has implemented a smart lighting system that uses sensors to adjust lighting levels based on pedestrian and vehicle traffic, reducing energy consumption and improving safety.
3. **Smart Irrigation in Hawke's Bay:** The Hawke's Bay region has implemented a smart irrigation system that uses soil moisture sensors and weather data to optimize irrigation schedules and reduce water usage, resulting in improved crop yields and reduced environmental impact.
4. **Smart Waste Management in Wellington:** Wellington has implemented a smart waste management system that uses sensors to optimize waste collection routes and reduce the amount of waste sent to landfills, resulting in cost savings and reduced greenhouse gas emissions.

5. **Smart Grid in Taupo:** Taupo has implemented a smart grid system that uses real-time data to monitor and optimize energy usage, reduce peak demand, and improve energy efficiency.
6. **Smart Water Management in Tauranga:** Tauranga has implemented a smart water management system that uses sensors to monitor water usage and detect leaks, optimizing water distribution and reducing water loss.
7. **Smart Parking in Dunedin:** Dunedin has implemented a smart parking system that uses real-time data to guide drivers to available parking spots, reducing congestion and improving the parking experience.
8. **Smart Public Transport in Hamilton:** Hamilton has implemented a smart public transport system that uses real-time data to optimize bus routes and schedules, reducing waiting times and improving the overall passenger experience.
9. **Smart Health in Otago:** Otago has implemented a smart health system that uses technology to enable remote consultations, telemedicine, and remote patient monitoring, improving access to healthcare services and reducing the burden on the healthcare system.
10. **Smart Tourism in Queenstown:** Queenstown has implemented a smart tourism system that uses technology to enhance the visitor experience, including interactive maps, augmented reality, and personalized recommendations for activities and attractions.

NORWAY

1. **FutureBuilt:** FutureBuilt is a program that aims to make urban development in Norway more sustainable. The initiative focuses on energy efficiency, carbon neutrality, and circular economy principles in new developments.
2. **+CityxChange:** +CityxChange is a smart city project that aims to transform the city of Trondheim into a sustainable, positive energy city. The initiative involves the development of a smart grid and the use of renewable energy sources.
3. **Urban Sharing:** Urban Sharing is a project that aims to develop more sustainable and shared mobility solutions in Norway. The initiative focuses on car-sharing, bike-sharing, and ride-sharing.
4. **Follo Line Project:** The Follo Line Project is a railway infrastructure project that will create a new high-speed train connection between

the cities of Oslo and Ski. The initiative will reduce travel times and ease congestion in the region.
5. **Halden Smart Grid:** The Halden Smart Grid is a project that aims to develop an intelligent energy system in the city of Halden. The initiative involves the use of smart meters, renewable energy sources, and demand response programs.
6. **Oslo Smart City:** The Oslo Smart City initiative aims to create a more sustainable and livable city through the use of technology. The initiative focuses on areas such as mobility, energy, and environmental sustainability.
7. **The Climate City:** The Climate City is a project that aims to develop a more sustainable and climate-friendly city in the municipality of Bærum. The initiative focuses on renewable energy, circular economy principles, and smart transportation solutions.
8. **Ruter Innovation Lab:** The Ruter Innovation Lab is a project that aims to develop innovative and sustainable transportation solutions for the Oslo region. The initiative focuses on electric buses, autonomous vehicles, and intelligent transport systems.
9. **ENOVA:** ENOVA is a government-funded agency that supports sustainable energy and climate measures in Norway. The initiative provides financial support and expertise for energy efficiency projects, renewable energy, and emission reduction programs.
10. **Drammen Smart City:** The Drammen Smart City initiative aims to create a more sustainable and livable city through the use of technology. The initiative focuses on areas such as mobility, energy, and environmental sustainability.

OMAN

1. **Muscat Smart City:** This project is located in Muscat, the capital city of Oman, and aims to create a smart city that will feature smart buildings, smart transportation, and sustainable energy solutions. The city will be a hub for technology, innovation, and entrepreneurship.
2. **Sohar Smart City:** This project is located in Sohar and aims to create a smart city that will feature smart buildings, smart transportation, and sustainable energy solutions. The city will be a hub for technology, innovation, and entrepreneurship.
3. **Salalah Smart City:** This project is located in Salalah and aims to create a smart and sustainable city that will feature smart homes,

smart transportation, and sustainable energy solutions. The city will be a hub for tourism and will attract international tourists.

4. **Al Seeb Smart City:** This project is located in Al Seeb and aims to create a smart and sustainable city that will feature smart homes, smart transportation, and sustainable energy solutions. The city will be a hub for technology, innovation, and entrepreneurship.
5. **Al Khuwair Smart City:** This project is located in Al Khuwair and aims to create a smart and sustainable city that will feature smart homes, smart transportation, and sustainable energy solutions. The city will be a hub for technology, innovation, and entrepreneurship.
6. **Al Hail Smart City:** This project is located in Al Hail and aims to create a smart city that will feature smart buildings, smart transportation, and sustainable energy solutions. The city will be a hub for technology, innovation, and entrepreneurship.
7. **Barka Smart City:** This project is located in Barka and aims to create a smart and sustainable city that will feature smart homes, smart transportation, and sustainable energy solutions. The city will be a hub for technology, innovation, and entrepreneurship.
8. **Nizwa Smart City:** This project is located in Nizwa and aims to create a smart and sustainable city that will feature smart homes, smart transportation, and sustainable energy solutions. The city will be a hub for technology, innovation, and entrepreneurship.
9. **Sur Smart City:** This project is located in Sur and aims to create a smart and sustainable city that will feature smart homes, smart transportation, and sustainable energy solutions. The city will be a hub for technology, innovation, and entrepreneurship.
10. **Ibra Smart City:** This project is located in Ibra and aims to create a smart and sustainable city that will feature smart homes, smart transportation, and sustainable energy solutions. The city will be a hub for technology, innovation, and entrepreneurship.

PHILIPPINES

1. **Cagayan de Oro River Basin Management, Cagayan de Oro:** A smart city project aimed at managing the Cagayan de Oro River Basin through the integration of smart technologies and infrastructure.
2. **Smart City Davao, Davao:** A smart city project aimed at transforming Davao into a leading smart city through the use of innovative technologies.

3. **Smart Energy Management System, Nationwide:** Implementing a smart energy management system to optimize energy usage and reduce costs.
4. **Smart Transportation System, Nationwide:** Implementing a smart transportation system to improve traffic flow, reduce congestion and improve road safety.
5. **Smart Water Management System, Nationwide:** Implementing a smart water management system to optimize water usage and reduce waste.
6. **E-Governance, Nationwide:** Implementing e-governance systems to provide citizens with online access to government services and information.
7. **Smart Health, Nationwide:** Implementing smart health systems to improve access to healthcare and reduce healthcare costs.
8. **Smart Agriculture, Nationwide:** Implementing smart agriculture systems to improve crop yields, reduce waste and increase sustainability.
9. **Smart Waste Management, Nationwide:** Implementing smart waste management systems to improve waste collection, recycling and disposal.
10. **Smart Buildings, Nationwide:** Implementing smart building technologies to optimize energy usage and reduce costs.

POLAND

1. **Smart Energy Management System, Warsaw:** Implementing a smart energy management system in the capital city of Warsaw to optimize energy usage and reduce costs.
2. **Intelligent Transportation System, Krakow:** Implementing an intelligent transportation system in the city of Krakow to improve traffic flow, reduce congestion and improve mobility.
3. **Smart Lighting, Wroclaw:** Implementing smart lighting systems in the city of Wroclaw to reduce energy consumption and improve public safety.
4. **E-Governance, Nationwide:** Implementing e-governance systems to provide citizens with online access to government services and information.
5. **Waste Management, Nationwide:** Implementing smart waste management systems to improve waste collection, recycling and disposal.

6. **Smart Grid, Nationwide:** Developing a smart grid system to improve energy distribution and reliability.
7. **Smart Health, Nationwide:** Implementing smart health systems to improve access to healthcare and reduce healthcare costs.
8. **Smart Agriculture, Nationwide:** Implementing smart agriculture systems to improve crop yields, reduce waste and increase sustainability.
9. **Intelligent Water Management, Poznan:** Implementing an intelligent water management system in the city of Poznan to optimize water usage and reduce waste.
10. **Smart Parks, Nationwide:** Developing smart park systems to improve park management, reduce waste and increase sustainability.

QATAR

1. **Lusail City, Lusail:** A fully integrated smart city development, incorporating cutting-edge technologies and sustainable practices.
2. **Al Daayen Smart City, Al Daayen:** A smart city project aimed at developing a sustainable and environmentally friendly city.
3. **Education City Smart Campus, Education City, Doha:** A smart campus project aimed at integrating technology into education and research.
4. **Hamad Port Smart Terminal, Hamad Port, Doha:** A smart port project aimed at optimizing operations and improving efficiency.
5. **Smart Grid, Nationwide:** Implementing a smart grid system to optimize energy distribution and reduce waste.
6. **Smart Transportation, Nationwide:** Implementing smart transportation systems to improve traffic flow and reduce congestion.
7. **Smart Water Management, Nationwide:** Implementing smart water management systems to improve water distribution and reduce waste.
8. **Smart Waste Management, Nationwide:** Implementing smart waste management systems to improve waste collection, recycling and disposal.
9. **Smart Building Management, Nationwide:** Implementing smart building management systems to optimize energy usage and reduce costs.
10. **Smart Health Care, Nationwide:** Implementing smart health care systems to improve patient outcomes and reduce costs.

ROMANIA

1. **Bucharest Smart City, Bucharest:** A smart city project aimed at transforming Bucharest into a smart city through the use of innovative technologies.
2. **Cluj-Napoca Smart City, Cluj-Napoca:** A smart city project aimed at transforming Cluj-Napoca into a smart city through the use of innovative technologies.
3. **Iasi Smart City, Iasi:** A smart city project aimed at transforming Iasi into a smart city through the use of innovative technologies.
4. **Smart Traffic Management System, Nationwide:** Implementing a smart traffic management system to improve traffic flow, reduce congestion and improve road safety.
5. **Smart Energy Management System, Nationwide:** Implementing a smart energy management system to optimize energy usage and reduce costs.
6. **Smart Buildings, Nationwide:** Implementing smart building technologies to optimize energy usage and reduce costs.
7. **Smart Water Management, Nationwide:** Implementing smart water management systems to improve water distribution and reduce waste.
8. **Smart Waste Management, Nationwide:** Implementing smart waste management systems to improve waste collection, recycling and disposal.
9. **Timisoara Smart City, Timisoara:** A smart city project aimed at transforming Timisoara into a smart city through the use of innovative technologies.
10. **Smart Agriculture, Nationwide:** Implementing smart agriculture systems to improve crop yields, reduce waste and increase sustainability.

RUSSIA

1. **Moscow's Digital City:** Moscow's Digital City project aims to make Moscow a more connected and efficient city by implementing smart technologies in areas such as transportation, energy, and public services.
2. **Kazan Smart City:** Kazan, the capital of the Republic of Tatarstan in Russia, has developed a comprehensive smart city plan that includes

a wide range of initiatives in areas such as transportation, healthcare, and education.
3. **Skolkovo Innovation Center:** Skolkovo is a high-tech business park located near Moscow that serves as a hub for innovation and entrepreneurship. It includes a number of smart city projects, such as a smart grid system and a platform for intelligent transportation.
4. **St. Petersburg Intelligent Transport System:** St. Petersburg has implemented an intelligent transport system that includes a variety of smart technologies to improve traffic flow and reduce congestion in the city.
5. **Smart Energy City:** The Smart Energy City project is a collaboration between the city of Ulyanovsk and the Skolkovo Foundation. The project aims to develop a smart energy system that integrates renewable energy sources and uses smart technologies to improve energy efficiency.
6. **Yandex Smart Home:** Yandex, a Russian technology company, has developed a smart home system that allows users to control their home appliances and devices using voice commands or a mobile app.
7. **Smart Lighting in Krasnodar:** The city of Krasnodar has implemented a smart lighting system that uses sensors to adjust the brightness of street lights based on the amount of pedestrian and vehicle traffic in the area.
8. **Nizhny Novgorod Intelligent Transport System:** Nizhny Novgorod has implemented an intelligent transport system that uses real-time data to optimize traffic flow and reduce congestion in the city.
9. **Sochi Smart City:** Sochi, the host city of the 2014 Winter Olympics, has implemented a number of smart city projects, including a smart traffic system and a smart lighting system.
10. **Tomsk Smart City:** Tomsk is a city in Siberia that has developed a comprehensive smart city plan aimed at improving the quality of life for its residents. The plan includes a number of smart initiatives in areas such as transportation, healthcare, and education.

SAUDI ARABIA

1. **NEOM:** NEOM is a new city development in Saudi Arabia that aims to be a sustainable and futuristic urban hub that incorporates smart city technologies. It is designed to be a global hub for innovation, entrepreneurship, and investment across various sectors.

2. **King Abdullah Economic City:** King Abdullah Economic City is a planned city development on the Red Sea coast of Saudi Arabia. It aims to be a sustainable and innovative city that incorporates smart city technologies to enhance the quality of life for residents.
3. **Jeddah Economic City:** Jeddah Economic City is another planned city development in Saudi Arabia. It aims to be a smart and sustainable city that leverages technology to enhance economic growth, environmental sustainability, and social well-being.
4. **Qiddiya:** Qiddiya is an entertainment and sports city development located outside Riyadh, Saudi Arabia. It is designed to be a sustainable and futuristic destination that incorporates smart city technologies to enhance the guest experience.
5. **Smart Riyadh:** Smart Riyadh is a smart city initiative that aims to enhance the quality of life for Riyadh's residents through various technology-driven services and solutions. It includes initiatives such as smart transportation, e-government services, and digital infrastructure.
6. **Smart Makkah:** Smart Makkah is a smart city initiative that aims to enhance the Hajj and Umrah experience for pilgrims by leveraging technology to improve transportation, crowd management, and public services.
7. **Smart Medina:** Smart Medina is a smart city initiative that aims to improve the quality of life for residents and visitors to Medina. It includes initiatives such as smart transportation, public services, and environmental sustainability.
8. **Knowledge Economic City:** Knowledge Economic City is a planned city development in Medina, Saudi Arabia. It aims to be a smart and sustainable city that leverages technology to foster innovation, knowledge-based industries, and economic growth.
9. **Al-Faisaliah City:** Al-Faisaliah City is a new urban development in Riyadh, Saudi Arabia. It aims to be a smart and sustainable city that leverages technology to enhance the quality of life for residents and promote economic growth.
10. **Al Widyan:** Al Widyan is a new urban development in Riyadh, Saudi Arabia that incorporates smart city technologies to promote sustainability and enhance the quality of life for residents. It includes initiatives such as smart transportation, energy-efficient buildings, and digital infrastructure.

SERBIA

1. **Smart Energy Management:** This project involves implementing smart energy management systems to monitor and optimize energy consumption, reducing energy costs and improving energy efficiency, for both the public and private sectors.
2. **Smart Traffic Management:** This project involves upgrading traffic management systems with real-time traffic monitoring, predictive analysis, and traffic flow optimization, to reduce congestion and improve road safety.
3. **Smart Water Management:** This project involves implementing smart water management solutions to monitor water quality, water distribution, and water consumption, improving water security and reducing water waste.
4. **Smart Public Services:** This project involves implementing smart solutions for public services, such as healthcare, education, and social services, to improve the quality and accessibility of these services, and enhance the well-being of citizens.
5. **Smart Environmental Monitoring:** This project involves implementing smart environmental monitoring systems to monitor air quality, noise levels, and other environmental factors, to improve public health and environmental sustainability.
6. **Smart Parking:** This project involves implementing smart parking solutions to manage and optimize parking spaces, reduce traffic congestion and improve accessibility for drivers.
7. **Smart Building Management:** This project involves implementing smart building management systems to optimize energy consumption, reduce maintenance costs and improve indoor air quality in buildings.
8. **Smart Waste Management:** This project involves implementing smart waste management solutions to monitor waste collection, reduce waste generation, and improve recycling and composting rates.
9. **Smart Emergency Services:** This project involves implementing smart emergency services to improve response times and coordination during emergencies, enhance public safety and reduce the impacts of natural disasters and other emergencies.
10. **Smart Security:** This project involves implementing smart security solutions to enhance public safety, monitor crime rates and improve emergency response times. These solutions may include smart

cameras, automated surveillance systems, and incident response systems.

SINGAPORE

1. **Smart Nation, Nationwide:** A national initiative aimed at transforming Singapore into a smart nation by leveraging technology to improve the quality of life for its citizens.
2. **Smart Grid, Nationwide:** Implementing a smart grid system to optimize energy usage and reduce costs.
3. **Smart Transportation, Nationwide:** Implementing a smart transportation system to improve traffic flow, reduce congestion and improve road safety.
4. **Smart Water Management, Nationwide:** Implementing a smart water management system to optimize water usage and reduce waste.
5. **E-Governance, Nationwide:** Implementing e-governance systems to provide citizens with online access to government services and information.
6. **Smart Health, Nationwide:** Implementing smart health systems to improve access to healthcare and reduce healthcare costs.
7. **Smart Waste Management, Nationwide:** Implementing smart waste management systems to improve waste collection, recycling and disposal.
8. **Smart Agriculture, Nationwide:** Implementing smart agriculture systems to improve crop yields, reduce waste and increase sustainability.
9. **Smart Buildings, Nationwide:** Implementing smart building technologies to optimize energy usage and reduce costs.
10. **Smart City Living Labs, Nationwide:** Implementing a series of smart city living labs to test and demonstrate innovative technologies and approaches to urban living.

SLOVAKIA

1. **Bratislava Smart City:** This project is located in Bratislava, the capital city of Slovakia, and aims to create a smart city that will feature smart buildings, smart transportation, and sustainable energy solutions.

The city will be a hub for technology, innovation, and entrepreneurship.
2. **Košice Smart City:** This project is located in Košice and aims to create a smart city that will feature smart buildings, smart transportation, and sustainable energy solutions. The city will be a hub for technology, innovation, and entrepreneurship.
3. **Trenčín Smart City:** This project is located in Trenčín and aims to create a smart and sustainable city that will feature smart homes, smart transportation, and sustainable energy solutions. The city will be a hub for tourism and will attract international tourists.
4. **Banská Bystrica Smart City:** This project is located in Banská Bystrica and aims to create a smart and sustainable city that will feature smart homes, smart transportation, and sustainable energy solutions. The city will be a hub for technology, innovation, and entrepreneurship.
5. **Nitra Smart City:** This project is located in Nitra and aims to create a smart and sustainable city that will feature smart homes, smart transportation, and sustainable energy solutions. The city will be a hub for technology, innovation, and entrepreneurship.
6. **Žilina Smart City:** This project is located in Žilina and aims to create a smart city that will feature smart buildings, smart transportation, and sustainable energy solutions. The city will be a hub for technology, innovation, and entrepreneurship.
7. **Poprad Smart City:** This project is located in Poprad and aims to create a smart and sustainable city that will feature smart homes, smart transportation, and sustainable energy solutions. The city will be a hub for tourism and will attract international tourists.
8. **Trnava Smart City:** This project is located in Trnava and aims to create a smart and sustainable city that will feature smart homes, smart transportation, and sustainable energy solutions. The city will be a hub for technology, innovation, and entrepreneurship.
9. **Prešov Smart City:** This project is located in Prešov and aims to create a smart and sustainable city that will feature smart homes, smart transportation, and sustainable energy solutions. The city will be a hub for technology, innovation, and entrepreneurship.
10. **Martin Smart City:** This project is located in Martin and aims to create a smart and sustainable city that will feature smart homes, smart transportation, and sustainable energy solutions. The city will be a hub for technology, innovation, and entrepreneurship.

SLOVENIA

1. **Smart Transportation for Slovenia:** This project aims to improve the transportation system in Slovenia, by implementing smart solutions for traffic management, public transportation, and parking, to reduce congestion and improve mobility.
2. **Green Energy for Slovenia:** This project aims to increase the use of renewable energy sources in Slovenia, such as solar and wind energy, to reduce the dependence on fossil fuels and improve energy efficiency.
3. **Intelligent Building Management System:** This project aims to implement smart solutions for heating, cooling, and lighting in buildings across Slovenia, in order to reduce energy consumption and improve comfort levels.
4. **E-Government for Slovenia:** This project aims to improve the delivery of government services to the citizens of Slovenia, by implementing electronic services and portals, to make it easier for citizens to access government services and information.
5. **Smart Waste Management for Slovenia:** This project aims to improve the waste management system in Slovenia, by implementing smart solutions for waste collection, treatment, and disposal, to reduce waste and improve the environment.
6. **Intelligent Street Lighting:** This project aims to improve the street lighting system in Slovenia, by implementing smart lighting solutions, to reduce energy consumption and enhance public safety.
7. **Water Management for Slovenia:** This project aims to improve the water management system in Slovenia, by implementing smart solutions for water quality, supply, and distribution, to ensure the availability of clean water and reduce waste.
8. **Smart Education for Slovenia:** This project aims to improve the education system in Slovenia, by implementing smart solutions for student assessment, teacher development, and educational resources, to enhance the quality of education.
9. **Smart Health for Slovenia:** This project aims to improve the health care system in Slovenia, by implementing smart solutions for telemedicine, disease prevention, and health monitoring, to provide better health care services to the citizens.
10. **Sustainable Development for Slovenia:** This project aims to promote sustainable development in Slovenia, by implementing smart solutions for energy, transportation, waste, and water

management, to reduce the impact of human activities on the environment and improve the quality of life.

SOUTH AFRICA

1. **Cape Town Smart City, Cape Town:** An initiative aimed at creating a smart and sustainable city through the integration of smart technologies and infrastructure.
2. **eThekwini Smart City, Durban:** A smart city initiative aimed at transforming Durban into a leading smart city through the use of innovative technologies.
3. **Johannesburg Smart City, Johannesburg:** An initiative aimed at transforming Johannesburg into a smart, sustainable and inclusive city.
4. **Smart Energy Management System, Nationwide:** Implementing a smart energy management system to optimize energy usage and reduce costs.
5. **Smart Transportation System, Nationwide:** Implementing a smart transportation system to improve traffic flow, reduce congestion and improve road safety.
6. **Waste Management, Nationwide:** Implementing smart waste management systems to improve waste collection, recycling and disposal.
7. **Smart Water Management System, Nationwide:** Implementing a smart water management system to optimize water usage and reduce waste.
8. **E-Governance, Nationwide:** Implementing e-governance systems to provide citizens with online access to government services and information.
9. **Smart Health, Nationwide:** Implementing smart health systems to improve access to healthcare and reduce healthcare costs.
10. **Smart Agriculture, Nationwide:** Implementing smart agriculture systems to improve crop yields, reduce waste and increase sustainability.

SOUTH KOREA

1. **Songdo International Business District:** A new city being built near Incheon, which is a smart city with a focus on sustainability, technology, and walkability.
2. **Smart Transportation System in Seoul:** A smart transportation system in Seoul that uses data analytics and advanced technology to manage traffic flow, reduce congestion, and improve public transportation.
3. **Busan Eco Delta City:** A smart city in Busan that aims to be eco-friendly and sustainable, with features like solar panels, green roofs, and efficient waste management systems.
4. **Gwanggyo New City:** A new city being built near Seoul, which is a smart city with a focus on green energy and sustainable design.
5. **U-City in Sejong:** A smart city in Sejong, which uses data and technology to improve the quality of life for residents, enhance public safety, and reduce environmental impact.
6. **Smart Grid Project in Jeju:** A project in Jeju, which aims to create a smart grid that can integrate renewable energy sources and improve energy efficiency.
7. **Smart Farming in South Korea:** A project that promotes the use of IoT and data analytics in agriculture, to optimize crop yield and minimize resource use.
8. **Smart Water Management System in Daegu:** A project that uses IoT and sensor technologies to monitor water quality and reduce waste in Daegu.
9. **Smart Tourism in Gangwon Province:** A project that uses technology to enhance the tourist experience in Gangwon province, through things like virtual reality tours and smart signage.
10. **Smart Factories in South Korea:** A project that promotes the use of Industry 4.0 technologies in manufacturing, to increase efficiency, reduce waste, and enhance product quality.

SPAIN

1. **Barcelona Smart City:** Barcelona's Smart City program aims to use technology to enhance the quality of life for residents and visitors while promoting sustainability and economic growth. Initiatives include smart transportation, urban planning, and open data.

2. **Madrid Smart City:** Madrid's Smart City program aims to use technology to improve the efficiency of public services, reduce energy consumption, and enhance the quality of life for residents. Initiatives include smart transportation, smart buildings, and citizen engagement.
3. **Valencia Smart City:** Valencia's Smart City program aims to use technology to promote economic growth, sustainability, and social innovation. Initiatives include smart mobility, sustainable energy, and open data.
4. **Malaga Smart City:** Malaga's Smart City program aims to use technology to improve the quality of life for residents and visitors, promote sustainability, and attract investment. Initiatives include smart transportation, public services, and urban innovation.
5. **Santander Smart City:** Santander's Smart City program aims to use technology to enhance the quality of life for residents and visitors, promote sustainability, and foster innovation. Initiatives include smart mobility, environmental monitoring, and citizen participation.
6. **Seville Smart City:** Seville's Smart City program aims to use technology to improve the efficiency of public services, reduce energy consumption, and enhance the quality of life for residents. Initiatives include smart transportation, sustainable energy, and urban planning.
7. **Bilbao Smart City:** Bilbao's Smart City program aims to use technology to promote economic growth, sustainability, and social innovation. Initiatives include smart mobility, sustainable energy, and citizen participation.
8. **Vitoria-Gasteiz Smart City:** Vitoria-Gasteiz's Smart City program aims to use technology to enhance the quality of life for residents, promote sustainability, and foster innovation. Initiatives include smart mobility, sustainable energy, and environmental monitoring.
9. **A Coruña Smart City:** A Coruña's Smart City program aims to use technology to improve the efficiency of public services, reduce energy consumption, and enhance the quality of life for residents. Initiatives include smart transportation, smart buildings, and citizen participation.
10. **Zaragoza Smart City:** Zaragoza's Smart City program aims to use technology to promote economic growth, sustainability, and social innovation. Initiatives include smart mobility, sustainable energy, and urban innovation.

SWEDEN

1. **Stockholm Royal Seaport:** The Stockholm Royal Seaport project aims to create a sustainable district in the city of Stockholm. The initiative involves the use of renewable energy, district heating and cooling, and smart transportation solutions.
2. **GrowSmarter:** GrowSmarter is a smart city project that aims to develop sustainable and innovative solutions in three cities in Sweden, including Stockholm, Malmö, and Gothenburg. The initiative involves the use of smart grids, renewable energy, and e-mobility solutions.
3. **ElectriCity:** ElectriCity is a project that aims to develop a sustainable public transport system in the city of Gothenburg. The initiative involves the use of electric buses, wireless charging, and smart transportation solutions.
4. **CIVIC:** CIVIC is a project that aims to develop an intelligent transport system in the city of Stockholm. The initiative involves the use of smart traffic management, e-mobility solutions, and connected and autonomous vehicles.
5. **Smart Housing Småland:** Smart Housing Småland is a project that aims to develop smart and sustainable housing solutions in the region of Småland in Sweden. The initiative involves the use of energy-efficient building design, renewable energy sources, and smart home automation.
6. **Viable Cities:** Viable Cities is a smart city project that aims to develop sustainable and livable urban areas in the city of Lund. The initiative involves the use of smart energy systems, renewable energy, and e-mobility solutions.
7. **Drive Sweden:** Drive Sweden is a national strategic innovation program that aims to develop sustainable and innovative mobility solutions in Sweden. The initiative focuses on autonomous vehicles, electric vehicles, and smart transportation systems.
8. **Gothenburg Green City Zone:** The Gothenburg Green City Zone project aims to create a sustainable and green district in the city of Gothenburg. The initiative involves the use of renewable energy, green roofs, and smart transportation solutions.
9. **Södermalm:** Södermalm is a project that aims to develop a sustainable and livable district in the city of Stockholm. The initiative involves the use of energy-efficient buildings, renewable energy sources, and smart transportation solutions.

10. **IRIS:** IRIS is a smart city project that aims to develop sustainable and innovative solutions in the city of Kristiansand in Sweden. The initiative involves the use of smart energy systems, renewable energy, and e-mobility solutions.

SWITZERLAND

1. **Smart City Zug:** Smart City Zug is a project that aims to develop a blockchain-based digital identity platform for the city of Zug. The initiative involves the use of blockchain technology to securely and efficiently manage the identity of residents and businesses.
2. **Future Cities Laboratory:** The Future Cities Laboratory is a research program that aims to develop sustainable urban solutions for cities in Southeast Asia, including Singapore, Jakarta, and Ho Chi Minh City. The initiative involves the use of smart energy systems, sustainable transportation, and green buildings.
3. **City of Things:** City of Things is a project that aims to develop a large-scale Internet of Things (IoT) infrastructure in the city of Antwerp, Belgium. The initiative involves the use of smart sensors and IoT devices to collect and analyze data in real-time.
4. **Smart City Solothurn:** Smart City Solothurn is a project that aims to develop a sustainable and livable city in Switzerland. The initiative involves the use of smart energy systems, sustainable transportation, and green buildings.
5. **Swiss Alps Energy:** Swiss Alps Energy is a project that aims to develop a sustainable and innovative mining infrastructure in the Swiss Alps. The initiative involves the use of renewable energy sources, such as hydro and solar power, to power mining operations.
6. **Smart City Basel:** Smart City Basel is a project that aims to develop a smart and sustainable city in Switzerland. The initiative involves the use of smart energy systems, sustainable transportation, and green buildings.
7. **EcoBus:** EcoBus is a project that aims to develop a sustainable and energy-efficient public transportation system in the city of Geneva. The initiative involves the use of electric buses, wireless charging, and smart transportation solutions.
8. **Smart City Geneva:** Smart City Geneva is a project that aims to develop a smart and sustainable city in Switzerland. The initiative involves the use of smart energy systems, sustainable transportation, and green buildings.

9. **Smart Energy Region Zurich:** The Smart Energy Region Zurich is a project that aims to develop a smart energy system for the region of Zurich. The initiative involves the use of renewable energy sources, energy storage systems, and smart grids.
10. **My Smart Life:** My Smart Life is a project that aims to develop sustainable and innovative solutions in the city of Nantes in France, as well as in several cities in Switzerland, including Zurich, Geneva, and Lausanne. The initiative involves the use of smart energy systems, sustainable transportation, and green buildings.

THAILAND

1. **Smart Traffic Management System, Bangkok:** Implementing a smart traffic management system to improve traffic flow and reduce congestion in the capital city of Bangkok.
2. **Energy-Efficient Building, Bangkok:** Promoting the construction of energy-efficient buildings in Bangkok to reduce energy consumption and improve sustainability.
3. **Smart Lighting, Bangkok:** Implementing smart lighting systems in Bangkok to reduce energy consumption and improve public safety.
4. **E-Governance, Nationwide:** Implementing e-governance systems to provide citizens with online access to government services and information.
5. **Intelligent Water Management, Chiang Mai:** Implementing an intelligent water management system to optimize water usage and reduce waste in Chiang Mai.
6. **Waste Management, Nationwide:** Implementing smart waste management systems to improve waste collection, recycling and disposal.
7. **Smart Grid, Nationwide:** Developing a smart grid system to improve energy distribution and reliability.
8. **Smart Health, Nationwide:** Implementing smart health systems to improve access to healthcare and reduce healthcare costs.
9. **Smart Agriculture, Nationwide:** Implementing smart agriculture systems to improve crop yields, reduce waste and increase sustainability.
10. **Smart Parks, Nationwide:** Developing smart park systems to improve park management, reduce waste and increase sustainability.

TURKEY

1. **Istanbul Smart City Project:** The project aims to provide an integrated digital infrastructure for Istanbul's transportation, energy, waste management, and security systems.
2. **Konya Smart City Project:** The project includes various initiatives such as smart transportation, smart health, smart energy management, and a smart city center.
3. **Izmir Smart City Project:** The project focuses on enhancing the quality of life of residents by improving public transportation, parking management, waste management, and water conservation.
4. **Antalya Smart City Project:** The project aims to make Antalya a more sustainable and livable city through initiatives like smart waste management, energy efficiency, and smart transportation.
5. **Bursa Smart City Project:** The project includes the implementation of a smart transportation system, smart parking system, and smart lighting system.
6. **Eskisehir Smart City Project:** The project involves initiatives such as smart energy management, smart water management, and smart transportation to make the city more sustainable and eco-friendly.
7. **Trabzon Smart City Project:** The project aims to improve public safety and security through the implementation of smart surveillance and emergency response systems.
8. **Sakarya Smart City Project:** The project focuses on sustainable development through initiatives such as smart energy management, smart transportation, and smart waste management.
9. **Adana Smart City Project:** The project aims to improve the quality of life of residents through initiatives like smart transportation, smart lighting, and smart waste management.
10. **Denizli Smart City Project:** The project focuses on the development of a smart transportation system, smart parking system, and smart waste management to improve the city's overall efficiency and sustainability.

UNITED ARAB EMIRATES

1. **Smart Dubai Platform, Dubai:** A comprehensive platform aimed at making Dubai the smartest and happiest city in the world by providing innovative digital solutions for city services.

2. **Dubai Electricity and Water Authority, DEWA Smart Grid, Dubai:** DEWA's smart grid project aims to integrate renewable energy sources and improve energy efficiency.
3. **Abu Dhabi Smart City, Abu Dhabi:** A smart city initiative aimed at creating a sustainable, efficient and connected city through the integration of smart technologies and infrastructure.
4. **Dubai Silicon Oasis Smart City, Dubai:** A smart city initiative aimed at creating a self-sufficient technology park in Dubai, providing residents with a high quality of life through the use of smart technologies.
5. **Masdar City, Abu Dhabi:** A sustainable smart city that leverages cutting-edge technology and renewable energy to create a model for sustainable urban living.
6. **Internet of Things, IoT Implementation, Nationwide:** Implementing IoT technology across the country to improve city services and enhance citizens' quality of life.
7. **Smart Parking System, Dubai:** Implementing a smart parking system in Dubai to improve parking efficiency and reduce traffic congestion.
8. **Dubai Data Initiative, Dubai:** A project aimed at transforming Dubai into a data-driven city through the collection and analysis of data from various sources.
9. **Smart Buildings, Nationwide:** Implementing smart building technology to improve building efficiency and reduce energy consumption.
10. **Smart Waste Management System, Nationwide:** Implementing smart waste management systems to improve waste collection, recycling and disposal.

UNITED KINGDOM

1. **Future Cities Catapult:** Future Cities Catapult is a government-backed organization that focuses on developing innovative solutions to urban challenges. It works with startups, universities, and other partners to create new smart city technologies and services.
2. **Bristol's Smart City Program:** Bristol's Smart City Program aims to create a sustainable, connected and innovative city that improves the quality of life for residents. The program includes initiatives such as the development of a smart energy grid, the implementation of

smart transportation systems, and the use of open data to inform policy-making.

3. **Manchester's CityVerve:** Manchester's CityVerve is a smart city demonstrator project that uses IoT technologies to improve public services, transportation, and healthcare. The project includes the deployment of smart streetlights, the use of sensors to monitor air quality, and the implementation of a connected healthcare system.

4. **London's Smart Infrastructure:** London is implementing a smart infrastructure system that includes the use of smart sensors to monitor traffic, air quality, and energy usage. The system also includes the development of a smart grid that incorporates renewable energy sources and energy storage.

5. **Glasgow's Future City:** Glasgow's Future City is a smart city demonstrator project that aims to use technology to enhance public services, promote economic growth, and improve quality of life. The project includes initiatives such as the deployment of smart streetlights, the use of data analytics to improve public transportation, and the implementation of smart energy systems.

6. **Newcastle's Urban Observatory:** Newcastle's Urban Observatory is a research project that uses sensors and data analytics to monitor the city's environment, infrastructure, and energy usage. The project aims to provide data-driven insights to inform policy-making and improve the quality of life for residents.

7. **Milton Keynes' Smart City Program:** Milton Keynes' Smart City Program is a long-term initiative that aims to create a sustainable and connected city that uses technology to improve public services and economic growth. The program includes initiatives such as the deployment of autonomous vehicles, the development of a smart energy grid, and the use of data analytics to inform policy-making.

8. **Leeds' Smart Cities Program:** Leeds' Smart Cities Program is a multi-year project that aims to use technology to improve public services, promote economic growth, and enhance quality of life for residents. The program includes initiatives such as the implementation of smart transportation systems, the use of data analytics to improve public health, and the development of a smart energy grid.

9. **Birmingham's Digital Districts:** Birmingham's Digital Districts is an initiative that aims to create new technology hubs in the city that promote innovation, entrepreneurship, and economic growth. The program includes the development of smart infrastructure, the use of data analytics to inform policy-making, and the creation of new digital training programs.

10. **Cambridge's Smart Cambridge:** Cambridge's Smart Cambridge is a project that uses IoT technologies to improve transportation, air quality, and public safety. The project includes initiatives such as the implementation of smart traffic management systems, the use of sensors to monitor air quality, and the deployment of connected emergency services.

UNITED STATES OF AMERICA

1. **LinkNYC in New York City:** LinkNYC is a network of digital kiosks that provide free Wi-Fi, phone calls, and device charging to the public. The kiosks also feature touchscreens that display maps, transit information, and local services, and provide emergency alerts and public safety information.
2. **Smart Columbus in Columbus, Ohio:** Smart Columbus is a smart city initiative that aims to transform the city's transportation system, promote sustainable energy, and enhance mobility options for residents. The project includes the deployment of electric vehicles, the development of smart transportation systems, and the implementation of data-driven decision-making processes.
3. **Smart Cities Challenge in the United States:** The Smart Cities Challenge is a competition sponsored by the U.S. Department of Transportation that provides funding to cities that propose innovative solutions to urban challenges. The competition has helped fund a variety of smart city projects, including the deployment of autonomous shuttles, the development of smart traffic management systems, and the creation of open data platforms.
4. **CityKey in Chicago:** CityKey is a digital ID card that provides access to a variety of city services, including public transportation, libraries, and cultural institutions. The card also functions as a payment card and can be used to access discounts at local businesses.
5. **Smart Parking Meters in San Francisco:** San Francisco has implemented smart parking meters that accept payment via mobile app and provide real-time data on parking availability. The meters use sensors to detect the presence of vehicles and adjust pricing based on demand.
6. **Living Lab in Seattle:** Seattle's Living Lab is a program that enables companies to test new technologies in a real-world urban environment. The program provides access to city data and

infrastructure, and encourages the development of innovative solutions to urban challenges.
7. **Smart Water Meters in Los Angeles:** Los Angeles has deployed smart water meters that use sensors to detect leaks and provide real-time data on water usage. The meters can also send alerts to customers in case of high usage or leaks.
8. **The Array of Things in Chicago:** The Array of Things is a network of sensors that collect data on the urban environment, including air quality, temperature, and noise levels. The data is used to inform urban planning and decision-making processes, and is made available to the public through an open data portal.
9. **Smart Lighting in San Diego:** San Diego has implemented a smart lighting system that uses sensors to adjust lighting levels based on pedestrian and vehicular traffic. The system can also detect faults and send alerts to maintenance crews.
10. **Smart911 in the United States:** Smart911 is a national emergency response system that enables users to create a profile with critical information, such as medical conditions, disabilities, and household details, which can be accessed by 911 operators in case of an emergency. The system can also receive text messages and photos, which can be helpful in situations where speaking is difficult or dangerous. Smart911 is currently available in more than 40 states across the US.

VIETNAM

1. **Da Nang Smart City, Da Nang:** A smart city project aimed at transforming Da Nang into a smart city through the use of innovative technologies.
2. **Ho Chi Minh City Smart City, Ho Chi Minh City: A** smart city project aimed at transforming Ho Chi Minh City into a smart city through the use of innovative technologies.
3. **Ha Noi Smart City, Ha Noi:** A smart city project aimed at transforming Ha Noi into a smart city through the use of innovative technologies.
4. **Smart Traffic Management System, Nationwide:** Implementing a smart traffic management system to improve traffic flow, reduce congestion and improve road safety.
5. **Smart Energy Management System, Nationwide:** Implementing a smart energy management system to optimize energy usage and reduce costs.

6. **Smart Buildings, Nationwide:** Implementing smart building technologies to optimize energy usage and reduce costs.
7. **Smart Water Management, Nationwide:** Implementing smart water management systems to improve water distribution and reduce waste.
8. **Smart Waste Management, Nationwide:** Implementing smart waste management systems to improve waste collection, recycling and disposal.
9. **Can Tho Smart City, Can Tho:** A smart city project aimed at transforming Can Tho into a smart city through the use of innovative technologies.
10. **Smart Agriculture, Nationwide:** Implementing smart agriculture systems to improve crop yields, reduce waste and increase sustainability.

Summary: Smart City projects are becoming increasingly important as urbanization continues to accelerate. These projects aim to use advanced technologies, data analytics, and citizen engagement to improve the delivery of public services and infrastructure. In this chapter, we will provide an overview of Smart City projects around the world, with a focus on initiatives launched in different countries. This will provide readers with a better understanding of how cities around the world are tackling the challenges of urbanization through Smart City projects.

Conclusion: Smart City projects are an important tool for addressing the challenges of urbanization. By leveraging advanced technologies, data analytics, and citizen engagement, these initiatives can help to improve the delivery of public services and infrastructure while also reducing the environmental impact of urbanization. This chapter has provided an overview of Smart City projects around the world, with a focus on the initiatives that have been launched in different countries. By examining these projects, readers can gain a better understanding of how cities around the world are using Smart City initiatives to build more efficient and sustainable urban environments.

Quote: "The true value of technological progress lies in its ability to enhance the quality of life for humanity." Emin Hasic

ONE SMART WORLD
ARE YOU READY FOR IT?

| X |
PROGRAMS BY INDUSTRY

Smart cities are the future of urban development, aimed at improving the quality of life for their citizens by leveraging cutting-edge technologies. Smart city programs are designed to tackle urban problems by using data-driven solutions that are integrated, efficient, and sustainable. The potential benefits of these programs range from increased citizen participation to improved public safety, better energy management, and enhanced transportation systems. Smart city programs are being implemented across a variety of industries, from transportation and energy to healthcare and education, in an effort to create a more connected and sustainable urban environment. In this chapter, we will explore the different types of smart city programs being implemented in various industries.

SMART CITY 5G NETWORK (SC5GN)

Using technology to enhance connectivity and communication infrastructure, such as 5G networks, which allows for faster data transmission and more connected devices.

The implementation of a SC5GN is a game-changer for any city. 5G networks are the next generation of wireless technology that promises to revolutionize the way we connect and communicate. With 5G, cities can experience faster data transmission and more connected devices, which is crucial for the smooth functioning of a smart city.

One of the main benefits of a 5G network is the increased speed of data transmission. With 5G, data can be transferred at much faster rates compared to previous generations of wireless technology. This means that a 5G network can handle more data, more devices, and more users, which is essential for a smart city that is constantly growing and evolving. With faster data transmission, smart city services such as traffic management, emergency response, and e-governance can be more efficient, providing an improved quality of life for citizens.

Another benefit of a 5G network is the increased connectivity of devices. 5G networks can support a much larger number of connected devices compared to previous generations of wireless technology. This means that a 5G network can connect more devices such as IoT devices, sensors, and cameras, which is crucial for a smart city that is constantly expanding its infrastructure. With more connected devices, smart city services such as public safety, energy management, and transportation can be more effective, providing an improved quality of life for citizens.

5G networks can support low latency, high-density connections, and high-speed data transfer. This is beneficial for industries such as healthcare, transportation, and manufacturing where real-time communication is vital. This can lead to faster decision-making, increased efficiency and cost savings.

GENERAL FRAMEWORK

The following framework is a general one and the actual implementation will depend on the specific requirements and constraints of the city.

1. Identify the specific cyber threats that are most likely to impact the community.
2. Conduct a risk assessment to determine the potential impact and likelihood of these threats.
3. Develop a comprehensive cybersecurity strategy that addresses the identified risks.

4. Implement encryption for all sensitive data stored and transmitted within the community.
5. Deploy intrusion detection systems to monitor for and detect potential security breaches.
6. Establish incident response protocols to quickly and effectively respond to security incidents.
7. Provide cybersecurity training and education for community members and employees.
8. Implement access controls to limit who can access sensitive data and systems.
9. Regularly update software and systems to ensure they are protected against known vulnerabilities.
10. Conduct regular security audits and vulnerability assessments to identify potential weaknesses.
11. Implement a disaster recovery plan to ensure the community can quickly recover from a security incident.
12. Establish partnerships with local law enforcement and other agencies to enhance incident response capabilities.
13. Develop a comprehensive incident reporting process to ensure all security incidents are tracked and investigated.
14. Implement multi-factor authentication for all system and network access.
15. Conduct regular penetration testing to identify and address vulnerabilities in the community's systems and networks.
16. Use firewalls and other network security tools to protect against cyber threats.
17. Use security information and event management (SIEM) systems to monitor for and analyze security-related data.
18. Implement data loss prevention (DLP) systems to prevent sensitive data from being inadvertently shared or stolen.
19. Utilize threat intelligence to stay informed about emerging cyber threats and how to protect against them.
20. Implement security best practices and guidelines, such as the NIST Cybersecurity Framework.
21. Develop an incident management plan that includes incident response and recovery procedures.
22. Regularly test incident response plan through simulated incident scenarios.
23. Regularly review and update the cybersecurity strategy to address new threats and vulnerabilities.

24. Develop a communication plan for notifying community members and stakeholders about security incidents and the actions being taken to address them.
25. Continuously monitor the effectiveness of the cybersecurity measures and make adjustments as needed.

FREQUENTLY ASKED QUESTIONS

What is a Smart City 5G Network, and how does it work? A Smart City 5G Network (SC5GN) is a wireless network that uses 5G technology to connect devices, sensors, and infrastructure within a city. It works by using small cells, which are low-power radio access nodes that can be placed on light poles, rooftops, or other city infrastructure. These small cells connect to a central network that can manage and process data from all the devices connected to the network.

What are the benefits of implementing an SC5GN in a city? There are several benefits to implementing an SC5GN in a city, including faster and more reliable connectivity, reduced latency, improved data collection and analysis, and enhanced public safety and security. It can also support the development of new smart city applications and services, such as traffic management, energy efficiency, and healthcare.

How will an SC5GN impact my daily life as a resident of a Smart City? As a resident of a Smart City with an SC5GN, you may experience faster and more reliable internet connectivity, improved public transportation systems, more efficient energy usage, and enhanced public safety and security. You may also have access to new services and applications that are designed to improve your quality of life.

What are the privacy and security concerns associated with an SC5GN? Privacy and security are important concerns when it comes to implementing an SC5GN. The vast amount of data collected by the network could potentially be used for nefarious purposes if not properly secured. As such, it's essential to implement strong security measures to protect both personal information and critical infrastructure from cyber-attacks.

What is the cost of implementing an SC5GN, and who pays for it? The cost of implementing an SC5GN can vary depending on the size and complexity of the project. Typically, it's a joint effort between city governments, private companies, and other stakeholders. Some cities may offer incentives to private companies to encourage them to invest in the network infrastructure.

What happens if the SC5GN fails or experiences downtime? Downtime on an SC5GN could potentially disrupt critical services and lead to significant economic losses. To minimize downtime, it's important to implement backup and redundancy systems, conduct regular maintenance and testing, and have contingency plans in place in case of emergencies.

Will an SC5GN create job opportunities for residents of the Smart City? The implementation of an SC5GN could create job opportunities in areas such as network installation, maintenance, and management. It could also support the growth of new businesses and services that rely on the network infrastructure.

How will the SC5GN impact the environment? The impact of an SC5GN on the environment can vary depending on the specific project. However, in general, the implementation of a Smart City with an SC5GN could help to reduce carbon emissions by improving energy efficiency in buildings and transportation systems. It could also help to reduce traffic congestion and air pollution.

What kind of devices can be connected to an SC5GN? Almost any device that has a wireless connection can be connected to an SC5GN. This includes smartphones, tablets, laptops, and IoT devices such as smart sensors, traffic lights, and security cameras.

What are the challenges associated with implementing an SC5GN? Some of the challenges associated with implementing an SC5GN include the need for significant infrastructure investment, regulatory hurdles, and the complexity of coordinating multiple stakeholders. Additionally, there may be public concerns regarding privacy and security that need to be addressed.

Conclusion: *The implementation of a SC5GN is an essential step towards building a smarter and more connected city. With faster data transmission and more connected devices, smart city services can be more efficient, providing an improved quality of life for citizens. 5G networks can support low latency, high-density connections and high-speed data transfer, which can lead to increased efficiency and cost savings for industries.*

SMART CITY AGRICULTURE (SCAG)

Using technology to improve agricultural productivity and sustainability, such as precision farming and vertical farming.

The implementation of SCAG is a vital step towards building a sustainable and resilient city. By using technology to improve agricultural productivity and sustainability, cities can ensure a stable food supply, reduce the environmental footprint, and create new economic opportunities.

One of the main benefits of SCAG is precision farming. Precision farming is a method of farming that uses technology such as sensors, drones, and GPS to optimize crop yields and reduce inputs such as water and fertilizer. With precision farming, farmers can target specific areas of the field that need attention, resulting in more efficient use of resources and higher crop yields. This means that cities can produce more food with less land, water, and fertilizer, which is crucial for a sustainable and resilient city.

Another benefit of SCAG is vertical farming. Vertical farming is a method of farming that uses technology such as LED lighting, hydroponics, and automation to grow crops in an indoor, stacked configuration. With vertical farming, cities can produce more food in less space, reduce the environmental footprint, and create new economic opportunities. This means that cities can produce fresh and healthy food year-round, regardless of weather or climate conditions, which is crucial for a sustainable and resilient city.

SCAG can also support the development of urban farming and community gardening, which can increase access to fresh and healthy food for citizens, particularly for those living in food deserts, and promote community cohesion.

GENERAL FRAMEWORK

The following framework is a general one and the actual implementation will depend on the specific requirements and constraints of the city.

1. Identify the specific agricultural challenges and opportunities in the city.
2. Conduct a feasibility study to assess the potential for implementing smart agriculture solutions.
3. Develop a comprehensive plan for implementing smart agriculture in the city.
4. Research and evaluate different smart agriculture technologies, such as precision farming and vertical farming.
5. Identify potential partners and stakeholders for implementing smart agriculture solutions.
6. Develop a strategy for financing the implementation of smart agriculture solutions.
7. Establish a monitoring and evaluation system to track progress and measure the impact of smart agriculture solutions.
8. Develop training and education programs for farmers, agricultural workers, and other stakeholders.
9. Implement precision farming technologies such as sensors, drones, and precision irrigation systems.
10. Develop vertical farming solutions, such as indoor and vertical hydroponic systems.
11. Incorporate sustainable practices such as water conservation and recycling, energy efficiency, and reducing chemical use.
12. Utilize big data and analytics to optimize crop yields and improve decision-making.
13. Incorporate IoT (Internet of Things) technology to monitor and control environmental conditions.

14. Develop urban agriculture projects, such as rooftop gardens and community gardens.
15. Use AI and machine learning to predict crop yields and optimize crop management.
16. Incorporate precision livestock management technologies to improve animal health and productivity.
17. Implement precision crop management technologies such as precision planting and precision harvesting.
18. Develop a strategy for the distribution and marketing of locally grown produce.
19. Develop a research and development program to continually improve smart agriculture technologies.
20. Create a platform for data sharing and collaboration among farmers and agricultural researchers.
21. Create a system for monitoring and reporting on food safety and quality.
22. Incorporate agricultural education in schools and community education programs.
23. Create a plan for the integration of smart agriculture solutions with other smart city initiatives.
24. Develop a strategy for engaging community members and stakeholders in smart agriculture initiatives.
25. Continuously evaluate and adjust the smart agriculture plan to address new challenges and opportunities.

FREQUENTLY ASKED QUESTIONS

What is Smart City Agriculture (SCAG), and how does it work? Smart City Agriculture (SCAG) is the practice of growing crops and raising livestock within urban environments. It works by using innovative farming techniques such as hydroponics, aquaponics, and vertical farming to maximize crop yield in limited space. This allows city dwellers to have access to locally grown, fresh produce.

What are the benefits of implementing SCAG in a city? There are several benefits of implementing SCAG in a city, including increased access to fresh produce, reduced carbon footprint from transport, improved food security, and the creation of green jobs. It also supports the development of a more sustainable and resilient food system for the city.

Can anyone participate in SCAG, or is it only for commercial use? SCAG can be practiced by both commercial and individual farmers. There are community gardens and urban farms that are open to the public and offer opportunities for city residents to participate in SCAG.

What kind of crops can be grown in SCAG, and how are they different from traditional farming? SCAG allows for a wide variety of crops to be grown, including leafy greens, herbs, fruits, and vegetables. These crops are grown using hydroponics or other innovative techniques, which require less water and space than traditional farming. Additionally, SCAG can provide fresh produce year-round, which is not possible with traditional farming in many areas.

What kind of technology is used in SCAG? SCAG utilizes a range of technologies, including sensors, automation, and data analytics. These technologies can be used to monitor plant growth, control the environment, and optimize crop yield.

How does SCAG impact the environment? SCAG can have a positive impact on the environment by reducing the carbon footprint associated with transportation and reducing the need for large-scale farming. Additionally, SCAG can help to promote green spaces in urban environments and reduce the urban heat island effect.

What are the challenges associated with implementing SCAG? Some of the challenges associated with implementing SCAG include the need for significant investment in infrastructure, the availability of suitable land for farming, and the need for skilled labour to operate the farms. There may also be regulatory challenges related to zoning and land use.

How can SCAG improve food security in a city? SCAG can improve food security by providing access to fresh produce for city residents, reducing the reliance on imported food, and creating a more resilient local food system. Additionally, SCAG can help to address food deserts and improve access to healthy food in underserved communities.

Is SCAG sustainable in the long term? SCAG has the potential to be sustainable in the long term if it is implemented in a way that is environmentally and economically viable. This requires careful planning and investment in sustainable practices such as renewable energy and water conservation.

Can SCAG help to address climate change? SCAG can help to address climate change by reducing the carbon footprint associated with transporting food and by sequestering carbon through plant growth. Additionally, SCAG can help to promote sustainable practices such as composting and reducing food waste.

Conclusion: *The implementation of SCAG is an essential step towards building a sustainable and resilient city. By using technology such as precision farming and vertical farming, cities can ensure a stable food supply, reduce the environmental footprint, and create new economic opportunities. Urban farming and community gardening can increase access to fresh and healthy food for citizens and promote community cohesion.*

SMART CITY AIR QUALITY MONITORING (SCAQM)

Using technology to monitor and manage air pollution, such as sensors and predictive analytics.

The implementation of SCAQM is a vital step towards building a healthy and livable city. By using technology to monitor and manage air pollution, cities can improve the air quality, protect citizens' health and well-being, and reduce environmental impacts.

One of the main benefits of SCAQM is the ability to measure and track air pollution in real-time. With sensors and predictive analytics, cities can measure and track the levels of pollutants such as particulate matter, nitrogen oxides, and sulphur dioxide. This means that cities can identify sources of air pollution and take action to reduce emissions, which is crucial for protecting citizens' health and well-being.

Another benefit of SCAQM is the ability to predict and forecast air pollution levels. With predictive analytics, cities can use historical data and weather patterns to predict air pollution levels and take action to reduce emissions. This means that cities can anticipate and prepare for air pollution events, which is crucial for protecting citizens' health and well-being.

SCAQM can also support the development of clean energy and sustainable transportation, which can reduce emissions and improve air quality. This can also lead to cost savings for both citizens and the city, in terms of healthcare costs, and also from the reduction of energy consumption.

GENERAL FRAMEWORK

The following framework is a general one and the actual implementation will depend on the specific requirements and constraints of the city.

1. Identify the specific air pollution challenges and opportunities in the city.
2. Conduct a feasibility study to assess the potential for implementing smart air quality monitoring solutions.
3. Develop a comprehensive plan for monitoring and managing air pollution in the city.
4. Research and evaluate different air quality monitoring technologies, such as sensors and predictive analytics.
5. Identify potential partners and stakeholders for implementing smart air quality monitoring solutions.
6. Develop a strategy for financing the implementation of smart air quality monitoring solutions.
7. Establish a monitoring and evaluation system to track progress and measure the impact of smart air quality monitoring solutions.
8. Develop training and education programs for city officials, residents, and other stakeholders.

9. Implement a network of air quality sensors throughout the city to measure various pollutants.
10. Use predictive analytics to forecast air pollution levels and inform management decisions.
11. Develop a system for real-time air quality monitoring and reporting.
12. Utilize big data and analytics to identify patterns and trends in air pollution.
13. Incorporate IoT (Internet of Things) technology to monitor and control emissions from buildings and vehicles.
14. Develop a strategy for reducing emissions from transportation, such as promoting electric vehicles and public transportation.
15. Use AI and machine learning to optimize traffic flow and reduce emissions from vehicles.
16. Develop a plan for addressing air pollution hotspots and non-compliance with air quality regulations.
17. Develop a system for reporting and addressing complaints related to air pollution.
18. Create a platform for data sharing and collaboration among city officials, researchers, and the public.
19. Create a system for monitoring and reporting on compliance with air quality regulations.
20. Develop a research and development program to continually improve smart air quality monitoring technologies.
21. Develop a strategy for engaging community members and stakeholders in smart air quality monitoring initiatives.
22. Incorporate air quality education in schools and community education programs.
23. Create a plan for the integration of smart air quality monitoring solutions with other smart city initiatives.
24. Develop a strategy for addressing the specific needs of vulnerable populations, such as children and the elderly.
25. Continuously evaluate and adjust the smart air quality monitoring plan to address new challenges and opportunities.

FREQUENTLY ASKED QUESTIONS

What is Smart City Air Quality Monitoring (SCAQM), and why is it important? Smart City Air Quality Monitoring (SCAQM) is a system that monitors the air quality in a city using a network of sensors.

It is important because air pollution is a major public health issue, and SCAQM can help to identify areas with poor air quality and take measures to reduce pollution.

How does SCAQM work? SCAQM works by using a network of sensors placed throughout a city to monitor air quality. These sensors can detect a range of pollutants, including particulate matter, ozone, and nitrogen dioxide. The data from the sensors is collected and analysed to provide information about air quality in real-time.

Who is responsible for implementing SCAQM in a city? The implementation of SCAQM in a city is typically the responsibility of the city government, with support from technology vendors and other stakeholders.

How can the public access the data collected by SCAQM? The data collected by SCAQM is typically made available to the public through a web portal or mobile app. This allows citizens to monitor air quality in real-time and make informed decisions about their activities.

What are the benefits of SCAQM for the general public? The benefits of SCAQM for the general public include improved public health, increased awareness of air pollution, and the ability to make informed decisions about outdoor activities.

How can SCAQM be used to reduce air pollution in a city? SCAQM can be used to identify areas with poor air quality and take measures to reduce pollution. For example, if the data shows that a particular road or industrial facility is a significant source of pollution, steps can be taken to reduce emissions from that source.

What kind of pollutants can SCAQM detect? SCAQM can detect a range of pollutants, including particulate matter, ozone, nitrogen dioxide, sulphur dioxide, and carbon monoxide.

How accurate are the sensors used in SCAQM? The sensors used in SCAQM are typically highly accurate, but there can be variability in the data depending on the location and environmental factors.

Therefore, it's important to use a network of sensors to ensure the accuracy of the data.

Can SCAQM be used to predict air quality in the future? SCAQM can be used to make predictions about air quality in the future based on historical data and environmental factors. This can help to anticipate areas with poor air quality and take measures to mitigate pollution.

What are the challenges associated with implementing SCAQM? Some of the challenges associated with implementing SCAQM include the cost of installing and maintaining the sensors, ensuring the accuracy and reliability of the data, and addressing privacy concerns related to the collection and use of personal data.

Conclusion: *The implementation of SCAQM is an essential step towards building a healthy and livable city. By using technology such as sensors and predictive analytics, cities can measure and track air pollution in real-time, predict and forecast air pollution levels, and take action to reduce emissions. Smart city air quality monitoring can also support the development of clean energy and sustainable transportation, which can reduce emissions and improve air quality, leading to cost savings for both citizens and the city.*

SMART CITY ANALYTICS (SCA)

Using technology to analyze and interpret data from smart city systems and services, such as data visualization and machine learning.

The implementation of SCA is a crucial step towards building a data-driven and efficient city. By using technology to analyze and interpret data from smart city systems and services, cities can gain valuable insights, make data-driven decisions, and improve the overall performance of the city.

One of the main benefits of SCA is the ability to visualize data and gain insights. With data visualization, cities can display large amounts of data in a clear and concise format, which allows for easy interpretation and understanding. This means that cities can identify patterns, trends, and anomalies in the data, which is crucial for making data-driven decisions.

Another benefit of SCA is the ability to use machine learning and artificial intelligence. With machine learning and AI, cities can analyze large amounts of data, identify patterns and make predictions. This means that cities can anticipate problems, optimize operations and make better decisions, which can lead to cost savings and improved services.

SCA can also support the development of citizen engagement, by providing citizens with access to data and insights related to the city's performance. This can help to build trust and transparency between the city and its citizens and also help citizens understand the impact of their actions on the city.

GENERAL FRAMEWORK

The following framework is a general one and the actual implementation will depend on the specific requirements and constraints of the city.

1. Identify the specific data and analytics challenges and opportunities in the city.
2. Conduct a feasibility study to assess the potential for implementing smart city analytics solutions.
3. Develop a comprehensive plan for analyzing and interpreting data from smart city systems and services.
4. Research and evaluate different data visualization and machine learning technologies.
5. Identify potential partners and stakeholders for implementing smart city analytics solutions.
6. Develop a strategy for financing the implementation of smart city analytics solutions.
7. Establish a monitoring and evaluation system to track progress and measure the impact of smart city analytics solutions.
8. Develop training and education programs for city officials and other stakeholders on data analysis and interpretation.

9. Implement a data management and warehousing system to store and organize data from smart city systems and services.
10. Use data visualization and dashboard tools to present data in an easily understandable format.
11. Utilize machine learning and AI to analyze data and uncover insights.
12. Develop a system for real-time data analysis and reporting.
13. Utilize big data and analytics to identify patterns and trends in city operations.
14. Incorporate IoT (Internet of Things) technology to gather data from connected devices.
15. Develop a strategy for data governance, including policies and procedures for data access and security.
16. Use predictive analytics to forecast future trends and inform decision-making.
17. Develop a system for reporting and addressing complaints related to city services.
18. Create a platform for data sharing and collaboration among city officials, researchers, and the public.
19. Create a system for monitoring and reporting on city performance metrics.
20. Develop a research and development program to continually improve smart city analytics technologies.
21. Develop a strategy for engaging community members and stakeholders in smart city analytics initiatives.
22. Incorporate data literacy education in schools and community education programs.
23. Create a plan for the integration of smart city analytics solutions with other smart city initiatives.
24. Develop a strategy for addressing the specific needs of vulnerable populations, such as children and the elderly.
25. Continuously evaluate and adjust the smart city analytics plan to address new challenges and opportunities.

FREQUENTLY ASKED QUESTIONS

What are Smart City Analytics? Smart City Analytics (SCA) refers to the use of technology to analyze and interpret data from various smart city systems and services. This includes the use of data visualization, machine learning, and other analytics techniques to make sense of the data.

What are the benefits of Smart City Analytics? The benefits of SCA include improved decision-making, enhanced resource allocation, increased efficiency, and better overall management of a city's resources and services. By analyzing data, city officials can identify trends, patterns, and potential areas of improvement.

What types of data can be analysed with Smart City Analytics? SCA can analyze a wide variety of data, including traffic patterns, energy usage, waste management, public transportation, and more. By analyzing this data, city officials can gain insights into how these systems are performing and identify potential areas for improvement.

How is Smart City Analytics different from traditional analytics? Smart City Analytics is different from traditional analytics in that it focuses specifically on analyzing data from smart city systems and services. This requires specialized tools and techniques, such as machine learning algorithms, that can handle the large and complex datasets that are typically generated by smart city systems.

What role does machine learning play in Smart City Analytics? Machine learning is a key component of Smart City Analytics, as it enables city officials to identify patterns and trends in the data that would be difficult or impossible to detect manually. Machine learning algorithms can also be used to make predictions about future trends and potential problems.

What are some examples of Smart City Analytics in action? Some examples of SCA in action include using traffic data to optimize traffic flow and reduce congestion, using energy usage data to identify potential areas for energy savings, and using public transportation data to improve the efficiency and reliability of bus and train schedules.

What are the challenges of implementing Smart City Analytics? One of the main challenges of implementing SCA is the need for specialized tools and expertise, which can be costly and difficult to obtain. In addition, there may be concerns about data privacy and security, as well as potential resistance from city officials and citizens who may be sceptical of the benefits of SCA.

What are some best practices for implementing Smart City Analytics? Best practices for implementing SCA include starting small and focusing on a few key areas of data analysis, building partnerships with other cities and organizations to share data and expertise, and involving citizens in the process to build trust and ensure that their needs and concerns are being addressed.

How can Smart City Analytics contribute to sustainability efforts? SCA can contribute to sustainability efforts by identifying areas of resource waste or inefficiency, such as excessive energy usage or traffic congestion. By addressing these issues, cities can reduce their environmental impact and improve their overall sustainability.

What is the future of Smart City Analytics? The future of SCA is likely to involve increased use of machine learning and artificial intelligence, as well as greater integration with other smart city technologies, such as Internet of Things (IoT) devices and smart sensors. As cities continue to grow and become more complex, SCA will play an increasingly important role in managing and optimizing these systems.

Conclusion: The implementation of SCA is a crucial step towards building a data-driven and efficient city. By using technology such as data visualization and machine learning, cities can gain valuable insights, make data-driven decisions, and improve the overall performance of the city. SCA can support the development of citizen engagement by providing citizens with access to data and insights related to the city's performance, which can help to build trust and transparency between the city and its citizens.

SMART CITY BIKE-SHARING (SCBS)

Using technology to improve the availability and accessibility of bike-sharing systems, such as real-time bike tracking and dockless systems. The implementation of SCBS is a vital step towards building a sustainable, healthy and accessible city. By using technology to improve the availability and accessibility of bike-sharing systems, cities can promote active transportation, reduce traffic congestion and emissions, and improve citizens' health and well-being.

One of the main benefits of SCBS is the ability to track bikes in real-time. With real-time bike tracking, cities can monitor the location, usage and availability of bikes. This means that cities can optimize bike distribution, reduce bike theft, and ensure that bikes are always available when and where they are needed, which is crucial for promoting active transportation.

Another benefit of SCBS is the ability to use dockless systems. Dockless systems allow users to pick up and drop off bikes anywhere, without the need for a fixed docking station. This means that cities can expand bike-sharing to areas that were previously unserved, and increase the accessibility of bike-sharing for citizens, particularly for those living in low-income areas.

SCBS can also support the development of multimodal transportation, by providing citizens with alternative transportation options, and reducing traffic congestion and emissions. This can also lead to cost savings for both citizens and the city, in terms of healthcare costs, and also from the reduction of energy consumption.

GENERAL FRAMEWORK

The following framework is a general one and the actual implementation will depend on the specific requirements and constraints of the city.

1. Conduct a study to assess the current demand and potential for bike-sharing in the city.
2. Research and evaluate different bike-sharing technologies and systems, such as dockless and real-time bike tracking.
3. Develop a comprehensive plan for implementing a bike-sharing system in the city.
4. Identify potential partners and stakeholders for the bike-sharing system, such as private companies and community organizations.
5. Develop a strategy for financing the implementation of the bike-sharing system.
6. Establish a monitoring and evaluation system to track progress and measure the impact of the bike-sharing system.
7. Develop a marketing and outreach strategy to promote the bike-sharing system to residents and visitors.
8. Create a system for real-time tracking and monitoring of bikes.
9. Develop a system for reporting and addressing complaints related to the bike-sharing system.
10. Utilize data analytics to inform decision-making and improve the bike-sharing system.
11. Develop a system for managing and maintaining the bikes, docks, and other equipment.
12. Create a system for managing and enforcing parking and usage rules for the bikes.
13. Use mobile and web-based technology to make it easy for people to find and rent bikes.
14. Incorporate payment systems such as credit card, mobile payment and smart card.
15. Create a program for recycling or disposing of bikes that are no longer in use.
16. Develop a system for incorporating bike-sharing into transportation planning and policy.
17. Incorporate bike-sharing into city's transportation network.
18. Develop a strategy for addressing the specific needs of vulnerable populations, such as children and the elderly.
19. Develop a plan for integrating bike-sharing with other smart city initiatives, such as public transportation and car-sharing.
20. Create a system for offering incentives for using bike-sharing, such as discounted public transportation fares.
21. Develop a program for educating users on safe biking practices and city biking laws.

22. Develop a program for encouraging bike-sharing usage among employees of large companies and organizations.
23. Develop a program for encouraging bike-sharing usage among students in schools and universities.
24. Create a system for providing bike-sharing access for low-income residents.
25. Continuously evaluate and adjust the bike-sharing plan to address new challenges and opportunities.

FREQUENTLY ASKED QUESTIONS

What is Smart City Bike-Sharing? Smart City Bike-Sharing (SCBS) is a technology-driven approach to improving the availability and accessibility of bike-sharing systems. This includes the use of real-time bike tracking, dockless systems, and other technologies to make bike-sharing more convenient and user-friendly.

How does SCBS work? SCBS typically involves the use of a mobile app that allows users to find and reserve bikes in real-time. The app may also provide information about bike availability, bike parking locations, and other useful features. Some SCBS systems use dockless bikes, which can be parked and locked anywhere, while others use traditional bike docks.

What are the benefits of SCBS? The benefits of SCBS include increased mobility, reduced traffic congestion, improved air quality, and enhanced overall quality of life. By providing easy access to bikes, SCBS can also promote healthy and active lifestyles.

What types of technology are used in SCBS? SCBS uses a variety of technologies, including GPS tracking, mobile apps, bike sensors, and other Internet of Things (IoT) devices. These technologies enable real-time tracking of bike availability, usage, and maintenance needs.

How does SCBS contribute to sustainability efforts? SCBS contributes to sustainability efforts by promoting the use of bicycles as a low-carbon mode of transportation. By providing easy access to bikes, SCBS can also help reduce the number of cars on the road, thereby reducing traffic congestion and air pollution.

What are the challenges of implementing SCBS? One of the main challenges of implementing SCBS is the need for a robust and reliable technology infrastructure. This includes the need for high-quality GPS tracking, reliable mobile networks, and secure data management systems.

How does SCBS benefit urban planning and design? SCBS can benefit urban planning and design by promoting more sustainable and active modes of transportation. This can help reduce the need for parking spaces and promote the development of more pedestrian- and bike-friendly urban spaces.

How can cities ensure that SCBS is accessible to all users? To ensure that SCBS is accessible to all users, cities may need to consider issues such as affordability, bike availability, and accessibility for people with disabilities. This may require the development of targeted outreach and education programs to promote the use of SCBS.

What are some best practices for implementing SCBS? Best practices for implementing SCBS include conducting a thorough needs assessment, developing a robust technology infrastructure, partnering with bike-sharing providers and other stakeholders, and involving community members in the planning and implementation process.

What is the future of SCBS? The future of SCBS is likely to involve increased use of advanced technologies such as autonomous bikes, electric bikes, and more sophisticated bike sensors. As SCBS continues to evolve, it will play an increasingly important role in promoting sustainable and active transportation in urban areas.

Conclusion: *The implementation of SCBS is a vital step towards building a sustainable, healthy and accessible city. By using technology such as* *real-time bike tracking and dockless systems, cities can promote active transportation, reduce traffic congestion and emissions, and improve citizens' health and well-being. SCBS can support the development of multimodal transportation, by providing citizens with alternative transportation options, and reducing traffic congestion and emissions, which can lead to cost savings for both citizens and the city.*

SMART CITY BUILDING MANAGEMENT (SCBM)

Using technology to optimize building performance and reduce energy consumption, such as building automation systems and smart HVAC systems.

The implementation of SCBM is a crucial step towards building a sustainable, efficient and comfortable city. By using technology to optimize building performance and reduce energy consumption, cities can improve the energy efficiency of buildings, reduce energy costs and carbon emissions, and create a more comfortable environment for citizens and building occupants.

One of the main benefits of SCBM is the ability to use building automation systems (BAS). Building automation systems use technology such as sensors, controllers, and software to monitor and control the various systems within a building, such as lighting, heating, ventilation, and air conditioning (HVAC). This means that cities can optimize the performance of buildings, reduce energy consumption and costs, and improve the comfort and safety of building occupants.

Another benefit of SCBM is the ability to use smart HVAC systems. Smart HVAC systems use technology such as sensors, controllers, and software to monitor and control the temperature, humidity, and air quality within a building. This means that cities can optimize the performance of buildings, reduce energy consumption and costs, and improve the comfort and safety of building occupants.

SCBM can also support the development of renewable energy, by providing buildings with the ability to generate, store and manage their own energy. This can lead to cost savings for both citizens and the city, in terms of energy costs and also in terms of reducing carbon emissions.

GENERAL FRAMEWORK

The following framework is a general one and the actual implementation will depend on the specific requirements and constraints of the city.

1. Conduct a study to assess the current energy consumption and performance of buildings in the city.
2. Research and evaluate different building automation and energy management technologies.
3. Develop a comprehensive plan for implementing smart building management systems in the city.
4. Identify potential partners and stakeholders for the smart building management systems, such as building owners and energy providers.
5. Develop a strategy for financing the implementation of the smart building management systems.
6. Establish a monitoring and evaluation system to track progress and measure the impact of the smart building management systems.
7. Develop a marketing and outreach strategy to promote the smart building management systems to building owners and tenants.
8. Create a system for real-time monitoring and control of building systems such as lighting, HVAC, security, and fire safety.
9. Develop a system for reporting and addressing complaints related to building systems.
10. Utilize data analytics to inform decision-making and improve building performance.
11. Develop a system for managing and maintaining the building automation systems.
12. Incorporate energy-efficient technologies such as LED lighting, smart thermostats, and smart HVAC systems.
13. Utilize building automation systems to optimize energy consumption and reduce costs.
14. Develop a plan to integrate building management systems with other smart city initiatives.

15. Develop a strategy for addressing the specific needs of vulnerable populations, such as elderly and disabled.
16. Develop a plan for incorporating smart building management into building codes and regulations.
17. Develop a program for educating building managers and tenants on the benefits and use of smart building management systems.
18. Develop a program for encouraging the adoption of smart building management systems among property owners.
19. Develop a plan for providing incentives for implementing smart building management systems.
20. Develop a program for providing technical assistance and training to building managers and tenants on smart building management systems.
21. Develop a program for incorporating smart building management systems into the curriculum of schools and universities.
22. Develop a system for providing access to smart building management systems for low-income building owners.
23. Continuously evaluate and adjust the smart building management systems to address new challenges and opportunities.
24. Develop a plan for integrating smart building management systems with the city's energy grid.
25. Develop a plan for integrating smart building management systems with the city's emergency response system.

FREQUENTLY ASKED QUESTIONS

What is Smart City Building Management (SCBM), and how does it work? SCBM is a system that uses technology to manage building operations and maintenance. It works by using sensors and data analytics to monitor and control various building systems, such as heating, ventilation, and lighting.

How can SCBM benefit building owners and managers? SCBM can benefit building owners and managers by improving building efficiency, reducing operating costs, and enhancing the occupant experience.

What kind of data can SCBM collect and analyze? SCBM can collect and analyze data on energy consumption, occupancy levels, temperature, humidity, and other factors that impact building performance.

Can SCBM help to improve building sustainability? Yes, SCBM can help to improve building sustainability by optimizing energy use and reducing waste.

How can building occupants benefit from SCBM? Building occupants can benefit from SCBM by experiencing improved indoor air quality, more comfortable temperatures, and better lighting conditions.

Who is responsible for implementing SCBM in a building? The implementation of SCBM in a building is typically the responsibility of the building owner or manager, with support from technology vendors and other stakeholders.

What kind of challenges are associated with implementing SCBM? Some of the challenges associated with implementing SCBM include ensuring the accuracy and reliability of data, managing the costs of implementing new technologies, and addressing privacy concerns related to the collection and use of personal data.

How can SCBM be used to improve building security? SCBM can be used to monitor building access and security systems, as well as to detect and respond to potential security threats.

Can SCBM be integrated with other smart city systems? Yes, SCBM can be integrated with other smart city systems, such as Smart City Energy Management (SCEM) and Smart City Waste Management (SCWM), to create a more comprehensive and effective smart city infrastructure.

What kind of skills and expertise are required to implement SCBM? Implementing SCBM requires a range of skills and expertise, including data analysis, programming, and facility management. In some cases, external consultants or vendors may be brought in to provide specialized expertise.

Conclusion: The implementation of SCBM is a crucial step towards building a sustainable, efficient and comfortable city. By using technology such as building automation systems and smart HVAC systems, cities can improve the energy efficiency of buildings, reduce energy costs and carbon emissions, and create a more comfortable environment for citizens and building occupants. SCBM can support the development of renewable energy, by providing buildings with the ability to generate, store and manage their own energy, which can lead to cost savings for both citizens and the city.

SMART CITY CITIZEN ENGAGEMENT (SCCE)

Using technology to improve communication and engagement between citizens and city government.

The implementation of SCCE is a vital step towards building a transparent, inclusive, and responsive city. By using technology to improve communication and engagement between citizens and city government, cities can increase citizen participation, build trust and transparency, and improve the overall performance of the city.

One of the main benefits of SCCE is the ability to improve communication between citizens and city government. With technology such as social media, mobile apps, and online platforms, citizens can easily access information about city services and provide feedback to city government. This means that cities can increase citizen participation, and improve communication and responsiveness, which is crucial for building trust and transparency.

Another benefit of SCCE is the ability to use data analytics to understand citizens' needs and preferences. With data analytics, cities can analyze data from citizen feedback, social media, and other sources, to understand citizens' needs and preferences. This means that cities can make data-driven decisions and improve the overall performance of the city.

SCCE can also support the development of citizen-led initiatives, by providing citizens with the tools and resources they need to develop and implement solutions to local problems. This can help to build community cohesion, and also can help citizens to take the initiative to solve the problems they are facing, which can lead to more efficient and effective solutions.

GENERAL FRAMEWORK

The following framework is a general one and the actual implementation will depend on the specific requirements and constraints of the city.

1. Develop a plan for improving communication and engagement between citizens and city government.
2. Research and evaluate different citizen engagement technologies, such as online platforms and mobile apps.
3. Identify potential partners and stakeholders for the citizen engagement initiatives, such as community organizations and advocacy groups.
4. Develop a comprehensive strategy for implementing citizen engagement technologies in the city.
5. Develop a marketing and outreach strategy to promote the citizen engagement technologies to citizens.
6. Establish a monitoring and evaluation system to track progress and measure the impact of the citizen engagement technologies.
7. Develop a system for managing and maintaining the citizen engagement technologies.
8. Develop a system for reporting and addressing complaints related to city services.
9. Utilize data analytics to inform decision-making and improve citizen engagement.
10. Develop a plan for incorporating citizen engagement technologies into existing city processes and procedures.

11. Develop a program for educating citizens on the benefits and use of citizen engagement technologies.
12. Develop a plan for providing incentives for using citizen engagement technologies.
13. Develop a program for providing technical assistance and training to citizens on citizen engagement technologies.
14. Develop a system for providing access to citizen engagement technologies for low-income citizens.
15. Continuously evaluate and adjust the citizen engagement technologies to address new challenges and opportunities.
16. Develop a system for providing real-time updates on city services and events.
17. Develop a system for providing personalized information and services to citizens.
18. Develop a system for providing feedback mechanisms for citizens to share their opinions and ideas.
19. Develop a system for providing transparency on city decision-making and budgeting.
20. Develop a system for providing easy-to-use interfaces for citizens to access information and services.
21. Develop a system for providing multilingual support for citizens.
22. Develop a system for providing access to city services and information through mobile devices.
23. Develop a system for providing easy-to-use interfaces for citizens to access information and services.
24. Develop a system for providing access to city services and information through mobile devices.
25. Develop a system for providing easy-to-use interfaces for citizens to access information and services, including online forms and online payments.

FREQUENTLY ASKED QUESTIONS

What is Smart City Citizen Engagement (SCCE), and how does it work? SCCE is an approach to smart city planning that emphasizes the active involvement of citizens in the development and implementation of smart city initiatives. It works by providing citizens with access to information, tools, and platforms that enable them to participate in the decision-making processes that affect their communities.

How can citizens get involved in SCCE initiatives? Citizens can get involved in SCCE initiatives by attending community meetings and events, participating in online forums and surveys, and providing feedback and suggestions to city officials and other stakeholders.

What are some of the benefits of SCCE for citizens? Some of the benefits of SCCE for citizens include increased transparency and accountability in local government, greater access to services and information, and the opportunity to have a greater impact on the development and implementation of smart city initiatives.

What kind of smart city initiatives can benefit from SCCE? A wide range of smart city initiatives can benefit from SCCE, including transportation, energy, waste management, public safety, and urban planning.

How can SCCE help to build trust and relationships between citizens and local government? SCCE can help to build trust and relationships between citizens and local government by providing opportunities for citizens to participate in decision-making processes and to contribute to the development of smart city initiatives.

What kind of challenges are associated with implementing SCCE? Some of the challenges associated with implementing SCCE include ensuring equitable access to information and participation opportunities, managing the costs of implementing engagement programs, and addressing the potential for conflict and disagreement between stakeholders.

How can technology be used to support SCCE? Technology can be used to support SCCE by providing online platforms for citizen participation and feedback, and by enabling the collection and analysis of data on citizen engagement and satisfaction.

Who is responsible for implementing SCCE initiatives? The responsibility for implementing SCCE initiatives typically falls on local government agencies, with support from community groups, technology vendors, and other stakeholders.

What kind of skills and expertise are required to implement SCCE initiatives? Implementing SCCE initiatives requires a range of skills and expertise, including community outreach, stakeholder engagement, communication, and technology management.

How can citizens measure the impact of SCCE initiatives? Citizens can measure the impact of SCCE initiatives by tracking changes in the quality and availability of services, improvements in community participation and engagement, and the achievement of specific goals and objectives related to smart city initiatives.

Conclusion: The implementation of SCCE is a vital step towards building a transparent, inclusive, and responsive city. By using technology to improve communication and engagement between citizens and city government, cities can increase citizen participation, build trust and transparency, and improve the overall performance of the city. SCCE can support the development of citizen-led initiatives, by providing citizens with the tools and resources they need to develop and implement solutions to local problems, which can help to build community cohesion, and lead to more efficient and effective solutions.

SMART CITY CITIZEN PARTICIPATION (SCCP)

Using technology to increase citizen participation in city decision-making, such as online voting and crowdsourcing platforms.

The implementation of SCCP is a crucial step towards building a transparent, inclusive, and responsive city. By using technology to increase citizen participation in city decision-making, cities can build trust and transparency, increase citizen engagement and improve the overall performance of the city.

One of the main benefits of SCCP is the ability to use online voting and crowdsourcing platforms. Online voting and crowdsourcing platforms allow citizens to participate in city decision-making from anywhere, at any time, using technology such as the internet and mobile devices. This means that cities can increase citizen participation and engagement, which is crucial for building trust and transparency.

Another benefit of SCCP is the ability to use data analytics to understand citizens' needs and preferences. With data analytics, cities can analyze data from citizen feedback, online voting, and crowdsourcing platforms, to understand citizens' needs and preferences. This means that cities can make data-driven decisions and improve the overall performance of the city.

SCCP can also support the development of citizen-led initiatives, by providing citizens with the tools and resources they need to develop and implement solutions to local problems. This can help to build community cohesion, and also can help citizens to take the initiative to solve the problems they are facing, which can lead to more efficient and effective solutions.

GENERAL FRAMEWORK

The following framework is a general one and the actual implementation will depend on the specific requirements and constraints of the city.

1. Develop a plan for increasing citizen participation in city decision-making through technology.
2. Research and evaluate different citizen participation technologies, such as online voting and crowdsourcing platforms.
3. Identify potential partners and stakeholders for the citizen participation initiatives, such as community organizations and advocacy groups.
4. Develop a comprehensive strategy for implementing citizen participation technologies in the city.

5. Develop a marketing and outreach strategy to promote the citizen participation technologies to citizens.
6. Establish a monitoring and evaluation system to track progress and measure the impact of the citizen participation technologies.
7. Develop a system for managing and maintaining the citizen participation technologies.
8. Develop a system for reporting and addressing complaints related to city services.
9. Utilize data analytics to inform decision-making and improve citizen participation.
10. Develop a plan for incorporating citizen participation technologies into existing city processes and procedures.
11. Develop a program for educating citizens on the benefits and use of citizen participation technologies.
12. Develop a plan for providing incentives for using citizen participation technologies.
13. Develop a program for providing technical assistance and training to citizens on citizen participation technologies.
14. Develop a system for providing access to citizen participation technologies for low-income citizens.
15. Continuously evaluate and adjust the citizen participation technologies to address new challenges and opportunities.
16. Develop a system for providing real-time updates on city services and events.
17. Develop a system for providing personalized information and services to citizens.
18. Develop a system for providing feedback mechanisms for citizens to share their opinions and ideas.
19. Develop a system for providing transparency on city decision-making and budgeting.
20. Develop a system for providing easy-to-use interfaces for citizens to participate in decision-making.
21. Develop a system for providing multilingual support for citizens.
22. Develop a system for providing access to city services and information through mobile devices.
23. Develop a system for providing easy-to-use interfaces for citizens to access information and services.
24. Develop a system for providing access to city services and information through mobile devices.

25. Develop a system for providing easy-to-use interfaces for citizens to access information and services, including online forms and online payments.

FREQUENTLY ASKED QUESTIONS

What is Smart City Citizen Participation (SCCP), and why is it important? SCCP is a key aspect of smart city planning that involves engaging citizens in the decision-making processes that affect their communities. It is important because it enables citizens to have a greater voice in the development and implementation of smart city initiatives.

How can citizens participate in SCCP initiatives? Citizens can participate in SCCP initiatives by attending community meetings and events, providing feedback and suggestions to city officials and other stakeholders, and participating in online forums and surveys.

What are some of the benefits of SCCP for citizens? Some of the benefits of SCCP for citizens include increased transparency and accountability in local government, greater access to services and information, and the opportunity to have a greater impact on the development and implementation of smart city initiatives.

What kind of smart city initiatives can benefit from SCCP? A wide range of smart city initiatives can benefit from SCCP, including transportation, energy, waste management, public safety, and urban planning.

How can SCCP help to build trust and relationships between citizens and local government? SCCP can help to build trust and relationships between citizens and local government by providing opportunities for citizens to participate in decision-making processes and to contribute to the development of smart city initiatives.

What kind of challenges are associated with implementing SCCP? Some of the challenges associated with implementing

SCCP include ensuring equitable access to information and participation opportunities, managing the costs of implementing participation programs, and addressing the potential for conflict and disagreement between stakeholders.

How can technology be used to support SCCP? Technology can be used to support SCCP by providing online platforms for citizen participation and feedback, and by enabling the collection and analysis of data on citizen participation and satisfaction.

Who is responsible for implementing SCCP initiatives? The responsibility for implementing SCCP initiatives typically falls on local government agencies, with support from community groups, technology vendors, and other stakeholders.

What kind of skills and expertise are required to implement SCCP initiatives? Implementing SCCP initiatives requires a range of skills and expertise, including community outreach, stakeholder engagement, communication, and technology management.

How can citizens measure the impact of SCCP initiatives? Citizens can measure the impact of SCCP initiatives by tracking changes in the quality and availability of services, improvements in community participation and engagement, and the achievement of specific goals and objectives related to smart city initiatives.

Conclusion: The implementation of SCCP is a crucial step towards building a transparent, inclusive, and responsive city. By using technology to increase citizen participation in city decision-making, cities can build trust and transparency, increase citizen engagement and improve the overall performance of the city. SCCP can support the development of citizen-led initiatives, by providing citizens with the tools and resources they need to develop and implement solutions to local problems, which can help to build community cohesion, and lead to more efficient and effective solutions.

SMART CITY CLIMATE RESILIENCE (SCCR)

Using technology to enhance the city's ability to withstand and adapt to the impacts of climate change, such as sea-level rise and extreme weather events.

The implementation of SCCR is a crucial step towards building a sustainable and resilient city. By using technology to enhance the city's ability to withstand and adapt to the impacts of climate change, cities can reduce the risks associated with extreme weather events and sea-level rise, and ensure the long-term sustainability of the city.

One of the main benefits of SCCR is the ability to use sensors and monitoring systems to detect and predict the impacts of climate change. These sensors and monitoring systems can collect data on sea-level, weather patterns, and other factors, which can be used to predict and prepare for the impacts of climate change. This means that cities can reduce the risks associated with extreme weather events and sea-level rise, and ensure the long-term sustainability of the city.

Another benefit of SCCR is the ability to use advanced simulation and modelling tools to analyze the impacts of climate change. These tools can be used to analyze the impacts of sea-level rise and extreme weather events, and to identify the most vulnerable areas of the city. This means that cities can develop and implement effective adaptation strategies that target the areas that are most at risk, which can reduce the impacts of climate change on citizens and the city.

SCCR can also support the development of renewable energy and energy efficiency, by providing buildings with the ability to generate, store and manage their own energy. This can lead to cost savings for both citizens and the city, in terms of energy costs, and also in terms of reducing carbon emissions, which will help to mitigate the impact of climate change.

GENERAL FRAMEWORK

The following framework is a general one and the actual implementation will depend on the specific requirements and constraints of the city.

1. Identify the specific climate risks and hazards that the city faces, such as sea-level rise, extreme weather events, and heat waves.
2. Develop a comprehensive climate resilience plan that addresses these risks and hazards, and includes measures such as green infrastructure, coastal protection, and emergency management.
3. Use GIS and other spatial analysis tools to map and model the potential impacts of climate change on the city's infrastructure and assets.
4. Implement early warning systems and other real-time monitoring systems to detect and respond to potential hazards and risks.
5. Use smart sensors, IoT devices, and other technology to gather real-time data on weather, air quality, and other environmental factors.
6. Use machine learning, predictive analytics, and other advanced techniques to analyze and interpret the data collected, and identify patterns and trends that can inform decision-making.
7. Develop and implement a comprehensive data management and governance plan to ensure that data is collected, stored, and shared in a secure and responsible manner.
8. Work with citizens, community groups, and other stakeholders to engage them in the planning and implementation of climate resilience measures.
9. Leverage public-private partnerships and other innovative financing mechanisms to fund climate resilience projects and programs.
10. Develop and implement a comprehensive communication and outreach plan to raise awareness of the risks and impacts of climate change, and the actions being taken to address them.
11. Develop and implement a comprehensive monitoring and evaluation program to track progress, measure outcomes, and identify areas for improvement.
12. Establish partnerships and collaborations with other cities and organizations to share best practices and learn from each other.
13. Develop and implement a comprehensive training and capacity-building program to ensure that city staff and other stakeholders have the skills and knowledge needed to effectively implement climate resilience measures.

14. Use big data analytics and other advanced techniques to track and analyze the city's greenhouse gas emissions and other environmental impacts.
15. Use Building Information Modelling (BIM) and other digital tools to design and construct new buildings and infrastructure that are more resilient to climate change.
16. Encourage and incentivize the use of low-carbon transportation options such as electric vehicles, bike-sharing, and public transit.
17. Develop and implement a comprehensive water management plan that includes measures such as rainwater harvesting, greywater reuse, and green roofs.
18. Use smart grid and other energy management technologies to optimize the city's energy consumption and reduce its carbon footprint.
19. Develop and implement a comprehensive waste management plan that includes measures such as recycling, composting, and reducing food waste.
20. Incorporate nature-based solutions, such as green roofs, urban forests, and wetlands, into the city's infrastructure and development plans.
21. Use smart city technologies, such as IoT and 5G, to enhance the city's ability to detect and respond to extreme weather events and other hazards.
22. Develop and implement a comprehensive land use and zoning plan that promotes compact, walkable, and resilient communities.
23. Use drones and other remote sensing technologies to monitor and assess the city's coastal and riverine areas, and identify areas that are vulnerable to sea-level rise.
24. Use smart city technologies, such as IoT and 5G, to enhance the city's ability to detect and respond to extreme weather events and other hazards.
25. Develop and implement a comprehensive disaster recovery and continuity of operations plan to ensure that the city can quickly and effectively respond to and recover from extreme weather events and other hazards.

FREQUENTLY ASKED QUESTIONS

What is Smart City Climate Resilience (SCCR) and why is it important? SCCR involves developing strategies and initiatives that enable cities to adapt to the impacts of climate change and to minimize the risks associated with extreme weather events. It is important because climate change poses significant threats to the safety, health, and well-being of city residents and infrastructure.

What are some of the key challenges associated with implementing SCCR initiatives? Some of the challenges associated with implementing SCCR initiatives include identifying the most effective strategies for adapting to specific climate risks, securing funding and resources for implementation, and engaging diverse stakeholders in the planning and decision-making process.

What kind of initiatives can help cities build climate resilience? Initiatives that can help cities build climate resilience include improving infrastructure and building design to withstand extreme weather events, developing emergency response plans, increasing the use of renewable energy, and implementing water management and conservation strategies.

How can citizens get involved in SCCR initiatives? Citizens can get involved in SCCR initiatives by participating in public meetings and events, providing feedback and suggestions to city officials and other stakeholders, and advocating for policies and initiatives that support climate resilience.

What role do technology and data play in SCCR? Technology and data can play an important role in SCCR by providing real-time information on weather conditions, enabling the tracking of energy and water usage, and supporting the development of predictive models for climate risks.

How can SCCR help to address social equity issues? SCCR can help to address social equity issues by ensuring that climate resilience initiatives are developed with the needs and perspectives of diverse communities in mind, and by providing resources and support to underserved and vulnerable populations.

Who is responsible for implementing SCCR initiatives? Responsibility for implementing SCCR initiatives typically falls on local government agencies, with support from community groups, technology vendors, and other stakeholders.

What kind of skills and expertise are required to implement SCCR initiatives? Implementing SCCR initiatives requires a range of skills and expertise, including climate science, emergency management, community engagement, and technology management.

How can cities measure the impact of SCCR initiatives? Cities can measure the impact of SCCR initiatives by tracking changes in infrastructure and building design, improvements in emergency response and preparedness, and reductions in the costs and impacts of extreme weather events.

What are some examples of cities that have successfully implemented SCCR initiatives? Cities such as Copenhagen, Amsterdam, and New York have implemented successful SCCR initiatives, including flood protection infrastructure, green roofs and walls, and community outreach and engagement programs.

Conclusion: The implementation of SCCR is a crucial step towards building a sustainable and resilient city. By using technology to enhance the city's ability to withstand and adapt to the impacts of climate change, cities can reduce the risks associated with extreme weather events and sea-level rise, and ensure the long-term sustainability of the city. SCCR can support the development of renewable energy and energy efficiency, by providing buildings with the ability to generate, store and manage their own energy, which can lead to cost savings for both citizens and the city and help to mitigate the impact of climate change.

SMART CITY COMMUNITY DEVELOPMENT (SCCD)

Using technology to promote social and community development, such as affordable housing and community engagement platforms.

The implementation of SCCD is a crucial step towards building a inclusive and equitable city. By using technology to promote social and community development, cities can address the needs of all citizens, and create opportunities for economic and social mobility.

One of the main benefits of SCCD is the ability to use technology to create affordable housing solutions. This can include the use of modular construction, 3D printing, and other innovative technologies to lower the cost of building and maintaining housing. This can help to address the issue of housing affordability, which is a significant concern for many citizens, particularly low-income families and individuals.

Another benefit of SCCD is the ability to use technology to promote community engagement and empowerment. This can include the use of social media, mobile apps, and online platforms to connect citizens with city services, resources, and each other. This can help to build community cohesion and empower citizens to take an active role in shaping their neighbourhoods and city.

SCCD can also support the development of community-based initiatives and programs, by providing citizens with the tools and resources they need to develop and implement solutions to local problems. This can help to build community cohesion, and also can help citizens to take the initiative to solve the problems they are facing, which can lead to more efficient and effective solutions.

GENERAL FRAMEWORK

The following framework is a general one and the actual implementation will depend on the specific requirements and constraints of the city.

1. Define the goals and objectives for the SCCD initiative.
2. Identify the specific community or communities that will be targeted for development.
3. Assess the current state of technology infrastructure in the targeted community.
4. Develop a comprehensive plan for improving the technology infrastructure, including internet access and connectivity.
5. Create a platform for community engagement, such as a website or mobile app, to facilitate communication and participation.
6. Develop a strategy for affordable housing development that incorporates technology, such as smart home systems and energy-efficient design.
7. Identify key stakeholders and partners to involve in the SCCD initiative, such as community organizations, government agencies, and private companies.
8. Develop partnerships and collaborations with local businesses and organizations to promote economic development and job creation.
9. Implement programs and services to support community development, such as education and job training programs.
10. Identify and address any barriers to technology access, such as lack of digital literacy or lack of affordable devices.
11. Develop a strategy for data collection and analysis to measure the effectiveness of the SCCD initiative.
12. Encourage the use of open data and open-source technologies to promote transparency and collaboration.
13. Develop a plan for community-wide internet access, such as public Wi-Fi hotspots and community technology centers.
14. Implement smart city technologies, such as smart lighting and smart transportation systems, to improve the quality of life in the community.
15. Develop a plan for community-wide energy efficiency, such as solar power and energy-efficient buildings.
16. Develop a plan for community-wide water management, such as rainwater harvesting and greywater recycling.
17. Develop a plan for community-wide waste management, such as recycling and composting programs.
18. Develop a plan for community-wide air quality management, such as monitoring and reducing emissions from transportation and industry.

19. Develop a plan for community-wide public safety, such as emergency preparedness and response.
20. Develop a plan for community-wide public health, such as promoting healthy lifestyles and reducing health disparities.
21. Develop a plan for community-wide mobility, such as bike-sharing and car-sharing programs.
22. Develop a plan for community-wide sustainability, such as green spaces and sustainable development.
23. Develop a plan for community-wide education, such as digital literacy programs and e-learning.
24. Develop a plan for community-wide civic engagement, such as voter registration and community service programs.
25. Develop a plan for community-wide resilience, such as disaster preparedness and emergency response plans.

FREQUENTLY ASKED QUESTIONS

What is Smart City Community Development (SCCD) and why is it important? SCCD involves developing initiatives and programs that support the economic, social, and environmental well-being of communities within a city. It is important because it can help to promote equity, access to resources, and overall quality of life for residents.

What kind of initiatives fall under SCCD? Initiatives that fall under SCCD can include affordable housing programs, community health initiatives, small business support programs, and access to public spaces and amenities.

How can SCCD promote equity within a city? SCCD can promote equity within a city by prioritizing investments and resources in underserved communities and by ensuring that all residents have access to quality housing, education, healthcare, and other essential resources.

Who is responsible for implementing SCCD initiatives? Responsibility for implementing SCCD initiatives typically falls on local government agencies, with support from community groups, non-profit organizations, and private sector partners.

How can residents get involved in SCCD initiatives? Residents can get involved in SCCD initiatives by participating in community meetings and events, providing feedback and suggestions to city officials and other stakeholders, and volunteering with local organizations and initiatives.

What role do technology and data play in SCCD? Technology and data can play an important role in SCCD by providing real-time information on community needs and assets, supporting community engagement and participation, and enabling data-driven decision-making.

How can SCCD help to address environmental issues? SCCD can help to address environmental issues by promoting sustainable and eco-friendly practices in housing, transportation, and other community development initiatives.

What kind of skills and expertise are required to implement SCCD initiatives? Implementing SCCD initiatives requires a range of skills and expertise, including community engagement, data analysis, program management, and knowledge of public policy and funding mechanisms.

How can SCCD initiatives benefit the local economy? SCCD initiatives can benefit the local economy by promoting small business development, creating job opportunities, and improving the overall quality of life for residents, which can attract new investment and development to the area.

What are some examples of cities that have successfully implemented SCCD initiatives? Cities such as Seattle, Portland, and Minneapolis have successfully implemented SCCD initiatives, including affordable housing programs, community health clinics, and small business support programs.

Conclusion: *The implementation of SCCD is a crucial step towards building an inclusive and equitable city. By using technology to promote* social and community development, cities can address the needs of all citizens, and create opportunities for economic and social mobility. SCCD can support the development of community-based initiatives and programs, by providing citizens with the tools and resources they need to develop and implement solutions to local problems, which can help to build community cohesion, and lead to more efficient and effective solutions.

SMART CITY CONNECTED AND AUTONOMOUS VEHICLES (SCCAV)

Using technology to improve the efficiency, safety and sustainability of urban transportation, such as connected cars, drones and self-driving vehicles.

The implementation of SCCAV is a crucial step towards building a more efficient, safe, and sustainable urban transportation system. By using technology to improve the efficiency, safety and sustainability of urban transportation, cities can reduce traffic congestion, improve air quality, and increase mobility for all citizens.

One of the main benefits of SCCAV is the ability to use technology to reduce traffic congestion and improve road safety. Connected cars and self-driving vehicles can communicate with each other and with traffic infrastructure to optimize traffic flow and avoid accidents. This means that cities can reduce traffic congestion, improve air quality, and increase mobility for all citizens.

Another benefit of SCCAV is the ability to use technology to promote sustainable transportation. Electric and autonomous vehicles can reduce emissions, and drones can be used for deliveries and cargo transportation, which can reduce the need for traditional road vehicles. This means that cities can reduce their carbon footprint and promote sustainable transportation.

SCCAV can also support the development of new and innovative transportation services, such as on-demand transportation and ride-sharing. This can increase mobility for all citizens, and also can help to reduce the number of vehicles on the road, which can lead to a reduction in traffic congestion, and an improvement in air quality.

GENERAL FRAMEWORK

The following framework is a general one and the actual implementation will depend on the specific requirements and constraints of the city.

1. Define the goals and objectives for the SCCAV initiative.
2. Assess the current state of transportation infrastructure in the targeted city or area.
3. Develop a comprehensive plan for improving transportation infrastructure, including roads, bridges, and public transportation.
4. Develop a strategy for implementing connected and SCCAV technology, such as vehicle-to-vehicle (V2V) communication and sensor-based navigation.
5. Identify key stakeholders and partners to involve in the SCCAV initiative, such as government agencies, private companies, and universities.
6. Develop partnerships and collaborations with local businesses and organizations to promote economic development and job creation.
7. Develop a plan for testing and piloting SCCAV technology on public roads.
8. Develop a plan for integrating SCCAV technology with existing transportation infrastructure, such as traffic lights and road signs.
9. Develop a plan for integrating SCCAV technology with public transportation systems, such as buses and trains.
10. Develop a plan for data collection and analysis to measure the effectiveness of the SCCAV initiative.
11. Develop a plan for addressing the cybersecurity challenges that come with connected vehicles.
12. Develop a plan for dealing with regulatory challenges that come with autonomous vehicles.
13. Develop a plan for dealing with the challenges of transitioning from human-driven to autonomous vehicles.

14. Develop a plan for dealing with privacy challenges that come with connected vehicles.
15. Develop a plan for dealing with the challenges of integrating drones and other unmanned aerial vehicles into the transportation system.
16. Develop a plan for dealing with the challenges of charging and maintaining electric vehicles.
17. Develop a plan for dealing with the challenges of integrating SCCAV technology into existing transportation networks.
18. Develop a plan for dealing with the challenges of dealing with the challenges of liability and insurance for autonomous vehicles.
19. Develop a plan for dealing with the challenges of dealing with the challenges of traffic management and routing in a world of autonomous vehicles.
20. Develop a plan for dealing with the challenges of dealing with the challenges of ensuring accessibility for persons with disabilities.
21. Develop a plan for dealing with the challenges of dealing with the challenges of ensuring the safety and security of passengers in autonomous vehicles.
22. Develop a plan for dealing with the challenges of dealing with the challenges of dealing with the challenges of managing traffic flow and congestion in a world of autonomous vehicles.
23. Develop a plan for dealing with the challenges of integrating SCCAV technology with existing transportation networks, such as public transportation and ride-sharing services.
24. Develop a plan for dealing with the challenges of ensuring public acceptance and understanding of connected and autonomous vehicle technology.
25. Develop a plan for scaling up and expanding the SCCAV initiative to other areas of the city or region.

FREQUENTLY ASKED QUESTIONS

What are SCCAV and how are they different from traditional vehicles? SCCAV are vehicles that are connected to the internet and other vehicles, as well as equipped with autonomous features that allow them to navigate without human intervention. They are different from traditional vehicles in that they have advanced sensors and communication technology that enables them to collect and analyze data in real time.

How will SCCAV improve traffic flow in cities? SCCAV can improve traffic flow in cities by communicating with each other and adjusting their speed and routes based on real-time traffic information, which can help to reduce congestion and minimize travel time.

How will SCCAV impact the environment? SCCAV have the potential to reduce greenhouse gas emissions and improve air quality by optimizing driving patterns and reducing congestion on the roads.

Are SCCAV safe for passengers and pedestrians? Safety is a top priority in the development of SCCAV, and they are designed to meet strict safety standards. Advanced sensors and communication technology allow SCCAV to quickly respond to potential hazards on the road, making them safer for passengers and pedestrians.

Who is responsible for regulating SCCAV on the roads? Regulation of SCCAV is typically the responsibility of local and national government agencies, who work closely with industry partners and stakeholders to develop and enforce safety standards and regulations.

Will SCCAV eliminate the need for human drivers? While SCCAV are designed to operate without human intervention, they will not eliminate the need for human drivers completely. There will still be situations that require human intervention, such as emergencies or unexpected road conditions.

How will SCCAV impact public transportation? SCCAV have the potential to complement and enhance existing public transportation systems by offering last-mile connectivity and reducing congestion on public transit routes.

How will SCCAV be powered? SCCAV can be powered by a range of energy sources, including electricity, hydrogen fuel cells, and traditional gasoline or diesel engines.

How will SCCAV impact the job market? The widespread adoption of SCCAV is expected to create new job opportunities in industries such as software development, data analysis, and maintenance and repair of the vehicles.

When will SCCAV become a reality in cities? SCCAV are already being tested in some cities, and they are expected to become more widespread in the coming years. The timeline for adoption will depend on factors such as regulatory approval, infrastructure development, and public acceptance.

Conclusion: *The implementation of SCCAV is a crucial step towards building a more efficient, safe, and sustainable urban transportation system. By using technology to improve the efficiency, safety and sustainability of urban transportation, cities can reduce traffic congestion, improve air quality, and increase mobility for all citizens. SCCAV can support the development of new and innovative transportation services, such as on-demand transportation and ride-sharing, which can increase mobility for all citizens, and help to reduce the number of vehicles on the road, leading to a reduction in traffic congestion and an improvement in air quality.*

SMART CITY CYBER-SECURITY (SCCS)

Using technology to protect Smart City systems and services from cyber threats, such as encryption and intrusion detection systems.

The implementation of SCCS is a crucial step towards building a secure and reliable smart city. By using technology to protect Smart City systems and services from cyber threats, cities can ensure the confidentiality, integrity, and availability of city data and services, and protect citizens and their personal information from cyber-attacks.

One of the main benefits of SCCS is the ability to use technology to protect city data and services from cyber threats. This can include the use of encryption, intrusion detection systems, and other security measures to protect city data and systems from unauthorized access, theft, and other cyber-attacks. This means that cities can ensure the confidentiality, integrity, and availability of city data and services, and protect citizens and their personal information from cyber-attacks.

Another benefit of SCCS is the ability to use technology to detect and respond to cyber threats in real-time. This can include the use of intrusion detection systems, firewall, and other security measures to detect and respond to cyber threats in real-time. This means that cities can detect and respond to cyber threats quickly, which can help to minimize the impact of cyber-attacks on citizens and the city.

SCCS can also support the development of incident response plans, and regular security assessments, as well as security awareness training for citizens, to protect citizens and their personal information from cyber-attacks. This can ensure that citizens are aware of the risks of cyber threats and can take steps to protect themselves and their personal information.

GENERAL FRAMEWORK

The following framework is a general one and the actual implementation will depend on the specific requirements and constraints of the city.

1. Identify critical assets and infrastructure in the city that must be protected from cyber threats.
2. Develop a comprehensive cyber security strategy that aligns with the city's overall mission and goals.
3. Establish a dedicated cyber security team, with clearly defined roles and responsibilities.
4. Conduct regular threat assessments to identify potential vulnerabilities and risks.
5. Implement encryption for all sensitive data and communication.
6. Implement intrusion detection systems (IDS) to detect and respond to potential cyber-attacks.

7. Regularly update and patch all software and systems to address known vulnerabilities.
8. Implement multi-factor authentication to protect against unauthorized access.
9. Train employees and citizens on cyber security best practices and how to recognize and report potential threats.
10. Implement a disaster recovery and business continuity plan in case of a cyber-attack.
11. Regularly conduct penetration testing to identify and remediate vulnerabilities.
12. Implement network segmentation to limit the potential impact of a cyber-attack.
13. Implement a security information and event management (SIEM) system to monitor and analyze security-related data.
14. Establish incident response protocols and regularly conduct drills to ensure readiness in case of a cyber-attack.
15. Develop an incident management process to respond to a security incident.
16. Establish an incident communication plan for both internal and external stakeholders.
17. Continuously monitor and adapt the cyber security strategy as new threats and technologies emerge.
18. Develop a Cybersecurity governance structure with clear roles, responsibilities and accountability.
19. Develop a Cybersecurity risk management framework.
20. Develop a Cybersecurity governance policy.
21. Develop a Cybersecurity standards, policies and procedures.
22. Develop a Cybersecurity awareness and training program.
23. Develop a Cybersecurity incident management and response plan.
24. Develop a Cybersecurity threat intelligence and sharing program.
25. Establish a Cybersecurity metrics and reporting system for measuring progress and effectiveness of the cyber security strategy.

FREQUENTLY ASKED QUESTIONS

What is Smart City Cyber-Security (SCCS)? SCCS refers to the measures taken to protect smart city infrastructure, networks, and data from cyber-attacks and ensure that they are secure and reliable.

What are the common cyber threats that smart cities face? Smart cities are vulnerable to a range of cyber threats, including malware, phishing, ransomware, DDoS attacks, and data breaches.

How can smart cities protect themselves from cyber threats? Smart cities can protect themselves from cyber threats by implementing security measures such as firewalls, encryption, intrusion detection, and access controls. They can also conduct regular security audits and train their employees on cyber-security best practices.

Who is responsible for cyber-security in a smart city? Cyber-security is a shared responsibility between the city government, service providers, and citizens. The city government is responsible for establishing cyber-security policies and regulations, while service providers are responsible for securing their systems and networks. Citizens are responsible for practicing safe online behaviour.

How can citizens contribute to smart city cyber-security? Citizens can contribute to smart city cyber-security by being aware of potential cyber threats and taking steps to protect their personal devices and data. This includes using strong passwords, avoiding suspicious links, and keeping software up-to-date.

What is the role of data privacy in SCCS? Data privacy is an important aspect of SCCS as smart city systems often collect and process sensitive personal data. SCCS should ensure that this data is collected and used in a transparent and ethical manner and that the privacy of citizens is protected.

Can SCCS be breached despite best efforts? While SCCS can significantly reduce the risk of cyber-attacks, it is impossible to completely eliminate the risk. However, by implementing effective security measures and responding quickly to security incidents, smart cities can minimize the impact of cyber-attacks.

What happens if there is a cyber-attack on a smart city? If a cyber-attack occurs, the smart city should have a response plan in place to quickly identify and mitigate the attack, prevent further damage, and restore normal operations as soon as possible.

How can SCCS be integrated into the planning of a smart city? SCCS should be considered at every stage of a smart city's development, from planning to implementation and ongoing management. This includes designing systems with security in mind, regularly testing and updating security measures, and training staff on cyber-security best practices.

Are there any international standards for SCCS? Yes, there are several international standards for SCCS, including ISO 27001, NIST Cybersecurity Framework, and IEC 62443. Smart cities can use these standards to guide their cyber-security efforts and ensure that they are implementing best practices.

Conclusion: The implementation of SCCS is a crucial step towards building a secure and reliable smart city. By using technology to protect Smart City systems and services from cyber threats, cities can ensure the confidentiality, integrity, and availability of city data and services, and protect citizens and their personal information from cyber-attacks. SCCS can support the development of incident response plans, and regular security assessments, as well as security awareness training for citizens, which can ensure that citizens are aware of the risks of cyber threats and can take steps to protect themselves and their personal information.

SMART CITY DATA MANAGEMENT (SCDM)

Using technology to collect, store, process and analyze data from Smart City systems and services, such as data lakes and analytics platforms.

The implementation of SCDM is a crucial step towards building a data-driven smart city. By using technology to collect, store, process and analyze data from Smart City systems and services, cities can make data-driven decisions, improve the efficiency of city services, and create new opportunities for economic and social development.

One of the main benefits of SCDM is the ability to use technology to collect and store large amounts of data from Smart City systems and services. This can include the use of data lakes and other data storage systems to collect and store data from a wide range of sources, such as traffic sensors, weather stations, and social media. This means that cities can collect and store large amounts of data, which can be used to make data-driven decisions and improve the efficiency of city services.

Another benefit of SCDM is the ability to use technology to process and analyze data. This can include the use of analytics platforms, machine learning and other data analysis tools to process and analyze data from Smart City systems and services. This means that cities can extract valuable insights from the data, which can be used to improve city services and create new opportunities for economic and social development.

SCDM can also support the development of data governance and data privacy policies, to ensure that citizens' data is collected, stored and used in a responsible and ethical way. This can ensure that citizens' data is protected and used in a way that respects citizens' privacy and rights.

GENERAL FRAMEWORK

The following framework is a general one and the actual implementation will depend on the specific requirements and constraints of the city.

1. Identify the types of data that will be collected from Smart City systems and services.
2. Develop a data governance structure with clear roles and responsibilities for managing the data.
3. Establish a data management strategy that aligns with the city's overall mission and goals.
4. Create a centralized data repository, such as a data lake, to store and process the data.
5. Implement data quality controls to ensure that the data is accurate and reliable.

6. Implement data security measures, such as encryption and access controls, to protect the data from unauthorized access or breaches.
7. Implement data analytics platforms, such as business intelligence (BI) tools, to analyze and visualize the data.
8. Develop data management policies and procedures to ensure compliance with relevant laws and regulations.
9. Identify key performance indicators (KPIs) to measure the effectiveness of the data management strategy.
10. Establish data sharing agreements with relevant stakeholders, such as other city departments and external partners.
11. Create a data catalogue to document the data assets and their metadata, including business definitions, lineage, and lineage.
12. Develop a data master plan that outlines data architecture, data integration, data quality, data security, data governance, and data management.
13. Develop data integration strategy that outlines how data will be integrated from different sources, and how it will be transformed to meet the city's needs.
14. Implement a data lineage process to track the flow of data from the source to the end-user.
15. Implement a data quality management process to ensure data is accurate, complete and timely.
16. Implement a data security management process to ensure data is protected from unauthorized access or breaches.
17. Implement a data governance process to ensure data is managed in accordance with the city's policies and procedures.
18. Implement a data analytics process to analyze and visualize the data to gain insights and make data-driven decisions.
19. Develop a data visualization strategy to present data in a clear and easy-to-understand format.
20. Develop a data management training program to ensure all employees understand the data management policies and procedures.
21. Develop a data management incident management and response plan to respond to data breaches or other data-related incidents.
22. Develop a data archiving and retention policy to ensure data is stored and retained in accordance with legal and regulatory requirements.
23. Establish data quality metrics to measure the quality of the data.
24. Establish data security metrics to measure the level of data protection.

25. Establish data governance metrics to measure the effectiveness of data management policies and procedures.

FREQUENTLY ASKED QUESTIONS

What is Smart City Data Management? Smart City Data Management (SCDM) refers to the collection, processing, storage, and analysis of data from various sources in a city, including sensors, social media, and other digital sources, to inform decision-making and improve city services and infrastructure.

Why is Smart City Data Management important? Smart City Data Management is important because it helps city governments to make informed decisions based on data analysis, leading to better planning and management of resources, and improved quality of life for citizens.

What kind of data is collected for Smart City Data Management? Data collected for Smart City Data Management can include a wide range of information, such as air quality, traffic patterns, energy consumption, waste management, crime rates, and social media activity.

Who is responsible for managing Smart City Data? The responsibility for managing Smart City Data typically lies with the city government, although it may also involve collaboration with private sector partners.

How is Smart City Data secured and protected? Smart City Data is secured and protected using a combination of encryption, access control, and other cyber security measures to prevent unauthorized access, data breaches, and other security threats.

What are the benefits of Smart City Data Management for citizens? The benefits of Smart City Data Management for citizens include improved public services, increased transparency and accountability, and a better quality of life through better urban planning and management.

What are some challenges of Smart City Data Management? Challenges of Smart City Data Management can include the need for secure and reliable data transmission, the high cost of implementing data management systems, and the need for skilled data analysts and management professionals.

How is Smart City Data used to improve city services? Smart City Data is used to improve city services by providing real-time data on service performance, enabling better resource allocation, and identifying areas where service improvements are needed.

How can citizens access Smart City Data? Citizens can access Smart City Data through city websites, mobile apps, and other digital platforms that provide access to data sets and real-time data feeds.

How can Smart City Data Management help to address urban challenges? Smart City Data Management can help to address urban challenges by providing insights into areas such as traffic management, energy efficiency, public safety, and waste management, leading to more efficient and effective city management.

Conclusion: The implementation of SCDM is a crucial step towards building a data-driven smart city. By using technology to collect, store, process and analyze data from Smart City systems and services, cities can make data-driven decisions, improve the efficiency of city services, and create new opportunities for economic and social development. SCDM can support the development of data governance and data privacy policies, which can ensure that citizens' data is collected, stored and used in a responsible and ethical way, protecting citizens' privacy and rights.

SMART CITY DIGITAL INCLUSION (SCDI)

Using technology to ensure that all citizens have access to digital services, such as digital literacy programs and community internet access.

The implementation of SCDI is a crucial step towards building an inclusive and equitable smart city. By using technology to ensure that all citizens have access to digital services, cities can promote social and economic inclusion, and bridge the digital divide between different segments of the population.

One of the main benefits of SCDI is the ability to use technology to provide digital services to all citizens, regardless of their socio-economic background or location. This can include the use of digital literacy programs, community internet access, and other digital inclusion initiatives to ensure that all citizens have access to digital services. This means that cities can promote social and economic inclusion, and bridge the digital divide between different segments of the population.

Another benefit of SCDI is the ability to use technology to provide access to digital services that can improve the quality of life for citizens. This can include access to online education, telemedicine, and other digital services that can improve citizens' access to healthcare, education, and other services. This means that cities can improve the quality of life for citizens, and provide access to services that can support social and economic development.

SCDI can also support the development of policies and programs to ensure that digital services are accessible to all citizens, including those with disabilities. This can ensure that digital services are inclusive and accessible to all citizens, regardless of their abilities.

GENERAL FRAMEWORK

The following framework is a general one and the actual implementation will depend on the specific requirements and constraints of the city.

1. Identify groups or communities within the city that have limited access to digital services and technology.

2. Develop a digital inclusion strategy that aligns with the city's overall mission and goals.
3. Establish partnerships with community organizations, non-profits, and private companies to expand access to digital services.
4. Implement digital literacy programs, such as computer training and workshops, to help citizens acquire the skills needed to access and use digital services.
5. Provide community internet access, such as Wi-Fi hotspots, to increase access to digital services for under-served communities.
6. Develop mobile apps and online portals that provide citizens with easy access to city services and information.
7. Increase access to technology, such as computers and tablets, for citizens who cannot afford to purchase their own.
8. Develop multilingual digital services to ensure that non-English speakers have access to city services.
9. Offer digital skills training for seniors, disabled and other groups with specific needs.
10. Develop a digital inclusion training program to ensure that city employees understand and support the city's digital inclusion goals.
11. Develop a digital inclusion metrics system to measure progress and effectiveness of the digital inclusion strategy.
12. Develop an outreach plan to educate and inform citizens of the available digital services and resources.
13. Develop a digital equity plan to ensure that all citizens have access to the same digital opportunities and resources.
14. Identify and remove barriers to digital inclusion such as lack of infrastructure and lack of digital literacy.
15. Develop a digital mentoring program to help citizens improve their digital skills.
16. Develop a digital volunteer program to promote digital inclusion.
17. Develop a digital inclusion policy to guide decision making and resource allocation.
18. Develop a digital inclusion governance structure with clear roles and responsibilities.
19. Develop a digital inclusion risk management framework.
20. Develop a digital inclusion standards, policies and procedures.
21. Develop a digital inclusion awareness and education program.
22. Develop a digital inclusion incident management and response plan.
23. Develop a digital inclusion threat intelligence and sharing program.

24. Establish a digital inclusion metrics and reporting system for measuring progress and effectiveness of the digital inclusion strategy.
25. Incorporate digital inclusion considerations into all Smart City initiatives and projects.

FREQUENTLY ASKED QUESTIONS

What is Smart City Digital Inclusion (SCDI)? SCDI refers to the efforts of Smart Cities to ensure that all residents, especially those from underserved communities, have access to and the skills to use digital technologies and platforms, including the internet.

Why is digital inclusion important for Smart Cities? Digital inclusion is important for Smart Cities because it helps to ensure that all residents have equal access to the benefits of technology, including access to online services, educational resources, job opportunities, and more.

What are some of the barriers to digital inclusion in Smart Cities? Barriers to digital inclusion in Smart Cities can include things like lack of internet access, affordability of devices and services, lack of digital skills, and language barriers.

How are Smart Cities addressing digital inclusion? Smart Cities are addressing digital inclusion in a variety of ways, such as investing in infrastructure to improve internet access, providing low-cost or free internet and devices to underserved communities, offering digital skills training programs, and providing multilingual resources and support.

How can residents get involved in promoting digital inclusion in Smart Cities? Residents can get involved in promoting digital inclusion in Smart Cities by advocating for policies and programs that promote access and affordability, volunteering with local organizations that offer digital skills training or support, and supporting initiatives that help bridge the digital divide.

What is the role of businesses and organizations in promoting digital inclusion in Smart Cities? Businesses and organizations can play a key role in promoting digital inclusion in Smart Cities by investing in infrastructure, offering low-cost or free services to underserved communities, providing training and support, and partnering with local governments and community organizations to address digital divides.

What are some examples of successful digital inclusion initiatives in Smart Cities? Examples of successful digital inclusion initiatives in Smart Cities include programs like the Chicago Connected initiative, which provides free high-speed internet to low-income families, and the San Francisco Digital Equity initiative, which offers digital skills training and support to underserved communities.

How can Smart Cities ensure that digital inclusion initiatives are sustainable and effective over the long term? To ensure that digital inclusion initiatives are sustainable and effective, Smart Cities can focus on building partnerships with community organizations and businesses, investing in long-term infrastructure, regularly assessing and updating programs and policies, and prioritizing equity and inclusion in all aspects of planning and implementation.

How can Smart Cities measure the impact of digital inclusion initiatives? Smart Cities can measure the impact of digital inclusion initiatives through a variety of metrics, such as the number of people who have gained access to the internet, the number of people who have received digital skills training, and the number of people who have successfully used digital resources to improve their lives.

How can Smart Cities learn from and collaborate with each other to promote digital inclusion? Smart Cities can learn from and collaborate with each other to promote digital inclusion by sharing best practices and success stories, participating in conferences and workshops, and engaging in collaborative projects and initiatives that help bridge the digital divide.

Conclusion: The implementation of SCDI is a crucial step towards building an inclusive and equitable smart city. By using technology to ensure that all citizens have access to digital services, cities can promote social and economic inclusion, and bridge the digital divide between different segments of the population. SCDI can support the development of policies and programs to ensure that digital services are accessible to all citizens, including those with disabilities, which can ensure that digital services are inclusive and accessible to all citizens, regardless of their abilities.

SMART CITY DISASTER MANAGEMENT (SCDIM)

Using technology to enhance the city's ability to prepare for, respond to, and recover from natural and man-made disasters.

The implementation of SCDIM is a crucial step towards building a resilient and safe smart city. By using technology to enhance the city's ability to prepare for, respond to, and recover from natural and man-made disasters, cities can minimize the impact of disasters on citizens and the city's infrastructure, and ensure a speedy recovery.

One of the main benefits of SCDIM is the ability to use technology to enhance the city's ability to prepare for and respond to disasters. This can include the use of sensors, drones, and other technologies to monitor and predict potential disasters, and provide early warning to citizens. It can also include the use of communication technologies, such as social media and mobile apps, to provide real-time information and guidance to citizens during a disaster. This means that cities can minimize the impact of disasters on citizens and the city's infrastructure, and ensure a speedy recovery.

Another benefit of SCDIM is the ability to use technology to improve the coordination and communication among different agencies and organizations during a disaster. This can include the use of shared situational awareness platforms, command and control systems, and other technologies to improve the coordination and communication among different agencies and organizations during a disaster. This means that cities can respond to disasters more effectively, and minimize the impact of disasters on citizens and the city's infrastructure.

SCDIM can also support the development of disaster recovery plans, and regular drills and exercises, which can ensure that cities are prepared to respond to disasters effectively. This can ensure that cities can respond to disasters quickly and effectively, and minimize the impact of disasters on citizens and the city's infrastructure.

GENERAL FRAMEWORK

The following framework is a general one and the actual implementation will depend on the specific requirements and constraints of the city.

1. Develop a comprehensive disaster management plan that aligns with the city's overall mission and goals.
2. Establish partnerships with emergency management agencies, community organizations, and private companies to enhance the city's disaster management capabilities.
3. Implement early warning systems, such as weather and seismic monitoring systems, to provide citizens with real-time information about potential disasters.
4. Develop a command and control center to coordinate the city's disaster management efforts.
5. Implement a crisis communication plan to ensure that citizens are informed and able to access important information during a disaster.
6. Develop a system for managing critical infrastructure, such as power and water supplies, during a disaster.
7. Implement a geographic information system (GIS) to provide real-time information about the location of emergency services, evacuation routes, and other important resources.
8. Develop a system for managing evacuation routes and shelters to ensure the safe and efficient movement of citizens during a disaster.
9. Implement a system for managing and tracking emergency resources, such as medical supplies and equipment.

10. Develop a system for managing and tracking the whereabouts of first responders and other key personnel during a disaster.
11. Develop a training and exercise program to ensure that city employees and citizens are prepared to respond to disasters.
12. Develop a disaster recovery plan to ensure that the city can quickly and effectively recover from a disaster.
13. Implement an incident management system (IMS) to track and manage the response to a disaster.
14. Develop a system for managing and tracking debris and waste during and after a disaster.
15. Develop a system for managing and tracking the needs and welfare of citizens during and after a disaster.
16. Develop a system for managing and tracking damage to city infrastructure and property during and after a disaster.
17. Develop a system for managing and tracking the restoration of essential services during and after a disaster.
18. Develop a system for managing and tracking volunteer and donations during and after a disaster.
19. Develop a system for managing and tracking the recovery of the local economy during and after a disaster.
20. Develop a system for managing and tracking the reconstruction and rebuilding of damaged infrastructure during and after a disaster.
21. Develop a system for managing and tracking the restoration of the environment during and after a disaster.
22. Develop a system for managing and tracking the recovery of the community during and after a disaster.
23. Develop a system for managing and tracking the recovery of the city's services during and after a disaster.
24. Develop a system for managing and tracking the recovery of the city's economy during and after a disaster.
25. Develop a system for managing and tracking the recovery of the city's environment during and after a disaster.

FREQUENTLY ASKED QUESTIONS

What is Smart City Disaster Management (SCDIM)? Smart City Disaster Management (SCDIM) is the use of technology and data to help cities prepare for and respond to natural and man-made disasters.

How does Smart City Disaster Management work? SCDIM works by using real-time data from sensors, cameras, and other devices to monitor conditions, predict events, and alert authorities and citizens. It also uses advanced analytics to help decision-makers plan and execute effective responses.

What kind of disasters can be managed with SCDIM? SCDIM can be used to manage a wide range of disasters, including natural disasters such as floods, earthquakes, and hurricanes, as well as man-made disasters such as terrorist attacks and industrial accidents.

How does SCDIM help citizens during a disaster? SCDIM helps citizens during a disaster by providing real-time alerts and information about the situation, including evacuation routes, shelter locations, and emergency services. It also allows citizens to communicate with authorities and each other, and to access resources such as food, water, and medical assistance.

How is privacy and data protection ensured with SCDIM? Privacy and data protection are ensured with SCDIM through the use of secure and encrypted data transmission and storage. Citizens' personal information is protected and used only for the purpose of disaster response.

What are the benefits of using SCDIM? The benefits of using SCDIM include improved preparedness and response, faster and more accurate information sharing, reduced risk to life and property, and more efficient use of resources.

Who is responsible for implementing SCDIM? The responsibility for implementing SCDIM is shared between government agencies, emergency services, technology companies, and citizens. Collaboration and partnership are key to the success of SCDIM.

How can citizens get involved in SCDIM? Citizens can get involved in SCDIM by participating in community preparedness programs, reporting potential hazards or risks, and volunteering during a disaster response.

What are the challenges of implementing SCDIM? The challenges of implementing SCDIM include the need for significant investment in technology and infrastructure, ensuring interoperability and integration of different systems, and addressing concerns around data privacy and security.

What are some successful examples of SCDIM in action? Successful examples of SCDIM in action include the use of sensors and predictive analytics to monitor and respond to floods in Jakarta, Indonesia, and the deployment of a smart emergency management platform in New York City, which integrates data from multiple agencies to improve coordination and response times during a disaster.

Conclusion: *The implementation of SCDIM is a crucial step towards building a resilient and safe smart city. By using technology to enhance the city's ability to prepare for, respond to, and recover from natural and man-made disasters, cities can minimize the impact of disasters on citizens and the city's infrastructure, and ensure a speedy recovery. SCDIM can support the development of disaster recovery plans, and regular drills and exercises, which can ensure that cities are prepared to respond to disasters effectively.*

SMART CITY EDUCATION (SCE)

Using technology to enhance the learning experience and improve educational outcomes, such as digital classrooms and online learning platforms.

The implementation of SCE is a crucial step towards building a smart city that promotes lifelong learning and educational opportunities for all citizens. By using technology to enhance the learning experience and improve educational outcomes, cities can improve access to education and training, and support the development of a highly skilled and educated workforce.

One of the main benefits of SCE is the ability to use technology to enhance the learning experience and improve educational outcomes. This can include the use of digital classrooms and online learning platforms, which can provide access to a wide range of educational resources and opportunities. This can also include the use of virtual and augmented reality technology, which can provide immersive and interactive learning experiences. This means that cities can improve access to education and training, and support the development of a highly skilled and educated workforce.

Another benefit of SCE is the ability to use technology to personalize the learning experience for individual students. This can include the use of data analytics and machine learning, which can provide personalized learning plans and resources for students, based on their individual needs and abilities. This means that cities can improve the effectiveness of education and training, and support the development of a highly skilled and educated workforce.

SCE can also support the development of programs and initiatives that promote lifelong learning, such as adult education and retraining programs. This can ensure that citizens have access to educational opportunities throughout their lives, and can acquire new skills and knowledge to adapt to changes in the economy and society.

GENERAL FRAMEWORK

The following framework is a general one and the actual implementation will depend on the specific requirements and constraints of the city.

1. Develop a comprehensive education technology plan that aligns with the city's overall mission and goals.

2. Establish partnerships with local schools, universities, and educational organizations to enhance the city's educational capabilities.
3. Implement digital classrooms equipped with interactive tools and technology to enhance the learning experience.
4. Develop and implement online learning platforms to provide students with flexible and alternative learning options.
5. Establish a system for managing and tracking student data, such as grades and attendance, to improve the effectiveness of educational programs.
6. Implement a system for monitoring and evaluating the effectiveness of educational programs to ensure that they are meeting the needs of students.
7. Develop a system for providing personalized learning experiences to students, such as adaptive learning software and digital tutoring.
8. Implement a system for providing professional development opportunities to teachers and educators to ensure they are equipped to use technology in the classroom effectively.
9. Develop a system for providing online resources and materials to students and teachers, such as digital textbooks and educational videos.
10. Implement a system for managing and tracking the use of technology in the classroom, such as the number of students using online resources and the number of digital classrooms.
11. Provide training and support to students, teachers and parents to ensure they have the necessary skills to effectively use technology in the classroom
12. Implement a system for providing technical support and troubleshooting for technology in the classroom
13. Develop a system for providing access to digital educational resources and materials to students and teachers, such as digital textbooks and educational videos.
14. Evaluate the needs of the students with disabilities, and provide necessary assistive technology to make the learning experience inclusive.
15. Invest in robust and reliable infrastructure to support the implementation of smart city education technology
16. Establish a dedicated budget for the implementation and maintenance of smart city education technology.
17. Develop a data security plan to protect student data and information.

18. Create a governance structure to oversee the implementation and management of smart city education technology.
19. Develop a communication plan to inform and educate students, teachers, and parents about the benefits and use of smart city education technology.
20. Develop a system for measuring and reporting on the success of smart city education technology.
21. Ensure that the privacy and security of student data is protected by complying with all relevant laws and regulations
22. Continuously evaluate and update technology used in the classrooms to ensure that it is current and effective
23. Encourage the use of open education resources (OERs) and develop a strategy for their implementation
24. Establish a feedback mechanism for students, teachers and parents to provide input and suggestions on the smart city education technology
25. Foster collaboration and sharing of best practices among educators and schools in the city to drive innovation and improvement in smart city education technology.

FREQUENTLY ASKED QUESTIONS

What is Smart City Education (SCE)? Smart City Education is an initiative to transform traditional education models by integrating technology and digital tools in the classroom. It aims to improve the quality of education, enhance learning outcomes and promote the use of digital skills among students.

What are the benefits of Smart City Education? Smart City Education has several benefits, including increased access to educational resources, improved learning outcomes, personalized learning experiences, and the development of digital skills. It also helps to bridge the digital divide and create equal opportunities for education.

How does Smart City Education work? Smart City Education uses technology to create an interactive and engaging learning experience for students. It involves the use of devices such as laptops, tablets, and interactive whiteboards, as well as software and applications that facilitate learning and student engagement.

What is the role of teachers in Smart City Education? Teachers play a vital role in Smart City Education. They are responsible for designing the curriculum, facilitating learning, and providing guidance and support to students. They also help to ensure that technology is used effectively in the classroom and that students develop the necessary digital skills.

How does Smart City Education promote equal opportunities for education? Smart City Education promotes equal opportunities for education by providing students with access to digital tools and resources, regardless of their socio-economic background. It also helps to bridge the digital divide and create a level playing field for all students.

What are some examples of Smart City Education initiatives? Some examples of Smart City Education initiatives include the use of e-learning platforms, digital textbooks, interactive whiteboards, and virtual learning environments. These initiatives are designed to enhance learning outcomes and promote the use of digital skills among students.

What is the impact of Smart City Education on the job market? Smart City Education has a positive impact on the job market by creating a workforce with the necessary digital skills to meet the demands of the modern economy. It helps to promote innovation, entrepreneurship and digital transformation, and opens up new opportunities for employment.

How does Smart City Education help to address the skills gap? Smart City Education helps to address the skills gap by providing students with the necessary digital skills to succeed in the modern economy. It focuses on developing skills in areas such as coding, data analytics, and digital marketing, which are in high demand in the job market.

What is the role of parents in Smart City Education? Parents play an important role in Smart City Education by supporting their children's learning and providing a safe and conducive environment for studying. They also play a role in encouraging their children to develop digital skills and making sure they have access to the necessary resources and tools.

How can cities implement Smart City Education initiatives? Cities can implement Smart City Education initiatives by investing in digital infrastructure and resources, partnering with educational institutions and technology companies, and promoting awareness and adoption of digital tools and resources among students, teachers, and parents. It also requires collaboration and coordination among different stakeholders to ensure that the initiatives are effective and sustainable.

Conclusion: The implementation of SCE is a crucial step towards building a smart city that promotes lifelong learning and educational opportunities for all citizens. By using technology to enhance the learning experience and improve educational outcomes, cities can improve access to education and training, and support the development of a highly skilled and educated workforce. SCE can support the development of programs and initiatives that promote lifelong learning, such as adult education and retraining programs, ensuring that citizens have access to educational opportunities throughout their lives, and can acquire new skills and knowledge to adapt to changes in the economy and society.

SMART CITY E-GOVERNANCE (SCEG)

Using technology to improve the efficiency and transparency of government operations and services.

Implementing SCEG involves using technology to improve the efficiency and transparency of government operations and services. This approach can lead to a number of benefits for citizens, businesses, and government agencies.

One of the key benefits is improved efficiency. With the use of technology, government services can be delivered in a more streamlined and timely manner. This can include services such as paying taxes, accessing government information, and applying for licenses and permits., technology can be used to automate repetitive and time-consuming tasks, freeing up government staff to focus on more important work.

Another benefit of SCEG is increased transparency. Technology can make it easier for citizens to access information about government operations, budgets, and decision-making processes. This can lead to greater accountability and trust in government, as citizens can see how their tax dollars are being spent and how decisions are being made. The use of technology can help government agencies better understand the needs and preferences of citizens, which can help inform policy decisions.

SCEG can also improve the delivery of services in urban areas, such as transportation and public safety. For example, technology can help cities better manage traffic flow and reduce congestion, which can improve air quality and save citizens time. Similarly, the use of technology can help enhance public safety by allowing for improved communication and coordination among first responders. Implementing SCEG can lead to cost savings for government agencies by reducing the need for manual labour, reducing errors and improving transparency.

GENERAL FRAMEWORK

The following framework is a general one and the actual implementation will depend on the specific requirements and constraints of the city.

1. Establish a dedicated team or department responsible for implementing the e-governance initiative.
2. Conduct a thorough assessment of current government operations and services to identify areas for improvement.
3. Develop a comprehensive e-governance strategy that outlines the goals and objectives of the initiative.

4. Identify and prioritize key services that will be transitioned to an online or digital format.
5. Establish a secure and reliable technology infrastructure to support the e-governance initiative.
6. Implement secure login and authentication systems to ensure the confidentiality and integrity of government data.
7. Develop and implement policies and procedures to govern the use and management of government data.
8. Create a user-friendly and accessible online portal for citizens to access government services and information.
9. Develop training programs for government employees to ensure they are equipped to deliver digital services to citizens.
10. Implement data analytics tools to monitor and evaluate the performance of e-governance systems and services.
11. Encourage citizen participation and feedback to improve the e-governance system.
12. Develop and implement a disaster recovery plan for e-governance systems and services.
13. Implement data security measures to protect government data from cyber threats.
14. Collaborate with other government agencies and organizations to share resources and best practices.
15. Establish standards and guidelines for the design and development of e-governance systems and services.
16. Continuously monitor and update e-governance systems and services to ensure they remain relevant and effective.
17. Develop a citizen engagement program to encourage citizens to adopt and use the e-governance systems and services.
18. Create an e-governance service level agreement to ensure that the services are delivered to the citizens on time.
19. Establish a complaint management system to ensure that citizens' complaints are addressed in a timely manner.
20. Implement a system for measuring the performance of the e-governance service and make adjustments as needed.
21. Collaborate with the private sector to develop innovative solutions and services that can improve e-governance.
22. Implement a system to monitor and analyze the usage of e-governance services to identify areas for improvement.
23. Establish a system for tracking the progress of e-governance initiatives to ensure that they are on track and within budget.

24. Encourage the use of open data and open source software for e-governance systems and services.
25. Continuously evaluate and improve e-governance systems and services to ensure they meet the evolving needs of citizens and the city.

FREQUENTLY ASKED QUESTIONS

What is Smart City e-governance? Smart City e-governance refers to the use of digital technologies to improve the efficiency and transparency of government services and decision-making in a city.

How does Smart City e-governance benefit the citizens? Smart City e-governance benefits the citizens by making government services more accessible, efficient, and transparent. It enables citizens to participate in decision-making and provides them with a platform to voice their concerns and feedback.

What are the key features of Smart City e-governance? The key features of Smart City e-governance include digital citizen services, online payment systems, integrated service delivery, real-time data monitoring, and transparent decision-making.

What are the challenges of implementing Smart City e-governance? The challenges of implementing Smart City e-governance include ensuring digital literacy and accessibility, data security and privacy concerns, lack of IT infrastructure, and resistance to change.

How can Smart City e-governance improve public service delivery? Smart City e-governance can improve public service delivery by making government services more accessible, efficient, and transparent. It also enables government departments to streamline their processes and improve their coordination.

How can citizens participate in Smart City e-governance? Citizens can participate in Smart City e-governance by using digital citizen services, providing feedback and suggestions through online platforms, participating in public consultations, and monitoring real-time data.

How can Smart City e-governance improve accountability? Smart City e-governance can improve accountability by enabling citizens to access information about government decisions and actions. It also provides a platform for citizens to hold government officials accountable for their actions.

How can Smart City e-governance improve decision-making? Smart City e-governance can improve decision-making by providing real-time data and insights to government officials. It also enables citizens to participate in decision-making and provides a platform for feedback and suggestions.

How can Smart City e-governance ensure data security and privacy? Smart City e-governance can ensure data security and privacy by implementing secure data storage and transmission protocols, restricting access to sensitive data, and ensuring that citizens' data is used only for the intended purposes.

What are some examples of Smart City e-governance initiatives? Examples of Smart City e-governance initiatives include digital citizen services, online payment systems, citizen feedback platforms, real-time data monitoring systems, and transparent decision-making platforms.

Conclusion: *Implementing SCEG can lead to a wide range of benefits for citizens, businesses and government agencies, including improved efficiency, increased transparency, better delivery of services and cost savings.*

SMART CITY EMERGENCY RESPONSE (SCER)

Using technology to enhance emergency response capabilities, such as real-time situational awareness and incident management systems.

Implementing SCER involves using technology to enhance emergency response capabilities, such as real-time situational awareness and incident management systems. This approach can lead to a number of benefits for citizens, first responders, and government agencies.

One of the key benefits is improved situational awareness. With the use of technology, first responders can access real-time information about an incident, such as the location of emergency personnel, the status of equipment and the condition of affected individuals. This can help first responders make more informed decisions and respond more quickly and effectively to an emergency.

Another benefit of SCER is enhanced incident management. Technology can be used to automate repetitive and time-consuming tasks, such as logging incident details and communicating with other agencies, freeing up first responders to focus on more important work. Technology can be used to better coordinate the actions of multiple agencies, such as police, fire, and emergency medical services, allowing for a more efficient and effective response to an emergency.

SCER can also improve the delivery of services to citizens, such as warning systems, emergency shelter and evacuation plans. For example, the use of technology can help cities better manage emergency warning system, which can help citizens prepare and protect themselves in case of emergency. Similarly, the use of technology can help enhance emergency shelter, evacuation plan and communication with citizens during emergency. Implementing SCER can lead to cost savings for government agencies by reducing the need for manual labour, reducing errors and improving transparency.

GENERAL FRAMEWORK

The following framework is a general one and the actual implementation will depend on the specific requirements and constraints of the city.

1. Developing a comprehensive emergency response plan that incorporates the use of technology.
2. Identifying and assessing potential emergency scenarios and hazards.
3. Establishing real-time situational awareness systems to monitor and respond to emergency situations.
4. Implementing incident management systems to streamline communication and coordination between emergency response agencies.
5. Providing emergency responders with access to real-time data and information, such as live video feeds and sensor data.
6. Integrating emergency response systems with existing city infrastructure, such as traffic management and public safety systems.
7. Developing and implementing training programs for emergency responders on the use of technology in emergency response.
8. Providing citizens with access to emergency information and alerts through mobile apps and other digital platforms.
9. Utilizing data analytics and machine learning to improve emergency response decision-making.
10. Implementing secure, reliable and resilient communication infrastructure for emergency responders.
11. Building a city-wide emergency alert system that can be activated in case of emergency.
12. Providing emergency responders with real-time access to information about critical infrastructure, such as power and water systems.
13. Providing real-time information to the public about emergency situations and evacuation routes.
14. Implementing geolocation-based emergency response systems to quickly locate and respond to incidents.
15. Establishing a system for collecting and analyzing feedback from citizens and emergency responders to continuously improve emergency response capabilities.
16. Building a centralized emergency response data repository for data analysis and reporting.
17. Implementing network-based security measures to protect emergency response systems from cyber threats.
18. Collaborating with other cities and regions to share best practices and technologies for emergency response.
19. Incorporating community involvement in emergency response planning and execution.

20. Building a dedicated team for the maintenance, management and modernization of emergency response systems.
21. Utilizing virtual and augmented reality technologies for emergency response training and simulation.
22. Implementing a system for tracking the location and status of emergency response assets and personnel.
23. Incorporating automation and robotics technologies in emergency response operations.
24. Providing emergency responders with access to real-time weather, traffic and other data to improve their decision-making.
25. Building a system for monitoring and analyzing social media and other online platforms for early warning of emergencies.

FREQUENTLY ASKED QUESTIONS

What is SMART CITY EMERGENCY RESPONSE (SCER)? SCER is a technology-enabled emergency response system designed to help cities respond quickly and efficiently to emergencies and natural disasters.

How does SCER work? SCER integrates multiple technologies such as sensors, cameras, drones, and AI to detect, monitor and respond to emergency situations in real-time. The system analyses data from various sources to provide real-time insights and alerts to emergency responders.

What types of emergencies can SCER help manage? SCER can help manage various types of emergencies, including natural disasters such as floods and earthquakes, public safety emergencies such as fires and accidents, and public health emergencies such as pandemics.

How does SCER help emergency responders? SCER provides emergency responders with real-time information on the location and severity of an emergency, helping them to respond more quickly and effectively. The system can also help prioritize resources and coordinate response efforts across multiple agencies.

How does SCER help citizens during emergencies? SCER can help citizens by providing real-time alerts and updates on emergency situations, enabling them to take appropriate actions to stay safe. The system can also help emergency responders locate and rescue citizens who may be in danger.

How is data collected and analysed in SCER? Data is collected through various sensors, cameras, and other devices deployed throughout the city. The data is then analysed using AI and other analytical tools to provide real-time insights on emergency situations.

Is SCER secure? Yes, SCER is designed with robust security features to ensure the confidentiality, integrity, and availability of data. The system uses encryption, authentication, and access control mechanisms to prevent unauthorized access to sensitive information.

How is SCER funded? SCER is typically funded by a combination of public and private sources, including government grants, corporate sponsorships, and philanthropic donations.

How is SCER different from traditional emergency response systems? SCER is different from traditional emergency response systems in that it uses advanced technologies such as AI, IoT, and big data analytics to provide real-time insights and alerts to emergency responders. The system is also designed to be more agile and adaptable to changing emergency situations.

Is SCER currently deployed in any cities? Yes, SCER is currently being deployed in various cities around the world, including Barcelona, Amsterdam, and New York City. The system is expected to become more widespread as cities increasingly adopt smart city technologies.

Conclusion: *Implementing SCER can lead to a wide range of benefits for citizens, first responders and government agencies, including improved situational awareness, enhanced incident management, improved delivery of services, and cost savings.*

SMART CITY EMERGENCY SERVICES (SCES)

Using technology to improve emergency response times, such as real-time tracking and location-based services.

Implementing SCES involves using technology to improve emergency response times, such as real-time tracking and location-based services. This approach can lead to a number of benefits for citizens, first responders, and government agencies.

One of the key benefits is faster emergency response times. With the use of technology, first responders can access real-time information about the location of an emergency, such as the location of emergency personnel and the status of equipment. This can help first responders respond more quickly and effectively to an emergency. Location-based services can be used to automatically dispatch the nearest emergency personnel to an incident, further reducing response times.

Another benefit of SCES is improved communication and coordination among first responders. Technology can be used to automate repetitive and time-consuming tasks, such as logging incident details and communicating with other agencies, freeing up first responders to focus on more important work. Technology can be used to better coordinate the actions of multiple agencies, such as police, fire, and emergency medical services, allowing for a more efficient and effective response to an emergency.

SCES can also improve the delivery of emergency services to citizens, such as warning systems, emergency shelter and evacuation plans. For example, the use of technology can help cities better manage emergency warning systems, which can help citizens prepare and protect themselves in case of emergency. Similarly, the use of technology can help enhance emergency shelter, evacuation plan and communication with citizens during emergency. Implementing SCES can lead to cost savings for government agencies by reducing the need for manual labour, reducing errors and improving transparency.

GENERAL FRAMEWORK

The following framework is a general one and the actual implementation will depend on the specific requirements and constraints of the city.

1. Identify key emergency services in the city such as fire, police, and medical services.
2. Develop a comprehensive emergency response plan that includes technology-based solutions.
3. Implement real-time tracking and location-based services for emergency vehicles and personnel.
4. Install emergency communication systems, such as radios and mobile devices, for emergency responders.
5. Develop a city-wide emergency notification system, such as text messaging or email alerts.
6. Implement a data management system for emergency services to share and access information in real-time.
7. Establish a centralized incident management system for emergency services to coordinate responses.
8. Implement a system for citizens to report emergencies and request assistance.
9. Use technology to improve emergency response times through traffic management and routing systems.
10. Develop a training program for emergency responders on the use of technology-based systems.
11. Establish a testing and maintenance schedule for emergency response technology systems.
12. Develop a data analytics system to monitor and improve emergency response performance.
13. Implement a system for emergency services to access building and infrastructure information, such as floor plans and emergency shut-off valves.
14. Develop a system for emergency responders to access real-time weather and environmental data.
15. Implement a system for remote monitoring of critical infrastructure, such as power plants and water treatment facilities.
16. Use technology to enhance public safety, such as cameras and license plate recognition systems.
17. Develop a system for emergency responders to access citizen information, such as medical conditions and emergency contacts.

18. Establish a data sharing agreement with neighbouring cities and emergency services.
19. Use technology to improve emergency response logistics, such as resource management and supply chain systems.
20. Develop a system for virtual and augmented reality training for emergency responders.
21. Use technology to improve emergency response coordination with other government agencies and non-governmental organizations.
22. Implement a system for real-time monitoring of emergency response activities through dashboards and visualization tools.
23. Use technology to improve emergency response resilience and recovery, such as disaster recovery plans and backup systems.
24. Develop a system for emergency responders to access real-time data from sensor networks and IoT devices.
25. Use technology to improve emergency response decision-making, such as predictive analytics and machine learning systems.

FREQUENTLY ASKED QUESTIONS

What are Smart City Emergency Services? Smart City Emergency Services (SCES) are advanced systems and technologies that enable emergency service providers to deliver quick and effective responses to emergency situations.

What are the components of SCES? SCES includes components such as emergency response centers, communication systems, location-based services, real-time data analysis, and remote monitoring.

How do SCES benefit emergency service providers? SCES help emergency service providers to respond more quickly and efficiently to emergency situations. With real-time data analysis and remote monitoring, emergency service providers can anticipate and respond to emergency situations before they escalate.

How do SCES benefit the public? SCES benefit the public by providing quick and effective emergency responses, reducing response times, and improving the safety and security of the community.

What are some examples of SCES in action? Examples of SCES include intelligent traffic management systems that reroute traffic in real-time to accommodate emergency vehicles, automated emergency response systems that provide location-based services, and remote monitoring systems that provide real-time data on the status of emergency situations.

What are some challenges with implementing SCES? Challenges with implementing SCES include funding, infrastructure requirements, data privacy and security concerns, and the need for interoperability between different systems and technologies.

How can communities get involved in implementing SCES? Communities can get involved in implementing SCES by working with emergency service providers, local government, and technology vendors to identify and implement solutions that address the unique needs of the community.

How do SCES help in natural disaster situations? SCES help in natural disaster situations by providing real-time data on the status of emergency situations, enabling emergency service providers to respond quickly and effectively to save lives and minimize damage.

How do SCES address the needs of vulnerable populations? SCES address the needs of vulnerable populations by providing location-based services and real-time data analysis that help emergency service providers to respond quickly to emergency situations that may impact these populations.

How are data privacy concerns addressed in SCES? Data privacy concerns are addressed in SCES by implementing secure data storage and transmission protocols, ensuring that personal data is only used for emergency response purposes, and providing transparent data governance policies that ensure data is used ethically and in compliance with data privacy regulations.

Conclusion: *Implementing SCES can lead to a wide range of benefits for citizens, first responders and government agencies, including faster emergency response times, improved communication and coordination, improved delivery of services and cost savings.*

SMART CITY ENERGY DISTRIBUTION (SMED)

Using technology to optimize the distribution of energy, such as microgrids and distributed energy resources (DERs).

Implementing SMED involves using technology to optimize the distribution of energy, such as microgrids and distributed energy resources (DERs). This approach can lead to a number of benefits for citizens, energy providers, and government agencies.

One of the key benefits of SMED is improved energy efficiency. With the use of technology, energy providers can access real-time information about energy usage and demand, allowing them to adjust energy production and distribution in real-time. This can help reduce energy waste, as well as improve the overall efficiency of the energy grid. The use of microgrids and DERs can help to balance local energy supply and demand, further increasing efficiency.

Another benefit of SMED is increased energy resilience and security. Microgrids and DERs can operate independently of the main energy grid, allowing them to continue providing energy in the event of an outage or disaster. This can help to ensure that critical services, such as hospitals and emergency services, continue to function during power outages. The use of SMED technology can help to detect and prevent potential security threats to the energy grid.

SMED can also help to reduce costs for energy consumers by enabling them to generate their own energy, for example, through the use of solar panels or other DERs. It can also help to reduce environmental impacts by increasing the use of renewable energy sources.

Implementing SMED can lead to cost savings for government agencies and energy providers by reducing the need for manual labour, reducing errors, and improving transparency.

GENERAL FRAMEWORK

The following framework is a general one and the actual implementation will depend on the specific requirements and constraints of the city.

1. Identify the specific energy distribution challenges in the city.
2. Research and evaluate different microgrid and DER technologies.
3. Develop a comprehensive plan for implementing microgrids and DERs in the city.
4. Secure funding for the project from government agencies and private investors.
5. Develop partnerships with local utilities and other stakeholders.
6. Conduct a feasibility study to determine the technical and economic viability of the project.
7. Assess the potential impact on the power grid and transmission infrastructure.
8. Develop a plan for integrating microgrids and DERs into the existing power grid.
9. Conduct an environmental impact assessment.
10. Develop a plan for managing and maintaining the microgrids and DERs.
11. Identify potential locations for microgrid deployment.
12. Develop a plan for connecting microgrids to the existing power grid.
13. Develop a plan for integrating renewable energy sources into the microgrids.
14. Develop a plan for energy storage systems.
15. Develop a plan for demand management and load balancing.
16. Develop a plan for integrating energy-efficient technologies into the microgrids.
17. Develop a plan for monitoring and controlling the microgrids and DERs.
18. Develop a plan for cybersecurity measures.
19. Develop a plan for educating and engaging the community.
20. Obtain necessary permits and approvals.
21. Procure equipment and materials for the project.

22. Begin construction of the microgrids and DERs.
23. Test and commission the microgrids and DERs.
24. Monitor and evaluate the performance of the microgrids and DERs.
25. Continuously improve the microgrids and DERs based on feedback and performance data.

FREQUENTLY ASKED QUESTIONS

What is Smart City Energy Distribution (SMED)? Smart City Energy Distribution (SMED) is a system that leverages advanced technologies to optimize energy generation, distribution, and consumption across a city.

How does SMED work? SMED leverages advanced sensors, analytics, and automation to intelligently monitor and optimize energy distribution and consumption across a city.

What benefits does SMED offer to the public? SMED offers several benefits, including reducing energy waste, improving energy efficiency, increasing the reliability and resilience of the energy grid, and reducing overall energy costs.

How does SMED impact the environment? SMED can significantly reduce a city's carbon footprint by reducing energy waste, promoting renewable energy adoption, and supporting other sustainability initiatives.

What kind of data does SMED collect? SMED collects a variety of data, including energy usage patterns, renewable energy generation, weather data, and other relevant information to optimize energy distribution and consumption across the city.

Is SMED expensive to implement? The cost of implementing SMED varies depending on the size of the city, the existing energy infrastructure, and the specific needs of the community. However, many cities have found that the long-term benefits of SMED outweigh the initial investment.

How can residents and businesses get involved in SMED? Residents and businesses can get involved in SMED by adopting energy-efficient practices, investing in renewable energy technologies, and participating in community initiatives to promote energy conservation.

How does SMED impact energy companies? SMED can help energy companies optimize energy generation and distribution, reduce energy waste, and promote renewable energy adoption. However, it may also disrupt traditional energy models and require energy companies to adapt to new business models.

Can SMED be customized to fit the needs of different cities? Yes, SMED can be customized to fit the unique needs of different cities, taking into account factors such as population density, energy demand, and existing energy infrastructure.

What are the potential future developments in SMED? Potential future developments in SMED include the integration of advanced AI and machine learning technologies, the expansion of renewable energy sources, and the development of new energy storage technologies to enhance energy reliability and resiliency.

Conclusion: Implementing SMED can lead to a wide range of benefits for citizens, energy providers and government agencies, including improved energy efficiency, increased energy resilience and security, cost savings for energy consumers and environmental benefits.

SMART CITY ENERGY MANAGEMENT (SCEM)

Using technology to optimize energy consumption and reduce costs, such as building automation and demand response programs.

Implementing SCEM involves using technology to optimize energy consumption and reduce costs, such as building automation and demand response programs. This approach can lead to a number of benefits for citizens, energy providers, and government agencies.

One of the key benefits of SCEM is improved energy efficiency. With the use of technology, building automation systems can be used to adjust lighting, heating, and cooling based on occupancy and weather conditions, reducing energy waste. Similarly, demand response programs can be used to encourage energy users to reduce their consumption during peak demand periods, further reducing energy waste.

Another benefit of SCEM is reduced costs for energy consumers. Building automation and demand response programs can help to reduce energy consumption, resulting in lower energy bills for consumers. The use of technology can help to detect and fix energy-wasting equipment, further reducing costs.

SCEM can also help to reduce the environmental impact of energy consumption by promoting the use of renewable energy sources and reducing the overall energy usage. Implementing SCEM can lead to cost savings for government agencies and energy providers by reducing the need for manual labour, reducing errors, and improving transparency.

GENERAL FRAMEWORK

The following framework is a general one and the actual implementation will depend on the specific requirements and constraints of the city.

1. Identify the specific energy management challenges in the city.
2. Research and evaluate different smart city energy management technologies and solutions.
3. Develop a comprehensive plan for implementing smart city energy management in the city.
4. Secure funding for the project from government agencies and private investors.
5. Develop partnerships with local utilities and other stakeholders.
6. Conduct a feasibility study to determine the technical and economic viability of the project.
7. Assess the potential impact on the power grid and transmission infrastructure.

8. Develop a plan for integrating smart city energy management solutions into the existing power grid.
9. Conduct an environmental impact assessment.
10. Develop a plan for managing and maintaining the smart city energy management systems.
11. Identify potential locations for smart city energy management deployment.
12. Develop a plan for connecting smart city energy management systems to the existing power grid.
13. Develop a plan for integrating renewable energy sources into smart city energy management systems.
14. Develop a plan for energy storage systems.
15. Develop a plan for demand management and load balancing.
16. Develop a plan for integrating energy-efficient technologies into smart city energy management systems.
17. Develop a plan for monitoring and controlling smart city energy management systems.
18. Develop a plan for cybersecurity measures.
19. Develop a plan for educating and engaging the community.
20. Obtain necessary permits and approvals.
21. Procure equipment and materials for the project.
22. Begin installation of the smart city energy management systems.
23. Test and commission the smart city energy management systems.
24. Monitor and evaluate the performance of the smart city energy management systems.
25. Continuously improve the smart city energy management systems based on feedback and performance data.

FREQUENTLY ASKED QUESTIONS

What is Smart City Energy Management? Smart City Energy Management (SCEM) is the use of technology and data analytics to optimize the use and distribution of energy in a city to ensure efficiency, cost savings, and sustainability.

How does SCEM work? SCEM uses various sensors and devices to collect data on energy usage in a city. This data is then analysed using machine learning algorithms to identify patterns, predict future usage, and optimize energy distribution in real-time.

How can SCEM benefit a city? SCEM can help a city reduce its energy consumption and costs, decrease carbon emissions, and improve the reliability and resiliency of the energy infrastructure. It can also enable more efficient and sustainable development of the city.

Who is involved in implementing SCEM in a city? SCEM requires collaboration between the city government, energy providers, technology companies, and citizens. It is typically spearheaded by the city government in partnership with relevant stakeholders.

What are some examples of SCEM technologies? Some examples of SCEM technologies include smart meters, energy storage systems, demand response systems, renewable energy generation systems, and energy management software.

How can citizens participate in SCEM? Citizens can participate in SCEM by adopting energy-efficient behaviors, installing renewable energy systems in their homes, and providing feedback to the city government on energy-related issues.

How does SCEM help in the transition to renewable energy? SCEM can help integrate renewable energy sources into the energy infrastructure of a city by optimizing energy storage, distribution, and usage. This can make it easier and more cost-effective to transition to renewable energy.

What are some challenges to implementing SCEM? Some challenges to implementing SCEM include the high upfront costs of deploying the necessary technologies, the need for skilled personnel to operate and maintain the systems, and ensuring data privacy and security.

What are some benefits of citizen engagement in SCEM? Citizen engagement in SCEM can lead to increased energy awareness and education, better energy efficiency, and more sustainable energy practices in the community.

How does SCEM align with the broader goals of a Smart City?
SCEM is an essential component of a Smart City as it promotes energy efficiency, sustainability, and resilience in the urban environment. By optimizing energy usage and reducing energy waste, SCEM can help cities become more livable, sustainable, and prosperous.

Conclusion: Implementing SCEM can lead to a wide range of benefits for citizens, energy providers, and government agencies, including improved energy efficiency, reduced costs for energy consumers, and environmental benefits.

SMART CITY ENERGY STORAGE (SCENS)

Using technology to store excess energy generated by renewable sources and make it available when needed, such as battery storage systems.

Implementing SCENS involves using technology to store excess energy generated by renewable sources and make it available when needed, such as battery storage systems. This approach can lead to a number of benefits for citizens, energy providers, and government agencies.

One of the key benefits of SCENS is improved energy reliability. With the use of energy storage systems, excess energy generated by renewable sources can be stored and used when needed, reducing the need for traditional fossil fuel-based power generation. This can help to ensure a consistent and reliable supply of energy, even during periods of low renewable energy generation, such as at night or during extended cloud cover.

Another benefit of SCENS is reduced costs for energy consumers. Energy storage systems can reduce the need for expensive peaker power plants, which are typically used to meet peak energy demand. Energy storage systems can be used to shift energy consumption to off-peak hours, reducing peak energy costs.

SCENS can also help to reduce the environmental impact of energy consumption by increasing the use of renewable energy sources and reducing the overall energy usage. Implementing SCENS can lead to cost savings for government agencies and energy providers by reducing the need for manual labour, reducing errors, and improving transparency.

GENERAL FRAMEWORK

The following framework is a general one and the actual implementation will depend on the specific requirements and constraints of the city.

1. Identify the specific energy storage challenges in the city.
2. Research and evaluate different smart city energy storage technologies and solutions.
3. Develop a comprehensive plan for implementing smart city energy storage in the city.
4. Secure funding for the project from government agencies and private investors.
5. Develop partnerships with local utilities and other stakeholders.
6. Conduct a feasibility study to determine the technical and economic viability of the project.
7. Assess the potential impact on the power grid and transmission infrastructure.
8. Develop a plan for integrating smart city energy storage solutions into the existing power grid.
9. Conduct an environmental impact assessment.
10. Develop a plan for managing and maintaining the smart city energy storage systems.
11. Identify potential locations for smart city energy storage deployment.
12. Develop a plan for connecting smart city energy storage systems to the existing power grid.
13. Develop a plan for integrating renewable energy sources into smart city energy storage systems.
14. Develop a plan for demand management and load balancing.
15. Develop a plan for integrating energy-efficient technologies into smart city energy storage systems.
16. Develop a plan for monitoring and controlling smart city energy storage systems.
17. Develop a plan for cybersecurity measures.
18. Develop a plan for educating and engaging the community.
19. Obtain necessary permits and approvals.

20. Procure equipment and materials for the project.
21. Begin construction of the smart city energy storage systems.
22. Test and commission the smart city energy storage systems.
23. Monitor and evaluate the performance of the smart city energy storage systems.
24. Continuously improve the smart city energy storage systems based on feedback and performance data.
25. Develop a plan for scalability and expansion of the energy storage systems as needed.

FREQUENTLY ASKED QUESTIONS

What is energy storage in a smart city? Energy storage is the process of capturing and storing excess energy produced by renewable energy sources like solar and wind turbines to be used when demand is high.

How does energy storage work in a smart city? Energy storage systems typically involve using batteries, flywheels, or other energy storage technologies to capture and store excess energy. This stored energy can then be used to power homes, businesses, and other buildings during peak demand times.

What are the benefits of energy storage in a smart city? Energy storage provides a number of benefits, including reduced energy costs, increased reliability and stability of the power grid, and improved environmental sustainability by reducing the need for fossil fuels.

What types of energy storage systems are used in smart cities? Smart cities may use a variety of energy storage technologies, including lithium-ion batteries, flow batteries, and compressed air energy storage systems, among others.

How can energy storage be integrated into a smart city? Energy storage can be integrated into a smart city in a number of ways, including using microgrids, demand response programs, and virtual power plants to manage energy supply and demand.

What role do government policies and regulations play in energy storage in smart cities? Government policies and regulations can play a significant role in incentivizing the adoption of energy storage technologies in smart cities. These policies may include tax incentives, subsidies, and mandates for utilities to use renewable energy sources and energy storage systems.

How can residents and businesses benefit from energy storage in a smart city? Residents and businesses can benefit from energy storage in smart cities by reducing their energy costs, improving the reliability and stability of their power supply, and contributing to a more sustainable energy future.

What are some challenges associated with energy storage in smart cities? Some challenges associated with energy storage in smart cities include the high upfront costs of energy storage systems, the need for effective management and coordination of energy storage systems, and the potential for environmental impacts associated with the production and disposal of batteries.

What is the future of energy storage in smart cities? The future of energy storage in smart cities is likely to involve increased adoption of energy storage technologies, the development of new and more efficient storage technologies, and continued innovation in the ways that energy storage is integrated into smart city infrastructure.

How can individuals and organizations get involved in promoting energy storage in smart cities? Individuals and organizations can get involved in promoting energy storage in smart cities by advocating for government policies and regulations that incentivize the adoption of energy storage technologies, supporting the development of renewable energy sources, and investing in energy storage technologies themselves.

Conclusion: *Implementing SCENS can lead to a wide range of benefits for citizens, energy providers, and government agencies, including improved energy reliability, reduced costs for energy consumers and environmental benefits.*

SMART CITY ENVIRONMENT (SCENV)

Using technology to improve the quality of life of citizens and the natural environment, such as green spaces and biodiversity monitoring.

Implementing SCENV involves using technology to improve the quality of life of citizens and the natural environment, such as green spaces and biodiversity monitoring. This approach can lead to a number of benefits for citizens, government agencies, and the environment.

One of the key benefits of SCENV is improved urban livability. With the use of technology, city officials can monitor and maintain green spaces, such as parks and public gardens, to ensure they are safe and accessible for citizens. Technology can be used to monitor the health of urban ecosystems, such as monitoring air and water quality, which can help to improve the overall livability of the city.

Another benefit of SCENV is improved sustainability. By monitoring and managing urban ecosystems, city officials can take steps to reduce the environmental impact of the city, such as reducing carbon emissions, reducing waste, and protecting biodiversity. Technology can be used to promote sustainable transportation, such as by providing real-time information on public transportation and bike-sharing options.

SCENV can also help to improve public health by monitoring and managing air and water quality, reducing exposure to pollutants and promoting active and healthy lifestyles. Implementing SCENV can lead to cost savings for government agencies by reducing the need for manual labour, reducing errors, and improving transparency.

GENERAL FRAMEWORK

The following framework is a general one and the actual implementation will depend on the specific requirements and constraints of the city.

1. Define the goals and objectives for a smart city environment.
2. Conduct a city-wide assessment of existing environmental infrastructure and resources.
3. Engage stakeholders, including citizens, government officials, and technology experts.
4. Develop a master plan for green spaces, including parks, gardens, and natural areas.
5. Implement technology solutions to monitor and manage waste and water resources.
6. Invest in renewable energy solutions, such as solar, wind, and geothermal.
7. Use smart sensors and data analytics to monitor air quality and reduce emissions.
8. Use technology to improve energy efficiency in buildings, transportation, and other systems.
9. Encourage sustainable transportation options, such as public transportation, biking, and walking.
10. Use technology to reduce greenhouse gas emissions and mitigate the effects of climate change.
11. Implement a smart grid system to manage electricity generation and distribution.
12. Use data analytics to monitor and manage the city's carbon footprint.
13. Develop a comprehensive recycling and waste management program.
14. Implement technology to monitor and protect biodiversity, including wildlife and plants.
15. Foster community gardens and urban agriculture programs.
16. Use technology to manage and preserve natural resources, such as water and forests.
17. Develop a smart city platform to integrate and manage environmental and sustainability initiatives.
18. Use technology to encourage citizen engagement and participation in environmental initiatives.
19. Invest in education and training programs to develop a skilled workforce in environmental technology.
20. Collaborate with regional and international organizations to share best practices and resources.
21. Use data and analytics to track and report on the city's progress in achieving its environmental goals.
22. Continuously evaluate and adjust the smart city environment plan based on changing conditions and feedback.

23. Encourage public-private partnerships to support investment in green infrastructure and sustainability initiatives.
24. Foster innovation and research to develop new and improved technologies for a sustainable environment.
25. Ensure legal and regulatory frameworks support and align with the smart city environment initiatives.

FREQUENTLY ASKED QUESTIONS

What is Smart City Environment (SCENV)? Smart City Environment (SCENV) is an approach to urban development that aims to make cities more sustainable and livable through the use of technology and data. It focuses on promoting environmental sustainability and reducing the carbon footprint of cities.

How does Smart City Environment help in reducing the carbon footprint of cities? SCENV uses a range of technologies such as smart grid systems, energy-efficient buildings, and smart transportation to reduce the energy consumption of cities. By reducing energy consumption, cities can significantly reduce their carbon footprint.

How does Smart City Environment promote sustainable transportation? Smart City Environment promotes sustainable transportation by encouraging the use of public transportation, biking, and walking. It also uses technologies such as electric vehicles and intelligent transportation systems to make transportation more efficient and sustainable.

How does Smart City Environment tackle waste management? SCENV employs advanced waste management systems that use smart technologies such as sensors and automation to improve waste collection and disposal. This reduces the amount of waste generated by cities and promotes recycling and waste reduction.

How does Smart City Environment address air pollution? SCENV uses air quality monitoring systems and advanced traffic management systems to reduce the amount of air pollution generated by cities. It also promotes the use of electric and hybrid vehicles and encourages the development of green spaces to improve air quality.

How does Smart City Environment ensure access to clean water? SCENV employs advanced water management systems that use sensors and data analytics to improve water quality and reduce water waste. This ensures that cities have access to clean and sustainable water sources.

How does Smart City Environment address climate change? SCENV addresses climate change by reducing greenhouse gas emissions and promoting sustainable practices such as energy-efficient buildings, green infrastructure, and sustainable transportation. It also prepares cities for the impacts of climate change such as sea-level rise and extreme weather events.

How does Smart City Environment promote green spaces? SCENV promotes green spaces by encouraging the development of parks, gardens, and other green infrastructure. This improves air quality, promotes biodiversity, and provides residents with access to outdoor spaces.

How does Smart City Environment address noise pollution? SCENV addresses noise pollution by using advanced noise monitoring systems and promoting noise-reducing technologies such as electric vehicles and noise barriers. This improves the quality of life of residents and reduces the negative impacts of noise pollution.

What are the benefits of Smart City Environment? The benefits of Smart City Environment include reduced carbon footprint, improved air quality, enhanced energy efficiency, better waste management, improved water quality, improved public health, and increased resilience to the impacts of climate change. It also promotes livability, sustainability, and economic growth in cities.

Conclusion: *Implementing SCENV can lead to a wide range of benefits for citizens, government agencies, and the environment, including* *improved urban livability, improved sustainability, and improved public health.*

SMART CITY ENVIRONMENTAL MONITORING (SCEVM)

Using technology to monitor and manage environmental factors, such as air and water quality, noise levels and weather conditions.

Implementing SCEVM involves using technology to monitor and manage various environmental factors, such as air and water quality, noise levels, and weather conditions. This approach can lead to a number of benefits for citizens, government agencies, and the environment.

One of the key benefits of SCEVM is improved public health. By monitoring air and water quality, city officials can identify areas of high pollution, which can help to reduce exposure to pollutants for citizens. Monitoring noise levels can help to reduce noise pollution, which can have a negative impact on mental and physical health. Real-time monitoring of weather conditions can also help citizens to be prepared for any potentially dangerous weather conditions.

Another benefit of SCEVM is improved environmental protection. By monitoring air and water quality, city officials can identify areas of high pollution, which can help to take steps to reduce it. Data analysis can be used to track pollution levels over time, which can help to identify trends and take steps to mitigate them. Monitoring weather conditions can also help in identifying environmental issues such as floods, heatwaves, and droughts, and take necessary actions.

SCEVM can also help to improve sustainability by identifying areas of high pollution, which can help to take steps to reduce it. Data analysis can be used to track pollution levels over time, which can help to identify trends and take steps to mitigate them. Implementing SCEVM can lead to cost savings for government agencies by reducing the need for manual labour, reducing errors, and improving transparency.

GENERAL FRAMEWORK

The following framework is a general one and the actual implementation will depend on the specific requirements and constraints of the city.

1. Define the goals and objectives for smart city environmental monitoring.
2. Conduct a city-wide assessment of existing environmental infrastructure and resources.
3. Engage stakeholders, including citizens, government officials, and technology experts.
4. Develop a master plan for monitoring and managing key environmental factors, such as air and water quality, noise levels, and weather conditions.
5. Implement smart sensors and data analytics systems to monitor and collect data on environmental factors.
6. Use data visualization tools to present data and insights on environmental factors in a user-friendly format.
7. Develop a data management and analysis system to store and analyze collected data.
8. Use machine learning algorithms to identify patterns and trends in environmental data.
9. Implement a warning system to alert stakeholders of potential environmental hazards.
10. Use data analytics to monitor and manage air pollution, including particulate matter and greenhouse gas emissions.
11. Use technology to monitor and manage water resources, including quality, availability, and distribution.
12. Monitor and manage noise levels in key locations to ensure community health and well-being.
13. Use technology to monitor weather conditions, including temperature, humidity, and precipitation.
14. Develop a comprehensive system for managing and analyzing environmental data.
15. Use technology to engage citizens in environmental monitoring and management activities.
16. Encourage public-private partnerships to support investment in environmental monitoring and management.
17. Foster innovation and research to develop new and improved technologies for environmental monitoring and management.

18. Ensure legal and regulatory frameworks support and align with the smart city environmental monitoring initiatives.
19. Collaborate with regional and international organizations to share best practices and resources.
20. Continuously evaluate and adjust the environmental monitoring plan based on changing conditions and feedback.
21. Use data analytics to track and report on the city's progress in monitoring and managing environmental factors.
22. Invest in education and training programs to develop a skilled workforce in environmental technology.
23. Encourage the use of open data and open source technologies to promote transparency and collaboration.
24. Use data and analytics to inform policy decisions and regulations related to environmental factors.
25. Develop and implement contingency plans for responding to environmental emergencies and crises.

FREQUENTLY ASKED QUESTIONS

What is Smart City Environmental Monitoring? Smart City Environmental Monitoring involves using technology to collect and analyze data on environmental factors like air quality, water quality, noise pollution, and more in urban areas.

Why is Smart City Environmental Monitoring important? Environmental monitoring is important because it helps cities understand the impact of urban development on the natural environment and public health. It can also help identify areas for improvement and guide decision-making for sustainable urban planning.

What types of environmental data can be collected using Smart City Environmental Monitoring? Smart City Environmental Monitoring can collect data on air quality, water quality, noise pollution, temperature, humidity, and more.

How is environmental data collected and analysed in a Smart City? Environmental data can be collected using sensors, cameras, and other devices that are installed in various locations throughout the city. The data is then transmitted to a central database, where it can be analysed and used to inform decision-making.

How is the public involved in Smart City Environmental Monitoring? The public can be involved in environmental monitoring by reporting environmental concerns, participating in citizen science projects, and accessing public data on environmental conditions in their community.

Can Smart City Environmental Monitoring help improve public health? Yes, Smart City Environmental Monitoring can help improve public health by identifying areas with high levels of pollution and informing policies to reduce harmful environmental impacts.

What are some examples of Smart City Environmental Monitoring projects? Examples of Smart City Environmental Monitoring projects include air quality monitoring in Beijing, water quality monitoring in New York City, and noise pollution monitoring in Barcelona.

How much does Smart City Environmental Monitoring cost? The cost of Smart City Environmental Monitoring varies depending on the scale of the project, the types of sensors used, and other factors. However, it is generally considered to be a cost-effective investment in sustainable urban development.

Can Smart City Environmental Monitoring help reduce carbon emissions? Yes, Smart City Environmental Monitoring can help reduce carbon emissions by identifying areas where energy efficiency improvements can be made, and informing policies to reduce carbon emissions.

Is Smart City Environmental Monitoring only for large cities? No, Smart City Environmental Monitoring can be implemented in communities of any size, from small towns to large cities. It can be adapted to the needs of any community and can help promote sustainable development at the local level.

Conclusion: *Implementing SCEVM can lead to a wide range of benefits for citizens, government agencies, and the environment, including improved public health, improved environmental protection, and improved sustainability. It can also help in better decision making and proactive actions in case of any environmental issues.*

SMART CITY FINANCING (SCF)

Using technology to attract investment and optimize the use of public and private funds for Smart City development, such as smart bonds and impact investing.

Implementing SCF involves using technology to attract investment and optimize the use of public and private funds for Smart City development. This approach can lead to a number of benefits for citizens, government agencies, and investors.

One of the key benefits of SCF is the ability to attract investment from a wide range of sources, including private investors, government agencies, and international organizations. By leveraging technology, cities can create investment opportunities that are transparent, efficient, and attractive to investors.

Another benefit of SCF is the ability to optimize the use of public and private funds. By using technology to track and manage the flow of funds, cities can ensure that they are being used effectively and efficiently to support Smart City development. Using smart bonds and impact investing can help to ensure that the funds are being used in a socially responsible manner.

SCF can also help to improve the overall financial stability of cities by creating new revenue streams and increasing the overall attractiveness of the city as a place to live, work, and invest.

Implementing SCF can lead to cost savings for government agencies by reducing the need for manual labour, reducing errors, and improving transparency.

GENERAL FRAMEWORK

The following framework is a general one and the actual implementation will depend on the specific requirements and constraints of the city.

1. Define the goals and objectives for smart city financing.
2. Conduct a city-wide assessment of existing financial resources and capacities.
3. Engage stakeholders, including citizens, government officials, and financial experts.
4. Develop a master plan for financing smart city initiatives, including smart bonds and impact investing.
5. Identify and prioritize smart city projects based on potential impact and financial feasibility.
6. Develop a comprehensive financing strategy that balances public and private funding sources.
7. Use technology, such as blockchain and digital currencies, to attract and manage investment.
8. Explore innovative financing mechanisms, such as smart bonds and impact investing, to fund smart city projects.
9. Foster public-private partnerships to support investment in smart city initiatives.
10. Use data analytics to assess and manage risk associated with investment in smart city projects.
11. Use technology to improve transparency and accountability in city financing processes.
12. Use data and analytics to inform investment decisions and track progress of financing initiatives.
13. Develop a system for managing and monitoring the flow of funds for smart city projects.

14. Encourage the use of open data and open source technologies to promote transparency and collaboration.
15. Foster innovation and research to develop new and improved technologies for smart city financing.
16. Ensure legal and regulatory frameworks support and align with the smart city financing initiatives.
17. Collaborate with regional and international organizations to share best practices and resources.
18. Continuously evaluate and adjust the smart city financing plan based on changing conditions and feedback.
19. Invest in education and training programs to develop a skilled workforce in smart city financing.
20. Use data analytics to track and report on the city's progress in financing smart city initiatives.
21. Encourage citizen engagement and participation in smart city financing decisions.
22. Use data and analytics to inform policy decisions and regulations related to smart city financing.
23. Develop and implement contingency plans for responding to financial emergencies and crises.
24. Foster partnerships with financial institutions, such as banks and investment firms, to support smart city financing.
25. Encourage investment in smart city initiatives that have a positive impact on the environment and social well-being.

FREQUENTLY ASKED QUESTIONS

What is Smart City Financing (SCF)? Smart City Financing refers to the methods and strategies used to fund smart city projects, such as public-private partnerships, grants, and municipal bonds.

How are Smart City Projects financed? Smart City Projects can be financed through a variety of means, such as private investment, government grants, or public-private partnerships.

What are the benefits of public-private partnerships for Smart City Financing? Public-private partnerships provide an opportunity for both public and private entities to work together to fund smart city projects, reducing the financial burden on governments while also bringing in private sector expertise and resources.

How can municipalities attract private investment for smart city projects? Municipalities can attract private investment for smart city projects by creating an attractive business environment, offering tax incentives, and providing a clear plan and timeline for the project's implementation.

What are the risks associated with Smart City Financing? Risks associated with Smart City Financing can include cost overruns, delays in project implementation, and the possibility that the technology being implemented may become obsolete.

What are some examples of successful Smart City Financing models? Successful Smart City Financing models include public-private partnerships, municipal bonds, and crowdfunding.

How can smart city projects be financed without burdening taxpayers? Smart city projects can be financed through private investment, grants, or other sources of funding that do not rely solely on taxpayer dollars.

How do Smart City Financing models differ from traditional infrastructure financing? Smart City Financing models differ from traditional infrastructure financing in that they often involve the use of new and emerging technologies, and may require different sources of funding.

Can Smart City Financing contribute to economic development in a city? Yes, Smart City Financing can contribute to economic development by attracting new businesses and talent to the city, and by creating new jobs in the technology sector.

How can citizens be involved in the Smart City Financing process? Citizens can be involved in the Smart City Financing process by providing input on project priorities and by participating in public meetings and forums where financing options are discussed.

Conclusion: *Implementing SCF can lead to a wide range of benefits for citizens, government agencies, and investors, including the ability to attract investment, optimize the use of public and private funds, improve overall financial stability and lead to cost savings. It can also help in creating new revenue streams and attracting more investors, which can lead to more development and growth in the city.*

SMART CITY FLOOD MANAGEMENT (SCFM)

Using technology to manage and reduce flood risks, such as real-time monitoring and early warning systems.

Implementing SCFM involves using technology to manage and reduce flood risks. This approach can lead to a number of benefits for citizens, government agencies, and the environment.

One of the key benefits of SCFM is the ability to reduce the risk of flooding to citizens and their property. By using real-time monitoring and early warning systems, cities can quickly detect and respond to potential flood hazards, helping to minimize damage and protect citizens.

Another benefit of SCFM is the ability to improve the overall efficiency of flood response and management. By using technology to track and manage flood-related data, cities can ensure that they are making data-driven decisions and taking the most effective actions to reduce flood risks.

SCFM can also help to improve the overall resilience of cities to flooding by reducing the impact of floods on infrastructure and the environment. Implementing SCFM can lead to cost savings for government agencies by reducing the need for manual labour, reducing errors, and improving transparency.

GENERAL FRAMEWORK

The following framework is a general one and the actual implementation will depend on the specific requirements and constraints of the city.

1. Assess the current state of flood management in the city.
2. Engage stakeholders, including citizens, government officials, and experts in flood management.
3. Develop a comprehensive flood management plan, incorporating technology solutions.
4. Conduct a risk assessment to identify potential flood hazards and prioritize areas for intervention.
5. Install real-time monitoring systems to gather data on water levels, rainfall, and other relevant indicators.
6. Implement early warning systems to alert residents and emergency responders of potential floods.
7. Invest in infrastructure, such as flood barriers and retaining walls, to reduce flood risks.
8. Use technology to manage and monitor water flow and drainage systems in real-time.
9. Implement green infrastructure solutions, such as rain gardens and green roofs, to manage rainwater and reduce flood risks.
10. Use predictive analytics to identify and respond to potential flood hazards proactively.
11. Engage citizens in flood management through public awareness campaigns and educational programs.
12. Use technology to improve communication and coordination between government agencies, emergency responders, and residents during flood events.
13. Develop and implement evacuation plans for residents in flood-prone areas.
14. Encourage investment in flood-resilient infrastructure and homes.
15. Use data and analytics to inform policy decisions and regulations related to flood management.
16. Collaborate with regional and international organizations to share best practices and resources.
17. Continuously evaluate and adjust the flood management plan based on changing conditions and feedback.
18. Invest in education and training programs to develop a skilled workforce in flood management.

19. Use data analytics to track and report on the city's progress in reducing flood risks.
20. Encourage citizen engagement and participation in flood management initiatives.
21. Develop contingency plans for responding to severe flood events and emergencies.
22. Foster partnerships with relevant organizations and businesses to support flood management efforts.
23. Use technology to manage and reduce non-structural flood risks, such as land use planning and zoning.
24. Encourage investment in research and innovation to develop new and improved technologies for flood management.
25. Collaborate with neighbouring cities and regions to develop a coordinated approach to flood management.

FREQUENTLY ASKED QUESTIONS

What is Smart City Flood Management (SCFM)? Smart City Flood Management (SCFM) is an approach that leverages technology and data to better manage and mitigate the risks of floods in urban areas.

How does Smart City Flood Management work? SCFM involves using data from sensors, satellites, and other sources to monitor water levels, rainfall, and other key factors. This information can be used to predict floods and alert citizens, emergency responders, and local authorities. SCFM also involves infrastructure improvements, such as building flood barriers or installing rain gardens to absorb excess water.

Who is responsible for implementing Smart City Flood Management? In most cases, the local government is responsible for implementing SCFM initiatives. However, it often involves collaboration with other stakeholders, such as utility companies, private developers, and citizens.

What are the benefits of Smart City Flood Management? SCFM can help prevent property damage, protect citizens' safety, and minimize the disruption of essential services. It can also help cities save money in the long term by avoiding the costs of flood damage.

How does Smart City Flood Management impact citizens? SCFM can help citizens prepare for and respond to floods, giving them more time to evacuate or protect their property. It can also help them access accurate and up-to-date information about the risks of floods and the actions they should take.

How can citizens contribute to Smart City Flood Management? Citizens can contribute to SCFM by reporting floods or potential flood risks, following guidance from local authorities, and participating in community flood preparedness programs.

What technologies are used in Smart City Flood Management? SCFM involves the use of a range of technologies, including sensors, weather forecasting models, geographic information systems (GIS), and machine learning algorithms.

What are some examples of Smart City Flood Management projects? Some examples of SCFM projects include the Thames Barrier in London, which protects the city from storm surges, and the smart rain gardens in Philadelphia, which absorb excess rainwater and prevent flooding.

How is Smart City Flood Management funded? SCFM projects are often funded through a combination of public and private financing, including grants, loans, and investments from utility companies, insurance providers, and other stakeholders.

How can Smart City Flood Management be integrated with other smart city initiatives? SCFM can be integrated with other smart city initiatives, such as transportation, energy management, and emergency response. For example, flood warnings can be used to reroute traffic, and renewable energy sources can be used to power flood control infrastructure.

Conclusion: *Implementing SCFM can lead to a wide range of benefits for citizens, government agencies, and the environment, including the ability to reduce flood risks, improve the efficiency of flood response and* *management, and improve the overall resilience of cities to flooding. It can also lead to cost savings for government agencies.*

SMART CITY GOVERNANCE (SCG)

Using technology to improve the efficiency, transparency and participation in city decision-making, such as e-governance and citizen engagement platforms.

Implementing SCG involves using technology to improve the efficiency, transparency, and participation in city decision-making. This approach can lead to a number of benefits for citizens, government agencies, and the community as a whole.

One of the key benefits of SCG is the ability to improve the efficiency of city decision-making. By using e-governance platforms, cities can streamline government operations and services, making it easier for citizens to access information and interact with government officials.

Another benefit of SCG is the ability to increase transparency in city decision-making. By using technology to make government information more accessible and transparent, cities can help to build trust and improve citizen engagement.

SCG can also help to improve citizen participation in city decision-making by providing citizens with more opportunities to have their voices heard. By using citizen engagement platforms, cities can encourage citizens to provide input and feedback on government policies and initiatives, helping to ensure that the needs and concerns of citizens are taken into account. Implementing SCG can lead to cost savings for government agencies by reducing the need for manual labour, reducing errors, and improving transparency.

GENERAL FRAMEWORK

The following framework is a general one and the actual implementation will depend on the specific requirements and constraints of the city.

1. Assess the current state of governance in the city.
2. Engage stakeholders, including citizens, government officials, and experts in governance and technology.
3. Develop a comprehensive governance plan, incorporating technology solutions.
4. Implement e-governance platforms to improve access to government services and information.
5. Use technology to improve transparency in decision-making, such as publishing data and records online.
6. Foster citizen engagement and participation in decision-making through online forums and platforms.
7. Use data analytics to inform and evaluate policy decisions.
8. Encourage public feedback and input through online portals and platforms.
9. Use technology to improve communication and collaboration between government agencies and departments.
10. Implement digital signatures and secure authentication systems to improve efficiency and reduce fraud.
11. Develop and implement measures to protect citizens' privacy and data security.
12. Encourage open data initiatives to increase transparency and accountability.
13. Provide training and support to government officials and staff to improve their digital literacy and competency.
14. Use technology to improve the efficiency of government processes and services.
15. Encourage public-private partnerships to support governance initiatives.
16. Develop and implement measures to promote digital inclusion and access for all citizens.
17. Use technology to monitor and evaluate the impact of governance initiatives.
18. Foster partnerships with relevant organizations and institutions to support governance initiatives.

19. Encourage citizen feedback and participation in governance initiatives.
20. Implement measures to address and prevent corruption and unethical behaviour in government.
21. Develop and implement measures to increase accountability and responsibility in government decision-making.
22. Use technology to improve the efficiency and effectiveness of public consultation and engagement.
23. Encourage collaboration and information sharing between cities to improve governance practices.
24. Invest in research and innovation to develop new and improved technologies for governance.
25. Continuously evaluate and adjust the governance plan based on changing conditions and feedback.

FREQUENTLY ASKED QUESTIONS

What is Smart City Governance, and how does it differ from traditional governance? Smart City Governance refers to the use of technology and data to improve governance and decision-making in cities. It differs from traditional governance in that it leverages data and technology to make more informed decisions, be more transparent, and engage citizens in the decision-making process.

How does Smart City Governance increase citizen participation in decision-making? Smart City Governance makes it possible for citizens to provide feedback and engage in decision-making through digital platforms, such as online forums, mobile apps, and social media. This technology enables real-time communication and feedback, which can help to inform decisions and improve outcomes.

What are the benefits of Smart City Governance for citizens? The benefits of Smart City Governance for citizens include greater transparency in decision-making, increased access to government services and information, and more opportunities for citizen engagement and feedback. It can also improve public safety, traffic management, and environmental sustainability.

How does Smart City Governance address issues of data privacy and security? Smart City Governance must prioritize the protection of citizen data and privacy. This includes implementing data protection policies, ensuring secure data storage and transmission, and making sure that citizens have control over how their data is used.

What role do local governments play in Smart City Governance? Local governments play a critical role in Smart City Governance, as they are responsible for implementing policies, providing infrastructure, and engaging with citizens. They must work with technology and data providers, as well as citizens, to develop and implement effective strategies.

How can Smart City Governance help address issues of social and economic inequality? Smart City Governance can help to address issues of social and economic inequality by providing greater access to information and services, promoting economic development and job creation, and improving access to education and healthcare.

How does Smart City Governance impact the relationship between citizens and government officials? Smart City Governance can improve the relationship between citizens and government officials by creating more opportunities for engagement and feedback. By using technology and data to improve decision-making, citizens can feel more connected to their government and more empowered to participate in the decision-making process.

What are some of the challenges associated with implementing Smart City Governance? Some of the challenges associated with implementing Smart City Governance include data privacy and security concerns, the need for effective collaboration between multiple stakeholders, and the potential for technological systems to perpetuate existing social and economic inequalities.

What are some examples of successful Smart City Governance initiatives? Some examples of successful Smart City Governance initiatives include Barcelona's use of technology to improve public transportation, Copenhagen's implementation of smart lighting systems, and Amsterdam's use of data to monitor and manage air quality.

How can citizens get involved in Smart City Governance initiatives? Citizens can get involved in Smart City Governance initiatives by providing feedback through digital platforms, attending public meetings and events, and participating in local government programs and initiatives. They can also engage with local officials and organizations to advocate for specific policies and initiatives.

Conclusion: Implementing SCG can lead to a wide range of benefits for citizens, government agencies, and the community as a whole, including the ability to improve the efficiency of city decision-making, increase transparency in city decision-making, and improve citizen participation in city decision-making. It can also lead to cost savings for government agencies.

SMART CITY GREEN ENERGY (SCGE)

Using technology to promote the use of green energy, such as solar and wind power, and reduce dependence on fossil fuels.

Implementing SCGE involves using technology to promote the use of renewable energy sources, such as solar and wind power, and reduce dependence on fossil fuels. This approach can lead to a number of benefits for citizens, the environment, and the economy.

One of the key benefits of SCGE is the ability to reduce the city's carbon footprint and help combat climate change. By increasing the use of renewable energy sources, cities can reduce their greenhouse gas emissions and contribute to the global effort to combat climate change.

Another benefit of SCGE is the ability to improve air and water quality by reducing the amount of pollution caused by fossil fuel use. This can lead to healthier living conditions for citizens and can help to preserve natural habitats and biodiversity.

SCGE can also lead to economic benefits, such as creating jobs in the renewable energy sector and reducing energy costs for citizens and businesses. It can improve energy security by reducing dependence on fossil fuel imports. SCGE can also lead to cost savings for government agencies by reducing the need to spend money on infrastructure and maintenance of traditional power plants.

GENERAL FRAMEWORK

The following framework is a general one and the actual implementation will depend on the specific requirements and constraints of the city.

1. Assess the current state of energy use in the city.
2. Develop a comprehensive plan for promoting and implementing green energy solutions.
3. Engage stakeholders, including citizens, government officials, energy providers, and experts in green energy technology.
4. Implement policies and incentives to encourage the use of green energy, such as tax credits and subsidies.
5. Use technology to monitor and manage energy usage in the city.
6. Develop and implement measures to promote energy efficiency and conservation.
7. Invest in research and development of green energy technologies.
8. Encourage the use of electric and hybrid vehicles to reduce dependence on fossil fuels.
9. Promote the use of renewable energy sources, such as wind and solar power, in city operations and facilities.
10. Develop and implement measures to integrate renewable energy into the city's energy grid.
11. Encourage the use of smart grid technologies to improve energy efficiency and reliability.
12. Invest in energy storage solutions to support the use of renewable energy.
13. Foster public-private partnerships to support green energy initiatives.
14. Develop and implement educational programs and awareness campaigns to promote green energy.
15. Encourage energy-efficient building design and construction practices.

16. Use technology to monitor and reduce greenhouse gas emissions.
17. Foster collaboration and information sharing between cities to improve green energy practices.
18. Encourage the use of green energy in public transportation systems.
19. Develop and implement measures to address and mitigate energy poverty.
20. Promote the use of renewable energy in rural and remote areas to improve access to energy.
21. Invest in research and innovation to develop new and improved green energy technologies.
22. Encourage the use of green energy in industries and businesses within the city.
23. Evaluate and adjust the green energy plan based on changing conditions and feedback.
24. Encourage citizen participation and feedback in green energy initiatives.
25. Continuously monitor and report on progress towards green energy goals and objectives.

FREQUENTLY ASKED QUESTIONS

What is Smart City Green Energy? Smart City Green Energy refers to the use of renewable and sustainable energy sources to power a city's infrastructure and operations, such as solar, wind, hydro, and geothermal energy.

What are the benefits of using green energy in a smart city? Using green energy in a smart city can reduce greenhouse gas emissions, improve air quality, and promote energy independence and security, while also creating new jobs and economic opportunities.

How does a smart city generate green energy? A smart city can generate green energy through various methods, such as installing solar panels on buildings, using wind turbines, or implementing hydropower systems.

How does a smart city distribute green energy to its residents and businesses? A smart city can distribute green energy through a smart grid system, which allows for the efficient and reliable distribution of electricity from renewable sources to consumers.

How can I contribute to the green energy movement in my smart city? You can contribute to the green energy movement in your smart city by using energy-efficient appliances, reducing energy consumption in your home or office, and supporting renewable energy projects in your community.

How much does it cost to implement green energy in a smart city? The cost of implementing green energy in a smart city varies depending on the specific technologies and infrastructure needed. However, over time, the use of green energy can result in cost savings and economic benefits.

What are some examples of successful green energy projects in smart cities? Some successful green energy projects in smart cities include Barcelona's Smart City Project, which uses solar panels and wind turbines to power the city's streetlights, and Copenhagen's carbon-neutral goal, which aims to eliminate all carbon emissions by 2025.

What are some challenges associated with implementing green energy in a smart city? Challenges associated with implementing green energy in a smart city include high initial costs, lack of infrastructure, and resistance from traditional energy *companies*.

How can a smart city ensure that it is using green energy sustainably? A smart city can ensure that it is using green energy sustainably by implementing energy storage solutions, promoting energy-efficient practices, and monitoring energy consumption to identify areas for improvement.

What is the future of green energy in smart cities? The future of green energy in smart cities is promising, as more cities and communities recognize the benefits of sustainable and renewable energy sources, and work to implement innovative solutions to reduce carbon emissions and promote energy independence.

Conclusion: Implementing SCGE can lead to a wide range of benefits for citizens, the environment, and the economy, including the ability to reduce the city's carbon footprint, improve air and water quality, create jobs, reduce energy costs, improve energy security, and save cost for government agencies.

SMART CITY GREEN INFRASTRUCTURE (SCGI)

Using technology to promote sustainable urban development, such as green roofs, rain gardens, and urban forestry.

SCGI refers to the integration of technology with sustainable urban development practices. This approach focuses on implementing green solutions such as green roofs, rain gardens, and urban forestry to create sustainable and livable urban environments.

Green roofs are roofs covered with vegetation, which help to reduce the heat island effect, lower energy consumption and improve air quality. Rain gardens are designed to manage stormwater runoff and help reduce the risk of flooding. Urban forestry involves planting and maintaining trees in urban areas to provide shade, improve air quality, and create a more livable environment.

The benefits of SCGI are numerous and wide-ranging. By implementing green solutions in urban areas, cities can help to mitigate the impacts of climate change and create more sustainable communities. For example, green roofs and rain gardens help to manage stormwater runoff, reducing the risk of flooding and preserving valuable water resources. Urban forestry can also help to improve air quality, reduce the heat island effect and provide shade, making urban areas more comfortable for residents.

The integration of technology with green infrastructure provides a unique opportunity to optimize the performance of these systems. For example, sensors and monitoring systems can be used to track the performance of green roofs, rain gardens, and urban forestry, allowing cities to optimize these systems and improve their overall performance. This results in more effective use of resources, improved sustainability and reduced costs.

GENERAL FRAMEWORK

The following framework is a general one and the actual implementation will depend on the specific requirements and constraints of the city.

1. Assess the current state of urban development in the city.
2. Develop a comprehensive plan for promoting and implementing green infrastructure solutions.
3. Engage stakeholders, including citizens, government officials, urban planners, and experts in green infrastructure technology.
4. Implement policies and incentives to encourage the use of green infrastructure, such as tax credits and subsidies.
5. Use technology to monitor and manage the implementation of green infrastructure projects.
6. Develop and implement measures to promote sustainable urban development practices.
7. Invest in research and development of green infrastructure technologies.
8. Promote the use of green roofs and rain gardens to improve water management and reduce runoff.
9. Implement urban forestry initiatives to improve air quality and provide shade.
10. Encourage the use of green infrastructure in public spaces and facilities.
11. Foster public-private partnerships to support green infrastructure initiatives.
12. Develop and implement educational programs and awareness campaigns to promote green infrastructure.
13. Encourage the integration of green infrastructure into urban planning and development practices.
14. Foster collaboration and information sharing between cities to improve green infrastructure practices.
15. Develop and implement measures to address and mitigate the impact of urban development on natural resources.

16. Encourage the use of green infrastructure in rural and remote areas to improve access to natural resources.
17. Invest in research and innovation to develop new and improved green infrastructure technologies.
18. Encourage the use of green infrastructure in industries and businesses within the city.
19. Evaluate and adjust the green infrastructure plan based on changing conditions and feedback.
20. Encourage citizen participation and feedback in green infrastructure initiatives.
21. Continuously monitor and report on progress towards green infrastructure goals and objectives.
22. Develop and implement measures to promote biodiversity and support wildlife habitats in urban areas.
23. Encourage the use of green infrastructure to address and mitigate the impacts of climate change.
24. Promote the integration of green infrastructure into disaster risk reduction and management planning.
25. Foster collaboration and information sharing with other sectors, such as tourism and agriculture, to promote sustainable urban development.

FREQUENTLY ASKED QUESTIONS

What is Smart City Green Infrastructure (SCGI)? SCGI refers to the use of technology to optimize and manage green infrastructure elements in urban environments, including parks, green roofs, green walls, and rain gardens.

How does SCGI benefit the environment? SCGI helps to improve air quality, reduce the urban heat island effect, reduce stormwater runoff, and support urban biodiversity by increasing the number of green spaces in cities.

What are some examples of SCGI technologies? Examples of SCGI technologies include green roofs that use sensors to manage moisture levels, smart irrigation systems that use weather data to optimize watering schedules, and green walls that use IoT sensors to monitor air quality.

How can SCGI help cities become more sustainable? SCGI can help cities become more sustainable by reducing energy consumption, improving water management, and reducing greenhouse gas emissions.

How can SCGI improve the quality of life in cities? SCGI can improve the quality of life in cities by providing access to green spaces, reducing noise pollution, and improving air quality.

How can the public get involved in SCGI projects? The public can get involved in SCGI projects by participating in public consultations, joining citizen science initiatives, and volunteering in local green infrastructure projects.

How does SCGI contribute to the economy? SCGI can contribute to the economy by creating jobs in areas such as green infrastructure installation and maintenance, and by increasing property values in neighbourhoods with improved green spaces.

How does SCGI help cities adapt to climate change? SCGI helps cities adapt to climate change by reducing the urban heat island effect, managing stormwater runoff, and increasing urban biodiversity.

How can SCGI improve public health in cities? SCGI can improve public health by reducing air pollution, providing spaces for physical activity, and improving mental health through access to green spaces.

How can cities measure the success of SCGI projects? Cities can measure the success of SCGI projects by monitoring changes in air and water quality, assessing the amount of stormwater runoff, and tracking changes in urban biodiversity.

Conclusion: SCGI offers a powerful tool for promoting sustainable urban development. By integrating technology with green solutions such as green roofs, rain gardens, and urban forestry, cities can create more livable, sustainable and efficient communities. The benefits of this approach are far-reaching, including improved air quality, reduced heat island effect, reduced risk of flooding and improved resource efficiency.

SMART CITY GRIDS (SCGR)

Using technology to improve the efficiency and reliability of the electricity grid, such as advanced metering infrastructure and demand response programs.

Implementing SCGR can bring many benefits to a city's electricity system. By incorporating technology such as advanced metering infrastructure and demand response programs, cities can improve the efficiency and reliability of their electricity grid. Advanced metering infrastructure (AMI) allows for more accurate and detailed measurement of electricity consumption, enabling the detection of outages and theft, and the ability to bill customers based on their actual consumption. This can also enable consumers to better understand and manage their energy consumption, resulting in cost savings and reduction of energy consumption.

Demand response programs also enables the grid to be more flexible by allowing customers to reduce or shift their electricity consumption during periods of peak demand. This can help to reduce the need for expensive peaker power plants and reduce the risk of blackouts. Implementing SCGR can also improve the integration of renewable energy sources such as solar and wind power, and enable cities to better manage the integration of electric vehicles into the grid.

GENERAL FRAMEWORK

The following framework is a general one and the actual implementation will depend on the specific requirements and constraints of the city.

1. Assess the current state of the electricity grid in the city.

2. Develop a comprehensive plan for improving the efficiency and reliability of the grid.
3. Engage stakeholders, including citizens, government officials, energy experts and utility companies.
4. Invest in advanced metering infrastructure (AMI) technology to improve data collection and management.
5. Implement demand response programs to encourage energy conservation and reduce peak demand.
6. Encourage the use of distributed energy resources, such as rooftop solar panels and microgrids, to improve grid resiliency.
7. Promote the integration of renewable energy sources into the grid to reduce dependence on fossil fuels.
8. Use technology to monitor and manage the grid in real-time to identify and address issues more quickly.
9. Develop and implement measures to improve grid security and protect against cyber threats.
10. Encourage the use of energy storage systems to improve grid reliability and stability.
11. Promote the use of electric vehicles and charging infrastructure to reduce dependence on fossil fuels.
12. Implement measures to address and mitigate the impacts of extreme weather events on the grid.
13. Develop and implement educational programs and awareness campaigns to promote energy conservation and efficiency.
14. Foster collaboration and information sharing between cities to improve grid efficiency and reliability practices.
15. Encourage the use of smart grid technologies in industries and businesses within the city.
16. Evaluate and adjust the grid improvement plan based on changing conditions and feedback.
17. Encourage citizen participation and feedback in grid improvement initiatives.
18. Continuously monitor and report on progress towards grid efficiency and reliability goals.
19. Foster public-private partnerships to support grid improvement initiatives.
20. Develop and implement measures to support low-income communities and ensure access to energy.
21. Encourage the use of technology to improve energy efficiency in buildings and homes.

22. Foster collaboration and information sharing with other sectors, such as transportation, to promote sustainable energy practices.
23. Promote the integration of grid improvement initiatives into urban planning and development practices.
24. Encourage the use of technology to monitor and manage carbon emissions and promote sustainability.
25. Invest in research and innovation to develop new and improved grid technologies.

FREQUENTLY ASKED QUESTIONS

What are Smart City Green Grids? Smart City Green Grids (SCGR) are advanced energy networks that rely on renewable energy sources and digital technology to make urban energy systems more efficient, resilient, and sustainable.

How do SCGR work? SCGR work by integrating renewable energy sources, such as solar, wind, and geothermal, into the city's existing energy grid. Advanced technologies, such as energy storage and smart sensors, are also used to optimize energy use and reduce waste.

What are the benefits of SCGR? The benefits of SCGR include reducing greenhouse gas emissions, increasing energy efficiency, promoting renewable energy sources, improving energy resilience, and reducing energy costs.

Who is responsible for implementing SCGR? SCGR are usually implemented by city governments, energy companies, and technology firms working in partnership.

How much do SCGR cost to implement? The cost of implementing SCGR varies depending on the size of the city, the existing energy infrastructure, and the desired level of integration. However, the long-term benefits of energy efficiency and sustainability often outweigh the initial costs.

What are the challenges of implementing SCGR? Challenges of implementing SCGR include the high initial costs, the need for significant infrastructure upgrades, potential regulatory hurdles, and the need for stakeholder collaboration.

What types of renewable energy sources are used SCGR? SCGR often rely on a mix of renewable energy sources, such as solar, wind, and geothermal energy, to power the city's energy needs.

Can SCGR be scaled up to power entire cities? Yes, SCGR can be scaled up to power entire cities. However, significant investment in energy infrastructure and renewable energy sources is necessary to achieve this goal.

How do SCGR contribute to a more sustainable future? SCGR help to reduce greenhouse gas emissions, promote renewable energy sources, and increase energy efficiency, which all contribute to a more sustainable future.

Are there any successful examples of SCGR in action? Yes, there are many successful examples of SCGR in action, including the city of San Diego's microgrid, the city of Stockholm's smart grid, and the city of Masdar's renewable energy grid.

Conclusion: *SCGR can help cities to improve the overall reliability and efficiency of their electricity system, reduce costs, and promote a more sustainable energy future.*

SMART CITY HEALTHCARE (SCHC)

Using technology to improve the delivery of healthcare services, such as telemedicine and remote monitoring of patients.

SCHC utilizes cutting-edge technology to enhance the provision of healthcare services. One of its most significant benefits is the introduction of telemedicine, which allows patients to consult with their doctors remotely. This eliminates the need for in-person visits and reduces the burden on healthcare facilities. Patients can access medical advice and consultation from the comfort of their own homes, reducing their exposure to infections and other risks associated with visiting healthcare facilities.

Another crucial benefit of SCHC is remote monitoring of patients. This feature enables healthcare providers to track the health status of their patients, even when they are not physically present. For example, wearable devices can be used to monitor vital signs, such as heart rate and blood pressure, and send real-time data to the healthcare provider. This information can help healthcare providers identify potential health issues early, enabling them to take proactive measures to prevent more serious problems from developing.

SCHC also offers improved accessibility to healthcare services. The use of technology helps to break down geographical barriers and provides people in remote areas with access to high-quality medical services. This is especially important for individuals who live in rural areas and may not have easy access to healthcare facilities. With SCHC, healthcare providers can offer telemedicine consultations, remote monitoring, and other services to patients anywhere and at any time.

SCHC helps to reduce healthcare costs by reducing the need for in-person visits and eliminating the costs associated with traveling to healthcare facilities. This makes healthcare services more affordable for patients and helps to reduce the financial burden on the healthcare system as a whole.

GENERAL FRAMEWORK

The following framework is a general one and the actual implementation will depend on the specific requirements and constraints of the city.

1. Assess the current state of healthcare delivery in the city.
2. Develop a comprehensive plan for improving healthcare delivery through technology.

3. Engage stakeholders, including healthcare providers, patients, government officials and technology experts.
4. Implement telemedicine and remote patient monitoring systems to increase access to care.
5. Develop and implement electronic medical record (EMR) systems to improve data management and patient care.
6. Encourage the use of wearable technology and smart devices to monitor patient health and well-being.
7. Implement real-time data analysis to identify and address public health concerns.
8. Foster collaboration and information sharing between healthcare providers to improve patient outcomes.
9. Implement measures to address and mitigate privacy and security concerns with regards to patient health data.
10. Encourage the use of technology to support disease prevention and health promotion initiatives.
11. Develop and implement educational programs and awareness campaigns to promote healthy lifestyles and behaviour.
12. Foster public-private partnerships to support healthcare technology initiatives.
13. Evaluate and adjust the healthcare technology plan based on changing conditions and feedback.
14. Encourage citizen participation and feedback in healthcare technology initiatives.
15. Continuously monitor and report on progress towards healthcare delivery goals.
16. Promote the integration of technology into medical education and training programs.
17. Encourage the use of technology to support mental and behavioural health services.
18. Implement measures to address and mitigate the impact of healthcare disparities and ensure access to care for all citizens.
19. Promote the use of technology to improve emergency response and disaster preparedness.
20. Foster collaboration and information sharing with other sectors, such as transportation and housing, to improve overall health outcomes.
21. Encourage the use of technology to support healthy aging initiatives.
22. Implement measures to encourage sustainable and environmentally friendly healthcare practices.
23. Promote the use of technology to support community health initiatives and address social determinants of health.

24. Invest in research and innovation to develop new and improved healthcare technologies.
25. Foster international cooperation and information sharing to improve global health outcomes.

FREQUENTLY ASKED QUESTIONS

What is Smart City Healthcare (SCHC)? Smart City Healthcare (SCHC) is a system that uses technology and data analytics to improve the efficiency, quality, and accessibility of healthcare services in a city.

How does SCHC work? SCHC works by using a variety of technologies such as IoT, AI, and data analytics to collect and analyze healthcare-related data. This data is used to improve the delivery of healthcare services, optimize resource allocation, and reduce costs.

What are the benefits of SCHC? Some of the benefits of SCHC include improved healthcare outcomes, reduced healthcare costs, improved patient experience, and more efficient healthcare delivery.

What types of healthcare services can be improved using SCHC? SCHC can be used to improve a variety of healthcare services such as emergency medical response, chronic disease management, patient monitoring, and preventive care.

How does SCHC ensure patient privacy and security? SCHC takes patient privacy and security very seriously and uses various measures such as data encryption, access controls, and authentication to ensure the privacy and security of patient data.

Who is responsible for implementing SCHC in a city? The implementation of SCHC is typically a collaborative effort between local government, healthcare providers, and technology companies.

What are the challenges associated with implementing SCHC? Some of the challenges associated with implementing SCHC include the cost of implementation, the need for interoperability between different healthcare systems, and ensuring patient privacy and security.

How can patients access healthcare services through SCHC? Patients can access healthcare services through SCHC by using various digital tools such as telemedicine, health apps, and wearables.

How can SCHC help in managing a pandemic such as COVID-19? SCHC can help in managing a pandemic by providing real-time data on the spread of the disease, enabling rapid response to outbreaks, and providing remote healthcare services to reduce the risk of infection.

Is SCHC only suitable for large cities? No, SCHC can be implemented in cities of any size, and the system can be scaled up or down depending on the size of the city and its healthcare needs.

Conclusion: *SCHC offers numerous benefits to both patients and healthcare providers. By utilizing technology to improve the delivery of healthcare services, SCHC provides patients with increased accessibility to medical advice, improved health outcomes, and reduced costs. At the same time, it enables healthcare providers to offer better quality services, more efficiently and effectively.*

SMART CITY INFRASTRUCTURE MANAGEMENT (SCIM)

Using technology to improve the management, maintenance and upgrade of city infrastructure, such as smart roads, bridges, and buildings.

SCIM is a technology-driven approach to managing the infrastructure of a city. One of its key benefits is improved maintenance and upgrade of city infrastructure. With SCIM, cities can use sensors and other technologies to monitor the condition of their roads, bridges, and buildings in real-time. This enables them to quickly identify potential issues and take proactive measures to address them, reducing the likelihood of major breakdowns and ensuring that city infrastructure remains in good working order.

Another benefit of SCIM is increased efficiency in infrastructure management. By using technology to automate many routine tasks, such as monitoring and maintenance, SCIM reduces the workload of city personnel and enables them to focus on more important tasks. This helps to improve the overall quality of infrastructure management, enabling cities to provide better services to their residents.

SCIM also helps to reduce the environmental impact of city infrastructure. By monitoring energy usage and optimizing energy-efficient systems, SCIM helps cities to conserve resources and reduce greenhouse gas emissions. This is particularly important in light of the growing global concern over climate change and the need for cities to become more environmentally sustainable.

SCIM enables cities to respond more quickly and effectively to emergencies. For example, in the event of a natural disaster, SCIM can help city authorities to quickly assess the damage to infrastructure and take the necessary steps to repair it. This helps to reduce the downtime caused by infrastructure failures, reducing the impact on city residents and businesses.

GENERAL FRAMEWORK

The following framework is a general one and the actual implementation will depend on the specific requirements and constraints of the city.

1. Assess existing infrastructure for the city and identify key areas for improvement.
2. Conduct a comprehensive analysis of the challenges and opportunities for smart infrastructure management.
3. Develop a detailed plan for implementing smart infrastructure management, including objectives, timelines, and budgets.

4. Engage with stakeholders and gather their requirements and feedback to inform the plan.
5. Define the key performance indicators (KPIs) for measuring the success of smart infrastructure management.
6. Identify and select technology partners who can provide the required hardware, software, and services.
7. Develop a communication plan to engage with citizens and stakeholders, provide regular updates, and ensure transparency.
8. Implement smart infrastructure management systems, such as smart roads, bridges, and buildings, using IoT and other advanced technologies.
9. Establish real-time monitoring and reporting systems to track the performance of infrastructure and quickly respond to any issues.
10. Automate maintenance schedules and work order management to improve the efficiency and speed of infrastructure repair and upgrade.
11. Implement predictive maintenance systems to identify potential infrastructure failures before they occur.
12. Implement smart grid systems to optimize the use of energy and reduce waste.
13. Implement asset management systems to track the condition and performance of infrastructure assets over time.
14. Establish a central data repository to store and manage infrastructure data and information.
15. Enable data sharing and integration between different systems and departments to ensure seamless coordination.
16. Develop dashboards and visualizations to provide real-time insights into the performance of infrastructure and help with decision-making.
17. Use simulation and modelling tools to optimize the design and management of infrastructure.
18. Foster collaboration and partnerships with other cities, research institutions, and private sector companies to learn from their experience and best practices.
19. Continuously evaluate and improve the smart infrastructure management system based on feedback and performance data.
20. Ensure data security and privacy in accordance with regulations and best practices.
21. Invest in workforce training and development to ensure the effective use and maintenance of smart infrastructure systems.

22. Provide regular performance reports and insights to citizens and stakeholders to demonstrate the benefits of smart infrastructure management.
23. Foster a culture of innovation and continuous improvement to drive the adoption of new technologies and best practices.
24. Participate in national and international smart city initiatives to share experience, knowledge, and best practices.
25. Continuously monitor and adapt to emerging trends and technologies to ensure the city remains at the forefront of smart infrastructure management.

FREQUENTLY ASKED QUESTIONS

What is Smart City Infrastructure Management (SCIM)? Smart City Infrastructure Management (SCIM) is a system of technologies and strategies that are used to optimize and manage urban infrastructure, including roads, buildings, utilities, and transportation systems, among others.

What benefits does SCIM provide to the city and its inhabitants? SCIM provides many benefits, including enhanced public safety, improved mobility, greater efficiency, reduced costs, and improved quality of life for residents.

Who is responsible for implementing SCIM in a city? SCIM is typically implemented by the city government, with support from private sector partners, including technology companies, utilities, and other service providers.

How does SCIM help manage traffic in a city? SCIM uses a range of technologies, including intelligent traffic management systems, sensors, and real-time data analysis to optimize traffic flow, reduce congestion, and improve safety.

How can SCIM improve the management of public utilities? SCIM can help utilities providers to monitor and manage infrastructure, identify potential issues before they become major problems, and optimize resource usage to improve efficiency and reduce costs.

Can SCIM help to reduce the carbon footprint of a city? Yes, SCIM can help to reduce the carbon footprint of a city by optimizing energy usage, improving public transportation systems, and promoting sustainable building practices.

How can SCIM improve public safety in a city? SCIM can improve public safety by providing real-time data on crime and emergency incidents, and enabling emergency responders to quickly and efficiently respond to incidents.

What are the challenges associated with implementing SCIM in a city? The challenges associated with implementing SCIM include data privacy and security concerns, the high cost of implementation, and the need for close collaboration between different government agencies and private sector partners.

How does SCIM help to promote economic growth in a city? SCIM can help to promote economic growth by improving the efficiency of transportation systems, making it easier for businesses to move goods and people around the city, and creating new opportunities for public-private partnerships.

What role do citizens play in the implementation of SCIM in a city? Citizens play an important role in the implementation of SCIM by providing feedback on the effectiveness of different technologies and strategies, and helping to ensure that the system is responsive to the needs and concerns of the community.

Conclusion: *SCIM offers numerous benefits to cities, including improved maintenance and upgrade of infrastructure, increased efficiency in infrastructure management, reduced environmental impact, and improved emergency response. By utilizing technology to manage city infrastructure, SCIM enables cities to provide better services to their residents, conserve resources, and become more sustainable.*

SMART CITY INNOVATION (SCIN)

Using technology to foster innovation and entrepreneurship in the city, such as incubators and accelerators for Smart City start-ups.

SCIN is an approach to using technology to encourage innovation and entrepreneurship in a city. One of the key benefits of SCIN is the creation of a supportive ecosystem for entrepreneurs and start-ups. By establishing incubators and accelerators, cities can provide aspiring entrepreneurs with the resources they need to turn their ideas into successful businesses. This includes access to funding, mentorship, training, and other support services.

Another significant benefit of SCIN is the generation of new jobs and economic growth. By fostering innovation and entrepreneurship, cities can attract and retain top talent and create new business opportunities. This, in turn, drives economic growth and creates new jobs, benefiting the city and its residents.

SCIN also helps cities to stay at the forefront of technological progress. By encouraging innovation, cities can become early adopters of new technologies and stay ahead of the curve in terms of technological advancement. This is particularly important in the rapidly changing technological landscape, where cities that are able to adopt new technologies quickly are likely to have a competitive advantage.

SCIN enables cities to address social and environmental challenges in innovative ways. For example, start-ups focused on sustainability or social impact can use technology to develop new solutions to some of the world's most pressing problems. This helps cities to become more socially and environmentally responsible, which is becoming increasingly important in the face of growing global challenges.

GENERAL FRAMEWORK

The following framework is a general one and the actual implementation will depend on the specific requirements and constraints of the city.

1. Identifying city challenges and opportunities for innovation.
2. Forming partnerships with local universities and research institutes.
3. Creating a dedicated innovation district or hub.
4. Providing access to funding and investment resources.
5. Establishing an innovation program office.
6. Developing an open data policy.
7. Encouraging collaboration and co-creation.
8. Hosting hackathons and innovation challenges.
9. Offering incubation and acceleration services.
10. Fostering a culture of innovation and risk-taking.
11. Building a network of mentors and advisors.
12. Creating a supportive regulatory environment.
13. Promoting entrepreneurship and start-up culture.
14. Building a skilled workforce.
15. Facilitating the transfer of technology.
16. Encouraging the creation of a Smart City community.
17. Supporting Smart City research and development.
18. Offering resources and facilities for product testing and market validation.
19. Building a Smart City technology ecosystem.
20. Establishing public-private partnerships.
21. Encouraging smart city entrepreneurship.
22. Developing a strategic plan for technology adoption.
23. Ensuring that the innovation process is inclusive and participatory.
24. Conducting regular evaluations of the innovation program.
25. Continuously evolving and adapting to changing needs and trends.

FREQUENTLY ASKED QUESTIONS

What is Smart City Innovation? Smart City Innovation refers to the development and implementation of advanced technologies, policies, and practices to address urban challenges and improve the quality of life for citizens.

What are some examples of Smart City Innovation? Examples of Smart City Innovation include advanced transportation systems, smart energy grids, connected public spaces, and enhanced public services such as health care, education, and emergency response.

Who is involved in Smart City Innovation? Smart City Innovation involves a range of stakeholders, including government agencies, technology companies, academic institutions, community organizations, and citizens.

What are the benefits of Smart City Innovation? The benefits of Smart City Innovation include improved quality of life for citizens, increased economic growth, enhanced sustainability, and improved public services and infrastructure.

How does Smart City Innovation address environmental sustainability? Smart City Innovation can address environmental sustainability by promoting energy efficiency, reducing waste, and enhancing the use of renewable energy sources.

How does Smart City Innovation address social inclusion? Smart City Innovation can address social inclusion by promoting access to public services and infrastructure for all citizens, regardless of their socioeconomic status or physical abilities.

What role does technology play in Smart City Innovation? Technology plays a central role in Smart City Innovation, as it provides the tools and platforms necessary to collect and analyze data, automate processes, and enhance communication and collaboration between stakeholders.

How does Smart City Innovation ensure data privacy and security? Smart City Innovation must prioritize data privacy and security by implementing robust security protocols, utilizing encryption and authentication methods, and adhering to strict data protection regulations.

How can citizens participate in Smart City Innovation? Citizens can participate in Smart City Innovation by providing feedback on public services and infrastructure, participating in community engagement activities, and contributing to data collection and analysis.

How does Smart City Innovation promote economic growth? Smart City Innovation can promote economic growth by attracting investment, promoting entrepreneurship, and creating job opportunities in emerging technology sectors.

Conclusion: *SCIN offers numerous benefits to cities, including the creation of a supportive ecosystem for entrepreneurs and start-ups, the generation of new jobs and economic growth, staying ahead of technological progress, and addressing social and environmental challenges in innovative ways. By utilizing technology to foster innovation and entrepreneurship, SCIN enables cities to become more vibrant, dynamic, and sustainable places to live, work, and do business.*

SMART CITY INTELLIGENT TRANSPORTATION SYSTEMS (SCITS)

Using data and technology to improve traffic flow, reduce congestion, and enhance public transportation.

SCITS is a technology-driven approach to improving transportation in cities. One of its key benefits is the reduction of traffic congestion. By using data and technology to optimize traffic flow, SCITS helps cities to reduce the time and cost associated with congestion. This results in improved mobility and reduced stress for city residents, enabling them to get to where they need to go more quickly and efficiently.

Another significant benefit of SCITS is the enhancement of public transportation. By using technology to improve the efficiency and reliability of public transportation systems, SCITS helps cities to provide a more convenient and attractive alternative to private vehicle use. This, in turn, helps to reduce traffic congestion, air pollution, and carbon emissions, making cities more environmentally sustainable.

SCITS also helps cities to respond more effectively to traffic incidents. By using real-time data and analytics, SCITS enables city authorities to quickly identify and respond to traffic incidents, reducing their impact on traffic flow and minimizing the time required for recovery. This helps to improve overall road safety and reduce the number of accidents on city roads.

SCITS enables cities to provide better transportation services to their residents. By using technology to optimize traffic flow and improve public transportation, SCITS helps cities to provide more accessible and convenient transportation options, which is particularly important for those who are unable to drive, such as seniors, children, and people with disabilities.

GENERAL FRAMEWORK

The following framework is a general one and the actual implementation will depend on the specific requirements and constraints of the city.

1. Identify the specific transportation challenges in the city.
2. Research and evaluate different smart city intelligent transportation systems (SCITS) and technologies.
3. Develop a comprehensive plan for implementing SCITS in the city.
4. Secure funding for the project from government agencies and private investors.
5. Develop partnerships with local transportation authorities and other stakeholders.
6. Conduct a feasibility study to determine the technical and economic viability of the project.
7. Assess the potential impact on the existing transportation infrastructure.
8. Develop a plan for integrating SCITS into the existing transportation infrastructure.
9. Conduct an environmental impact assessment.

10. Develop a plan for managing and maintaining the SCITS systems.
11. Identify potential locations for SCITS deployment.
12. Develop a plan for data collection, analysis and management.
13. Develop a plan for integrating traffic control and management systems into SCITS.
14. Develop a plan for integrating public transportation systems into SCITS.
15. Develop a plan for integrating real-time traffic and transportation information for public use.
16. Develop a plan for integrating dynamic routing and scheduling systems into SCITS.
17. Develop a plan for integrating parking management systems into SCITS.
18. Develop a plan for cybersecurity measures.
19. Develop a plan for educating and engaging the community.
20. Obtain necessary permits and approvals.
21. Procure equipment and materials for the project.
22. Begin installation of the SCITS systems.
23. Test and commission the SCITS systems.
24. Monitor and evaluate the performance of the SCITS systems.
25. Continuously improve the SCITS systems based on feedback and performance data.

FREQUENTLY ASKED QUESTIONS

What is a Smart City Intelligent Transportation System (SCITS)? A Smart City Intelligent Transportation System (SCITS) is a system that uses advanced technologies such as sensors, communication devices, and information processing systems to improve the efficiency and safety of transportation.

What is the goal of SCITS? The goal of SCITS is to improve the efficiency, safety, and sustainability of transportation in urban areas through the use of intelligent transportation technologies.

What are the components of SCITS? The components of SCITS include traffic management systems, smart parking systems, public transportation management systems, traveler information systems, and vehicle-to-infrastructure (V2I) communication systems.

How can SCITS improve traffic flow? SCITS can improve traffic flow by providing real-time traffic information to drivers, optimizing traffic signal timing, and facilitating the coordination of public transportation systems.

How does SCITS enhance safety? SCITS enhances safety by providing drivers with real-time safety alerts, detecting hazardous driving conditions, and improving emergency response times.

How can SCITS reduce air pollution? SCITS can reduce air pollution by optimizing traffic flow, reducing congestion, and facilitating the use of electric and hybrid vehicles.

How does SCITS improve public transportation systems? SCITS can improve public transportation systems by providing real-time information on the availability and location of public transportation services, optimizing transit schedules, and improving the efficiency of transit operations.

What are the challenges of implementing SCITS? The challenges of implementing SCITS include the high cost of deploying advanced technologies, the need for significant infrastructure upgrades, and the need for strong public-private partnerships.

Who pays for the deployment of SCITS? The deployment of SCITS is typically funded through a combination of public and private sources, including government grants, private investments, and user fees.

How can the public get involved in the development of SCITS? The public can get involved in the development of SCITS by providing feedback on transportation issues, participating in community meetings and forums, and supporting local initiatives to improve transportation.

Conclusion: *SCITS offers numerous benefits to cities, including the reduction of traffic congestion, the enhancement of public transportation, improved incident response, and better transportation services. By utilizing technology to improve transportation, SCITS enables cities to become more livable, sustainable, and accessible places to live, work, and do business.*

SMART CITY IoT (SCIoT)

Using technology to connect and control various Smart City systems and services, such as sensors, actuators and communication networks.

Implementing the Internet of Things (IoT) in a smart city can bring a wide range of benefits to the community. By connecting various systems and services with IoT technology, cities can improve the efficiency and effectiveness of their operations, as well as provide new and improved services to citizens.

One of the key benefits of using IoT in a smart city is the ability to gather and analyze large amounts of data from sensors and other devices. This data can be used to optimize city services, such as traffic flow, energy consumption, and emergency response. For example, by using IoT-enabled sensors to monitor traffic patterns, cities can adjust traffic lights in real-time to reduce congestion and improve traffic flow.

IoT technology can also be used to improve the management and maintenance of city infrastructure. Sensors can be used to monitor the condition of roads, bridges, and buildings, allowing for proactive maintenance and repairs. This can help to extend the life of infrastructure and reduce costs associated with repairs and replacements.

Another benefit of IoT in smart cities is the ability to provide new and improved services to citizens. For example, IoT-enabled smart parking systems can help drivers find available parking spots, while smart waste management systems can optimize garbage collection routes and reduce costs.

GENERAL FRAMEWORK

The following framework is a general one and the actual implementation will depend on the specific requirements and constraints of the city.

1. Identify the specific IoT challenges in the city.
2. Research and evaluate different smart city IoT (SCIoT) technologies and solutions.
3. Develop a comprehensive plan for implementing SCIoT in the city.
4. Secure funding for the project from government agencies and private investors.
5. Develop partnerships with local technology companies and other stakeholders.
6. Conduct a feasibility study to determine the technical and economic viability of the project.
7. Develop a plan for integrating SCIoT into the existing infrastructure and systems in the city.
8. Conduct an environmental impact assessment.
9. Develop a plan for managing and maintaining the SCIoT systems.
10. Identify potential locations for SCIoT deployment.
11. Develop a plan for data collection, analysis, and management.
12. Develop a plan for integrating communication networks into SCIoT.
13. Develop a plan for integrating sensors and actuators into SCIoT.
14. Develop a plan for integrating real-time data and information for public use.
15. Develop a plan for integrating SCIoT with other Smart City systems and services.
16. Develop a plan for integrating machine learning and artificial intelligence in SCIoT
17. Develop a plan for cybersecurity measures.
18. Develop a plan for educating and engaging the community.
19. Obtain necessary permits and approvals.
20. Procure equipment and materials for the project.
21. Begin installation of the SCIoT systems.
22. Test and commission the SCIoT systems.
23. Monitor and evaluate the performance of the SCIoT systems.

24. Continuously improve the SCIoT systems based on feedback and performance data.
25. Develop a plan for scalability and expansion of the SCIoT systems as needed.

FREQUENTLY ASKED QUESTIONS

What is SMART CITY IoT (SCIoT) and how does it work? SCIoT is a system of interconnected devices and sensors that collect data to improve the efficiency and quality of life in cities. It works by using various sensors and devices to gather data on things like traffic, energy consumption, and air quality, then using that data to make informed decisions about how to improve the city.

How does SCIoT benefit residents? SCIoT can benefit residents in many ways, including reducing traffic congestion, improving air quality, and providing real-time information about events and public transportation. It can also help cities to respond more quickly to emergencies and improve overall quality of life.

Who is responsible for implementing SCIoT? Implementing SCIoT typically requires collaboration between government agencies, private companies, and other stakeholders in the community. In many cases, local governments will take the lead in implementing SCIoT projects, but private companies and community groups may also be involved.

How secure is SCIoT? Security is a top priority for SCIoT systems. They use various security measures, including encryption and authentication, to protect against data breaches and unauthorized access.

How much does it cost to implement SCIoT? The cost of implementing SCIoT can vary widely depending on the size and scope of the project. Some projects may cost only a few thousand dollars, while others may cost millions or even billions of dollars.

How does SCIoT impact privacy? SCIoT can collect a lot of data about individuals, so privacy is a concern. However, most SCIoT systems are designed to protect privacy by collecting only anonymous data and ensuring that personal information is kept secure.

What are some examples of SCIoT in action? Some examples of SCIoT in action include smart traffic systems that use real-time data to manage traffic flow, smart energy systems that use data to optimize energy usage, and smart waste management systems that use data to improve waste collection.

How can I get involved in SCIoT? There are many ways to get involved in SCIoT, including volunteering with local organizations that are working on SCIoT projects, attending community meetings and events, and advocating for SCIoT initiatives.

What are some challenges that SCIoT faces? SCIoT faces several challenges, including the high cost of implementing these systems, concerns about privacy and security, and the need for ongoing maintenance and updates.

How will SCIoT evolve in the future? As technology continues to advance, SCIoT is likely to become even more sophisticated and widespread. In the future, we can expect to see more advanced sensors, more sophisticated data analysis tools, and even greater collaboration between different stakeholders in the community.

Conclusion: *The implementation of SCIoT can help to improve the quality of life for citizens, increase the efficiency of city operations, and reduce costs. With the ability to connect and control various systems and services, smart cities can become more responsive and adaptable to the changing needs of the community.*

SMART CITY IRRIGATION (SCIR)

Using technology to optimize watering schedules for crops and lawns, reducing water waste and conserving resources.

Implementing SCIR technology can provide a wide range of benefits for both the city and its residents. One of the main advantages is the ability to optimize watering schedules for crops and lawns, which can help to reduce water waste and conserve resources. This can be achieved through the use of sensors and data analysis to monitor soil moisture levels, weather conditions, and other factors that affect plant growth.

Another benefit of SCIR technology is the ability to improve the efficiency of irrigation systems. By using IoT-enabled devices, such as smart valves and sprinklers, cities can remotely monitor and control irrigation systems in real-time. This can help to reduce water consumption and costs, while also ensuring that plants receive the right amount of water at the right time.

SCIR technology can also help to improve the health of plants and the overall environment. By using data and technology to optimize watering schedules and reduce water waste, cities can promote healthier plant growth, reduce the risk of disease, and improve air and water quality.

SCIR technology can also support the development of sustainable urban agriculture and community gardening projects, which can provide fresh produce for the community and create green spaces in the city.

GENERAL FRAMEWORK

The following framework is a general one and the actual implementation will depend on the specific requirements and constraints of the city.

1. Identify the specific irrigation challenges in the city.
2. Research and evaluate different smart city irrigation (SCIR) technologies and solutions.
3. Develop a comprehensive plan for implementing SCIR in the city.
4. Secure funding for the project from government agencies and private investors.

5. Develop partnerships with local farmers, irrigation suppliers, and other stakeholders.
6. Conduct a feasibility study to determine the technical and economic viability of the project.
7. Develop a plan for integrating SCIR into the existing irrigation infrastructure and systems in the city.
8. Conduct an environmental impact assessment.
9. Develop a plan for managing and maintaining the SCIR systems.
10. Identify potential locations for SCIR deployment, such as public parks, green spaces, and agricultural land.
11. Develop a plan for data collection, analysis and management.
12. Develop a plan for integrating weather data into SCIR.
13. Develop a plan for integrating soil moisture sensors into SCIR.
14. Develop a plan for integrating automatic valves and controllers into SCIR.
15. Develop a plan for integrating real-time data and information for public use.
16. Develop a plan for integrating SCIR with other Smart City systems and services.
17. Develop a plan for integrating machine learning and artificial intelligence in SCIR
18. Develop a plan for cybersecurity measures.
19. Develop a plan for educating and engaging the community.
20. Obtain necessary permits and approvals.
21. Procure equipment and materials for the project.
22. Begin installation of the SCIR systems.
23. Test and commission the SCIR systems.
24. Monitor and evaluate the performance of the SCIR systems.
25. Continuously improve the SCIR systems based on feedback and performance data.

FREQUENTLY ASKED QUESTIONS

What is SMART CITY IRRIGATION (SCIR) and how does it work? SCIR is a system of interconnected sensors and devices that collect data on weather conditions, soil moisture levels, and other factors that affect plant growth. It works by using this data to automatically adjust the amount and timing of irrigation, optimizing water use and improving crop yields.

How does SCIR benefit farmers? SCIR can benefit farmers in many ways, including reducing water usage and costs, improving crop yields, and reducing the risk of crop failure due to drought or other environmental factors.

How does SCIR impact the environment? SCIR can help to reduce the impact of agriculture on the environment by reducing water usage and minimizing the use of fertilizers and pesticides.

Who is responsible for implementing SCIR? Implementing SCIR typically requires collaboration between farmers, technology providers, and government agencies. In many cases, local governments or agricultural cooperatives will take the lead in implementing SCIR projects.

How secure is SCIR? Security is a top priority for SCIR systems. They use various security measures, including encryption and authentication, to protect against data breaches and unauthorized access.

How much does it cost to implement SCIR? The cost of implementing SCIR can vary widely depending on the size and scope of the project. Some projects may cost only a few thousand dollars, while others may cost millions or even billions of dollars.

What are some examples of SCIR in action? Some examples of SCIR in action include systems that use real-time weather data to adjust irrigation schedules, sensors that measure soil moisture levels to optimize irrigation, and remote monitoring systems that allow farmers to monitor their crops from a distance.

How can I get involved in SCIR? There are many ways to get involved in SCIR, including volunteering with local organizations that are working on SCIR projects, attending community meetings and events, and advocating for SCIR initiatives.

What are some challenges that SCIR faces? SCIR faces several challenges, including the high cost of implementing these systems, the need for ongoing maintenance and updates, and the need to integrate with existing irrigation systems.

How will SCIR evolve in the future? As technology continues to advance, SCIR is likely to become even more sophisticated and widespread. In the future, we can expect to see more advanced sensors, more sophisticated data analysis tools, and even greater collaboration between different stakeholders in the agricultural community.

Conclusion: *SCIR technology can help to improve the efficiency, sustainability, and health of urban environments, while also promoting conservation of resources.*

SMART CITY LIGHTING (SCL)

Using sensors and data analysis to optimize street lighting and reduce energy consumption.

The implementation of SCL can bring a wide range of benefits to cities and their residents. By using sensors and data analysis to optimize street lighting, cities can significantly reduce energy consumption, which in turn can lead to cost savings for both the city and its residents.

One of the key benefits of SCL is that it allows cities to better control and manage the lighting on their streets. With the use of sensors and data analysis, cities can monitor the level of light on their streets in real-time and make adjustments as necessary. This can help to ensure that the lighting is always at the appropriate level for the conditions and that it is not wasting energy by being too bright or too dim.

SCL also helps to improve the safety and security of cities. Properly lit streets can deter crime and make it easier for people to see and navigate their surroundings. SCL can also be used to help improve visibility at intersections and other areas where visibility is poor. By reducing energy consumption and costs, the smart lighting systems can also help to save tax dollars and provide more budget for other public needs.

Another benefit of SCL is that it can help to improve the overall sustainability of cities. By reducing energy consumption and costs, SCL can help to reduce greenhouse gas emissions and other pollutants. This can help to improve air and water quality, as well as reduce the overall environmental impact of cities.

GENERAL FRAMEWORK

The following framework is a general one and the actual implementation will depend on the specific requirements and constraints of the city.

1. Identify the current state of street lighting in the city, including the type and number of lights, the energy consumption, and the maintenance costs.
2. Define the goals and objectives for the smart city lighting project, such as reducing energy consumption, improving safety and security, and promoting sustainability.
3. Conduct a thorough survey of the city's lighting infrastructure, including mapping all the lights and their locations, and assessing their condition and performance.
4. Develop a plan for the deployment of sensors and other monitoring equipment, including the selection of appropriate technologies and the design of the data collection and analysis system.
5. Install the sensors and monitoring equipment, and test the system to ensure that it is functioning correctly.
6. Develop a data analysis and visualization system to process the data collected by the sensors and present it in a useful and actionable format.
7. Implement a control system to adjust the lighting based on the data collected by the sensors, such as dimming lights when they are not needed or increasing their brightness in response to increased pedestrian or vehicular traffic.
8. Develop a maintenance and support plan for the system, including regular inspections and repairs, software updates, and data backups.
9. Train city staff on how to use and maintain the system, and educate the public about the benefits of smart city lighting.
10. Continuously monitor and analyze the system's performance and make adjustments as needed.

11. Integrate the smart lighting system with other city systems and services, such as traffic management and public transportation.
12. Develop a communication plan to inform and educate the public on the smart lighting system and its benefits
13. Conduct a cost-benefit analysis of the smart lighting system, to track the energy savings and other cost reductions over time.
14. Develop a plan for the expansion of the smart lighting system to additional areas of the city.
15. Conduct a safety and security assessment to identify potential risks and implement appropriate measures to mitigate them.
16. Ensure that the smart lighting system is compliant with all relevant laws, regulations, and standards.
17. Develop a plan for the integration of renewable energy sources, such as solar power, into the smart lighting system.
18. Partner with other cities and organizations to share data and best practices for smart lighting.
19. Develop a plan for the integration of the smart lighting system with other smart city systems and services, such as smart transportation and smart energy management.
20. Develop a plan for the integration of the smart lighting system with existing city infrastructure, such as CCTV cameras.
21. Develop a plan for the integration of the smart lighting system with smart home and IoT devices.
22. Develop a plan for the integration of the smart lighting system with emergency response systems.
23. Develop a plan for the integration of the smart lighting system with public Wi-Fi networks.
24. Develop a plan for the integration of the smart lighting system with city-wide analytics systems.
25. Develop a plan for the integration of the smart lighting system with the city's 5G network infrastructure.

FREQUENTLY ASKED QUESTIONS

What is SMART CITY LIGHTING (SCL) and how does it work? SMART CITY LIGHTING is a system of connected lighting fixtures that can be remotely controlled and managed. It works by using sensors and data analysis to adjust lighting levels based on factors such as foot traffic, weather conditions, and time of day.

What are the benefits of SCL? SCL can help to reduce energy usage and costs, improve public safety by increasing visibility at night, and reduce light pollution by adjusting lighting levels when and where they are needed.

How can I report a broken streetlight in my neighbourhood? In many cases, you can report a broken streetlight in your neighbourhood by contacting your local government's public works department or utility company. Some cities also have online reporting systems or mobile apps that allow residents to report broken streetlights.

How does SCL impact the environment? SCL can have a positive impact on the environment by reducing energy usage and light pollution, as well as by using renewable energy sources like solar power.

Who is responsible for maintaining SCL systems? SCL systems are typically maintained by local governments or utility companies. In some cases, private companies may also be responsible for maintaining SCL systems.

How does SCL improve public safety? SCL can improve public safety by increasing visibility at night, reducing crime, and providing a sense of security for residents and visitors.

How much does it cost to implement SCL? The cost of implementing SCL can vary widely depending on the size and scope of the project, as well as the type of lighting technology used. However, many cities have found that the long-term cost savings from reduced energy usage make SCL a worthwhile investment.

How does SCL improve quality of life? SCL can improve quality of life by providing better lighting for public spaces and reducing light pollution. This can lead to more attractive and welcoming public spaces, as well as improved mental and physical health for residents and visitors.

What are some challenges that SCL faces? SCL faces several challenges, including the high cost of implementing these systems, the need to integrate with existing infrastructure, and concerns about data privacy and security.

How can I get involved in SCL projects in my community? There are many ways to get involved in SCL projects in your community, including attending local government meetings, participating in public forums and community meetings, and joining advocacy groups that are working on SCL initiatives.

Conclusion: *The implementation of SCL can bring a wide range of benefits to cities and their residents. From cost savings to improved safety and sustainability, SCL has the potential to greatly improve the quality of life in cities. The use of sensors and data analysis to optimize street lighting can help cities to reduce energy consumption and costs, improve safety and security, and promote sustainability. It is a smart, cost-effective and environmentally friendly solution for cities of all sizes.*

SMART CITY MAINTENANCE (SCM)

Using technology to improve the maintenance of city infrastructure, such as predictive maintenance and condition monitoring.

Implementing SCM through technology can have numerous benefits for a city. One of the main benefits is the ability to predict when maintenance is needed, rather than waiting for equipment or infrastructure to break down. This can be achieved through predictive maintenance, which uses data and analytics to identify potential issues before they occur. By proactively addressing maintenance needs, cities can reduce downtime and save money on costly repairs.

Another benefit is the ability to monitor the condition of equipment and infrastructure in real-time. This is achieved through condition monitoring, which uses sensors and other technologies to detect changes in equipment performance. By continuously monitoring the condition of city infrastructure, cities can identify potential issues early on, and take action to prevent them from becoming major problems.

SCM can also improve the efficiency of maintenance operations. By using technology to track equipment and infrastructure, cities can better coordinate maintenance activities and optimize the use of resources. This can lead to more effective and cost-efficient maintenance operations, and ultimately save the city money.

SCM can also improve the safety of city infrastructure. By monitoring equipment and infrastructure, cities can identify potential safety hazards and take steps to address them before they become a problem. This can help to reduce the risk of accidents and injuries, and improve the overall safety of the city.

GENERAL FRAMEWORK

The following framework is a general one and the actual implementation will depend on the specific requirements and constraints of the city.

1. Identify key infrastructure assets that require maintenance and prioritize them based on criticality and potential cost savings.
2. Develop a plan for implementing smart city maintenance technology, including the selection of appropriate sensors and communication networks.
3. Install sensors on infrastructure assets to collect data on their condition and performance.
4. Utilize data analysis techniques to identify patterns and trends in the collected data, and use this information to predict when maintenance will be needed.
5. Develop a system for communicating maintenance needs to appropriate staff or contractors, and for tracking the completion of maintenance tasks.
6. Establish a system for monitoring the condition of infrastructure assets on an ongoing basis, using techniques such as vibration analysis and ultrasonic testing.
7. Implement a predictive maintenance program to perform maintenance tasks before problems arise, reducing downtime and prolonging the life of the assets.

8. Implement a condition-based maintenance program to perform maintenance tasks based on the actual condition of the assets, rather than a predetermined schedule.
9. Develop a program for monitoring the energy consumption of street lighting and other city-owned lighting systems.
10. Use data analysis to identify patterns in energy consumption and opportunities for energy savings.
11. Implement a lighting control system to optimize street lighting and reduce energy consumption, such as sensor-controlled dimming or adaptive lighting.
12. Continuously monitor and evaluate the performance of the smart city maintenance system, looking for opportunities to improve efficiency and reduce costs.
13. Develop a plan for maintaining and updating the smart city maintenance system over time.
14. Consider integrating smart city maintenance with other smart city systems, such as transportation or energy management, to maximize efficiency and cost savings.
15. Establish a system for reporting on the status and performance of smart city maintenance to city officials and other stakeholders.
16. Identify key performance indicators (KPIs) to track the effectiveness of smart city maintenance and use this data to make informed decisions.
17. Develop and implement a plan for training staff and contractors on the use of smart city maintenance technology.
18. Consider working with third-party experts to design, implement, and maintain smart city maintenance systems.
19. Establish a budget for smart city maintenance technology and related expenses.
20. Prioritize and plan the funding of smart city maintenance systems based on the costs and benefits.
21. Develop a system for reporting and communicating the benefits of smart city maintenance systems to citizens and stakeholders.
22. Identify and assess the potential risks associated with smart city maintenance systems and develop a plan for mitigating these risks.
23. Continuously monitor and evaluate the performance of the smart city maintenance system, looking for opportunities to improve efficiency and reduce costs.
24. Plan for the scalability and adaptability of smart city maintenance systems to support future growth.

25. Continuously evaluate and improve the smart city maintenance systems to ensure that they remain relevant and effective over time.

FREQUENTLY ASKED QUESTIONS

What is SMART CITY MAINTENANCE (SCM) and what does it entail? SMART CITY MAINTENANCE involves using technology to manage and maintain public infrastructure such as roads, buildings, and utilities. This includes monitoring for damage and wear, scheduling repairs, and optimizing maintenance schedules to reduce costs.

How can SCM benefit my community? SCM can benefit your community by improving the condition and functionality of public infrastructure, reducing maintenance costs over time, and improving the overall quality of life for residents.

Who is responsible for SCM in my community? SCM is typically managed by local government agencies or contractors who are responsible for maintaining public infrastructure.

How does technology help with SCM? Technology such as sensors and data analysis can be used to monitor infrastructure and identify potential issues before they become major problems, which can help reduce costs and improve efficiency.

How does SCM impact sustainability and the environment? SCM can help reduce waste and energy usage by optimizing maintenance schedules and reducing the need for frequent repairs.

How can I report a maintenance issue in my community? You can report a maintenance issue in your community by contacting your local government agency responsible for maintenance, or by using online reporting systems or mobile apps.

What are some challenges faced by SCM initiatives? Challenges faced by SCM initiatives include the cost of implementing new technology and the need for infrastructure upgrades, as well as the potential for privacy and security concerns related to the use of data and sensors.

How can community members get involved with SCM initiatives? Community members can get involved with SCM initiatives by attending local government meetings, participating in public forums and community meetings, and joining advocacy groups that are working on SCM initiatives.

How does SCM impact economic development? SCM can help improve the condition and functionality of public infrastructure, which can lead to increased economic development by attracting new businesses and investments.

What are some examples of SCM projects in other cities? Examples of SCM projects in other cities include using sensors to monitor traffic and optimize traffic signals, using data analysis to prioritize road repairs, and using drones to inspect infrastructure such as bridges and towers.

Conclusion: *Implementing SCM through technology can improve the efficiency, safety and cost-effectiveness of maintaining a city's infrastructure. This can ultimately lead to a better quality of life for citizens, and a more sustainable and livable city for all.*

SMART CITY MOBILITY (SCMO)

Using technology to improve the accessibility and affordability of mobility options, such as bike-sharing and electric vehicles.

Implementing SCMO solutions can have numerous benefits. One of the key benefits is improved accessibility to transportation options. For example, by providing bike-sharing programs, residents have a low-cost and eco-friendly option for getting around the city. The implementation of electric vehicle charging stations can make it easier for residents to switch to electric cars, reducing the city's dependence on fossil fuels.

Another benefit of SCMO is reducing congestion on the roads. By using technology such as real-time traffic monitoring and dynamic routing, transportation officials can improve traffic flow and reduce delays. This can not only make it easier for residents to get around but also reduce the environmental impact of vehicle emissions.

SCMO solutions can also improve affordability for residents. For example, by implementing a payment system for public transportation that is integrated with other transportation options, such as bike-sharing, it can make it more affordable for residents to use a variety of transportation options. By providing real-time information on transportation options and pricing, residents can make more informed decisions about how to get around.

GENERAL FRAMEWORK

The following framework is a general one and the actual implementation will depend on the specific requirements and constraints of the city.

1. Identify the current state of mobility in the city, including transportation infrastructure and modes of transportation used by residents.
2. Conduct a needs assessment to identify areas where technology can improve mobility and accessibility.
3. Develop a vision and set of goals for the smart city mobility project.
4. Research and evaluate different technology options, such as bike-sharing systems and electric vehicle charging infrastructure.
5. Develop a detailed plan for implementation, including timelines, budgets, and key stakeholders.
6. Secure funding for the project through grants, partnerships, or other means.
7. Identify and engage key stakeholders, including city officials, transportation providers, and community organizations.
8. Develop a communication plan to inform the public about the smart city mobility project and gather feedback.

9. Implement the bike-sharing or electric vehicle charging infrastructure and accompanying technology, such as mobile apps or payment systems.
10. Monitor and evaluate the effectiveness of the technology in improving mobility and accessibility.
11. Identify and address any issues or challenges that arise during implementation.
12. Continuously gather data and feedback to improve the smart city mobility project.
13. Use data and analytics to improve the efficiency and effectiveness of the bike-sharing or electric vehicle charging infrastructure.
14. Develop a plan to maintain and upgrade the technology over time.
15. Encourage the use of low-emission vehicles and alternative modes of transportation.
16. Implement smart parking management solutions to improve accessibility and reduce congestion
17. Invest in inter-modal transportation infrastructure to connect different modes of transportation seamlessly.
18. Invest in transportation infrastructure that caters to the differently-abled and the elderly.
19. Invest in technology that can help in traffic management and reducing congestion.
20. Encourage carpooling and car-sharing
21. Implement a transportation demand management program to reduce the number of cars on the road
22. Encourage the use of public transportation by making it more convenient, accessible and affordable
23. Implement a fare integration system to make it easier for passengers to switch between different modes of transportation
24. Invest in technology that can help in incident management and emergency response
25. Continuously engage with the community to gather feedback and evaluate the effectiveness of the smart city mobility project.

FREQUENTLY ASKED QUESTIONS

What is SMART CITY MOBILITY (SCMO) and how does it impact the community? SMART CITY MOBILITY is a concept that focuses on improving transportation and mobility in urban areas through the use of technology. This can include things like smart traffic management, connected vehicles, and public transportation systems. The impact on the community is generally positive, as it can improve accessibility, reduce congestion, and enhance overall quality of life.

How does SCMO address environmental concerns such as air pollution and greenhouse gas emissions? SCMO can address environmental concerns by promoting the use of low-emission vehicles and public transportation, optimizing traffic flow to reduce congestion and idling, and integrating sustainable energy sources such as electric vehicle charging stations and solar-powered streetlights.

What role do local government and private companies play in SCMO? Local government plays a major role in implementing SCMO initiatives, but private companies such as ride-sharing and technology firms can also contribute through partnerships and innovation.

How does SCMO impact public transportation systems? SCMO can improve public transportation systems through the use of technology such as real-time tracking and route optimization, as well as the integration of other mobility options such as bike-sharing and car-sharing services.

What are some of the benefits of SCMO for individual commuters? Benefits of SCMO for individual commuters can include reduced travel time, improved safety, and increased access to transportation options.

How does SCMO impact urban planning and design? SCMO can impact urban planning and design by prioritizing the development of pedestrian and bicycle infrastructure, promoting mixed-use development to reduce the need for travel, and reducing the amount of space dedicated to parking.

What are some potential challenges associated with implementing SCMO initiatives? Challenges can include the cost of implementing new technology and infrastructure, potential privacy concerns related to data collection and analysis, and the need to coordinate with multiple stakeholders such as transportation providers, local government agencies, and private companies.

How can the public provide feedback or get involved with SCMO initiatives? The public can provide feedback and get involved by attending public meetings and participating in community forums, submitting comments and suggestions online, and joining advocacy groups that support SCMO initiatives.

What role do emerging technologies such as autonomous vehicles play in SCMO? Emerging technologies such as autonomous vehicles have the potential to significantly impact SCMO by improving safety and reducing traffic congestion, but also present challenges related to infrastructure and regulation.

What are some successful examples of SCMO initiatives in other cities? Successful examples of SCMO initiatives in other cities include the implementation of bike-sharing programs, the development of pedestrian-friendly zones, and the use of smart traffic management systems to reduce congestion and improve traffic flow.

Conclusion: *Implementing SCMO solutions can lead to a more sustainable, accessible, and efficient transportation system, improving the quality of life for residents.*

SMART CITY PARKING MANAGEMENT (SCPM)

Using technology to optimize parking availability and reduce congestion, such as dynamic pricing and real-time parking guidance.

Implementing SCPM can bring a wide range of benefits for a city and its inhabitants. By using technology such as dynamic pricing and real-time parking guidance, cities can optimize parking availability and reduce congestion. This can lead to a reduction in traffic and air pollution, as well as increased accessibility for drivers. It can also help to improve the overall livability of the city by making it easier for people to find a place to park, which can lead to increased foot traffic in commercial areas and support local businesses.

SCPM can also help to increase revenue for the city by implementing dynamic pricing and other revenue-generating strategies. This can be achieved by adjusting the price of parking spots based on demand, which can help to ensure that parking spots are always available and reduce the need for drivers to circle around looking for a spot. By implementing real-time parking guidance, cities can also help to reduce the number of cars on the road, which can lead to a reduction in congestion and travel times.

SCPM can also help to improve the overall sustainability of the city. By reducing the number of cars on the road, it can help to reduce the carbon footprint of the city and decrease the amount of air pollution. By making it easier for people to find a place to park, it can also help to encourage the use of sustainable modes of transportation such as biking and walking.

GENERAL FRAMEWORK

The following framework is a general one and the actual implementation will depend on the specific requirements and constraints of the city.

1. Identify key areas of high parking demand and congestion within the city.
2. Develop a comprehensive parking management strategy that addresses the needs of both residents and visitors.
3. Implement parking sensors and other monitoring technologies to gather data on parking usage and occupancy.
4. Utilize data analysis to identify patterns and trends in parking usage, and make adjustments to parking policies and pricing accordingly.
5. Implement dynamic pricing for parking spaces, based on demand and occupancy levels.
6. Develop and implement a real-time parking guidance system, using digital signs and mobile apps to guide drivers to available parking spaces.

7. Develop a parking reservation system to allow drivers to reserve parking spaces in advance.
8. Implement a permit-based parking system for residents and frequent visitors.
9. Invest in technologies such as electric vehicle charging stations to support the transition to electric vehicles.
10. Encourage the use of alternative modes of transportation, such as biking and public transit, to reduce the overall demand for parking.
11. Develop partnerships with private parking operators to expand the available supply of parking spaces.
12. Implement a parking enforcement system that utilizes technology, such as license plate recognition, to improve compliance and reduce violations.
13. Develop a parking citation and appeals process that is easy to understand and navigate.
14. Utilize data and analytics to measure the effectiveness of parking management strategies and make adjustments as needed.
15. Encourage the use of carpooling and ride-sharing to reduce the number of vehicles on the road and the demand for parking.
16. Develop a parking management system that is integrated with other smart city systems, such as transportation and traffic management.
17. Implement a system for reporting and addressing parking-related issues, such as illegal parking and broken meters.
18. Utilize technology such as drones for monitoring parking lots and garages.
19. Develop a parking management system that can be easily accessed and operated by the city and citizens.
20. Develop a system for providing parking information to drivers through digital signs, mobile apps and websites
21. Encourage the use of parking cashless payments options.
22. Develop a system for providing parking information to drivers through digital signs, mobile apps and websites.
23. Invest in technologies such as Automatic Number Plate Recognition (ANPR) to improve parking management
24. Develop a system for managing parking for people with disabilities.
25. Utilize data and analytics to measure the effectiveness of parking management strategies and make adjustments as needed.

FREQUENTLY ASKED QUESTIONS

What is SMART CITY PARKING MANAGEMENT (SCPM) and how does it work? SCPM is a system that uses technology to optimize parking in urban areas, including the use of sensors to monitor parking availability, real-time data analysis, and the ability to reserve and pay for parking spaces using a mobile app.

How does SCPM benefit the public? SCPM benefits the public by reducing the time and frustration associated with finding parking, reducing traffic congestion, and promoting more efficient use of parking spaces.

What are the costs associated with implementing SCPM? Costs can vary depending on the scale of the project, but typically include the cost of installing sensors and other technology, software development, and ongoing maintenance.

How does SCPM impact local businesses? SCPM can benefit local businesses by increasing the availability of parking spaces and making it easier for customers to find parking, which can lead to increased foot traffic and revenue.

What impact does SCPM have on public safety? SCPM can improve public safety by reducing congestion and the number of vehicles circling the block looking for parking, which can reduce the risk of accidents and improve emergency response times.

What role do local government and private companies play in SCPM? Local government is typically responsible for implementing SCPM initiatives, but private companies such as parking lot operators and technology firms can also play a role through partnerships and innovation.

How does SCPM address concerns related to sustainability and the environment? SCPM can address concerns related to sustainability by reducing the amount of time and fuel spent looking for parking, promoting the use of alternative transportation such as public transit and bikes, and integrating sustainable energy sources such as electric vehicle charging stations.

Can SCPM be customized to meet the needs of different communities? Yes, SCPM can be customized to meet the unique needs and characteristics of different communities, including factors such as parking demand, traffic patterns, and land use.

How can the public provide feedback or get involved with SCPM initiatives? The public can provide feedback and get involved by attending public meetings and participating in community forums, submitting comments and suggestions online, and joining advocacy groups that support SCPM initiatives.

What are some examples of successful SCPM initiatives in other cities? Successful SCPM initiatives in other cities have included the implementation of variable pricing based on demand, the use of mobile apps to reserve and pay for parking spaces, and the integration of sensors to monitor parking availability in real-time.

Conclusion: Implementing SCPM can bring a wide range of benefits for a city and its inhabitants. From reducing congestion and increasing revenue, to improving sustainability and livability, it can help to create a more efficient and livable city for everyone.

SMART CITY PLATFORM (SCP)

Using technology to integrate and optimize various Smart City systems and services, such as IoT (Internet of Things) and data analytics.

Implementing a SCP can bring a wide range of benefits to a city, through the integration and optimization of various systems and services. By utilizing technology such as the Internet of Things (IoT) and data analytics, a SCP can provide a centralized and interconnected system for managing and controlling various aspects of a city's infrastructure and services.

One of the main benefits of a SCP is improved efficiency. By integrating various systems and services, a SCP can streamline communication and data sharing between different departments and organizations, reducing the need for manual data entry and increasing the speed of decision making. This can lead to more effective and efficient management of city resources, such as energy and water, and improved delivery of services to citizens.

Another benefit of a SCP is the ability to collect and analyze large amounts of data, which can provide valuable insights into city operations and help identify areas for improvement. By using data analytics, a SCP can monitor and analyze key performance indicators, such as traffic flow, air quality, and energy consumption, and provide real-time information to decision-makers. This can enable them to make data-driven decisions, and optimize the performance of various systems and services.

A SCP can improve citizen engagement and participation, by providing easy access to city services and information, and enabling citizens to share their feedback and ideas. This can help to build trust and engagement between citizens and city officials, and contribute to the overall livability and sustainability of the city.

A SCP can also help to attract investment and promote economic development, by providing a foundation for the development of new technologies and services, and creating opportunities for innovation and entrepreneurship.

GENERAL FRAMEWORK

The following framework is a general one and the actual implementation will depend on the specific requirements and constraints of the city.

1. Define the scope and objectives of the Smart City Platform (SCP) project.

2. Identify and assess the current systems and services in place in the city.
3. Create a detailed plan for integrating various Smart City systems and services, including IoT and data analytics.
4. Identify and evaluate potential technology providers and platforms.
5. Establish a governance structure for the SCP project.
6. Develop a data management plan for the SCP project.
7. Identify and acquire the necessary hardware and software for the SCP project.
8. Develop and implement security measures for the SCP project.
9. Develop and implement a training program for all stakeholders involved in the SCP project.
10. Create a communication plan for the SCP project.
11. Create a budget and financial plan for the SCP project.
12. Develop and implement a system for monitoring and evaluating the SCP project.
13. Develop and implement a system for continuous improvement of the SCP project.
14. Design and implement an interface for the SCP project.
15. Design and implement a dashboards and visualization tools for the SCP project.
16. Design and implement data integration and storage system for the SCP project.
17. Design and implement a system for data analysis and mining for the SCP project.
18. Design and implement a system for data reporting and sharing for the SCP project.
19. Design and implement a system for data governance for the SCP project.
20. Develop and implement a system for data quality assurance and data validation for the SCP project.
21. Develop and implement a system for data archiving and retention for the SCP project.
22. Develop and implement a system for data security and privacy for the SCP project.
23. Develop and implement a system for data recovery and disaster recovery for the SCP project.
24. Develop and implement a system for data integration and migration for the SCP project.
25. Develop and implement a system for data integration and migration for the SCP project.

FREQUENTLY ASKED QUESTIONS

What is the SMART CITY PLATFORM (SCP) and how does it work? SCP is a software platform that integrates data from different smart city technologies and systems to provide a unified view of city operations, enabling better decision-making and resource allocation.

How does SCP benefit the public? SCP benefits the public by enabling more efficient and effective city management, including better traffic management, improved emergency response times, and enhanced public services.

What are the costs associated with implementing SCP? Costs can vary depending on the scale of the project, but typically include the cost of software development, hardware installation, and ongoing maintenance.

How does SCP impact local businesses? SCP can benefit local businesses by enabling more efficient use of city resources and enhancing public services, which can promote economic development and a better business environment.

What impact does SCP have on public safety? SCP can improve public safety by providing real-time data and insights to support emergency response teams and law enforcement agencies, as well as enabling more efficient and effective management of traffic and other public safety concerns.

What role do local government and private companies play in SCP? Local government is typically responsible for implementing SCP initiatives, but private companies such as technology firms and data analytics providers can also play a role through partnerships and innovation.

How does SCP address concerns related to sustainability and the environment? SCP can address concerns related to sustainability by enabling more efficient use of city resources, promoting sustainable transportation options, and supporting the integration of renewable energy sources.

Can SCP be customized to meet the needs of different communities? Yes, SCP can be customized to meet the unique needs and characteristics of different communities, including factors such as population, geography, and economic activity.

How can the public provide feedback or get involved with SCP initiatives? The public can provide feedback and get involved by attending public meetings and participating in community forums, submitting comments and suggestions online, and joining advocacy groups that support SCP initiatives.

What are some examples of successful SCP initiatives in other cities? Successful SCP initiatives in other cities have included the integration of traffic management systems with public transit data to improve mobility, the use of data analytics to optimize waste management and recycling, and the integration of air quality sensors with weather data to support public health initiatives.

Conclusion: *Implementing a SCP can greatly improve the efficiency, sustainability, and livability of a city, while providing valuable insights to decision-makers and promoting economic growth.*

SMART CITY PUBLIC ART (SCPA)

Using technology to enhance public art and create interactive experiences, such as digital murals and augmented reality installations.

Implementing SCPA can bring a number of benefits to a city and its residents. One major benefit is the ability to enhance public art and create interactive experiences through the use of technology. Digital murals and augmented reality installations can provide a new level of engagement and excitement for residents and visitors, creating a unique and memorable experience for all. Technology can be used to create a more inclusive and accessible art experience for people of all ages and abilities.

For example, digital murals can be designed to be interactive, allowing people to engage with the art in a more hands-on way, and augmented reality installations can be designed to be accessible to people with visual impairments through audio descriptions.

Another benefit of using technology to enhance public art is the ability to create a more connected and cohesive community. Digital murals and augmented reality installations can be used to create a sense of place and belonging, by telling the stories and highlighting the unique features of the community. Technology can be used to create a more sustainable and efficient public art experience. For example, digital murals can be designed to be energy-efficient and require less maintenance than traditional murals, and augmented reality installations can be designed to be interactive and engaging, reducing the need for physical signage and brochures.

SCPA can also bring economic benefits to a city. By creating a unique and memorable experience for residents and visitors, SCPA can attract more tourists and increase economic activity in the area. The use of technology in public art can also create opportunities for local businesses and entrepreneurs, such as technology companies and digital artists, to participate in the development and maintenance of public art projects.

GENERAL FRAMEWORK

The following framework is a general one and the actual implementation will depend on the specific requirements and constraints of the city.

1. Identify key areas in the city where public art can be enhanced with technology.
2. Research and gather information on the latest technology and trends in digital art and interactive installations.
3. Create a vision and strategy for the implementation of smart city public art.
4. Develop a budget and secure funding for the implementation.

5. Identify and select artists and designers to create digital murals and augmented reality installations.
6. Work with local government and community organizations to obtain necessary approvals and permits.
7. Install sensors and other technology to enable interactive features in the public art installations.
8. Develop and implement a maintenance plan for the technology and art installations.
9. Create a website and mobile app to provide information and enhance the user experience of the public art.
10. Partner with businesses and organizations to create sponsored and branded installations.
11. Create a program for public participation in the design and creation of art installations.
12. Develop a system for data collection and analysis of the impact and usage of the art installations.
13. Create a plan for expanding the program to other areas of the city.
14. Provide training and support for maintenance and management of the technology.
15. Develop a system for monitoring and troubleshooting technical issues.
16. Create a system for evaluating and selecting new technology and installations.
17. Develop a plan for retiring older installations and technology.
18. Develop a plan for integrating the public art installations with other smart city systems and services
19. Create a plan for promoting and marketing the smart city public art program.
20. Develop a plan for measuring and reporting on the success of the program.
21. Create a plan for ongoing evaluation and improvement of the program.
22. Develop a plan for collaboration and sharing of resources with other cities and organizations.
23. Create a plan for emergency response and security for the technology and art installations.
24. Develop a plan for community engagement and education related to the smart city public art program.
25. Create a plan for building a sustainable and long-term program for smart city public art.

FREQUENTLY ASKED QUESTIONS

What is SMART CITY PUBLIC ART (SCPA) and how is it different from traditional public art installations? SCPA is a program that integrates technology into public art installations to create interactive experiences and enhance community engagement with the artwork.

Who decides what kind of art is installed through SCPA? The selection of SCPA installations is typically made by a committee of artists, community leaders, and city officials, with input from the public.

How does SCPA benefit the community? SCPA benefits the community by fostering a sense of identity and community pride, promoting public engagement with the arts, and providing a unique and memorable experience for visitors.

What technologies are used in SCPA installations? Technologies used in SCPA installations can include interactive displays, augmented reality, and mobile apps that enhance the visitor experience and allow for real-time interaction with the artwork.

How is SCPA funded? Funding for SCPA can come from a variety of sources, including public grants, private donations, and corporate sponsorships.

Can SCPA installations be temporary or permanent? SCPA installations can be either temporary or permanent, depending on the scope and goals of the project.

How can the public provide input on SCPA initiatives? The public can provide input on SCPA initiatives through public forums, community meetings, and online surveys.

How does SCPA promote cultural diversity and inclusivity? SCPA can promote cultural diversity and inclusivity by featuring works from diverse artists and highlighting the unique cultural identities and histories of different communities.

How can SCPA help promote tourism and economic development? SCPA can promote tourism and economic development by creating a unique and memorable visitor experience that attracts new visitors to the area and promotes local businesses and cultural institutions.

What are some successful SCPA installations in other cities? Successful SCPA installations in other cities have included interactive sculptures, murals that use augmented reality to create a 3D experience, and installations that use light and sound to create an immersive experience for visitors.

Conclusion: *Implementing SCPA can bring a number of benefits to a city and its residents, including enhanced public art experiences, a more connected and cohesive community, and economic benefits. By using SCPA technology to enhance public art, cities can create a unique and memorable experience for residents and visitors, and foster a sense of place and belonging for the community.*

SMART CITY PUBLIC HEALTH (SCPH)

Using technology to improve the health and wellbeing of citizens, such as telemedicine, remote monitoring and health data analytics.

Implementing SCPH initiatives can bring a range of benefits to citizens and the city as a whole. One major benefit is the ability to improve access to healthcare services through the use of telemedicine and remote monitoring. This can allow individuals to receive medical consultations and treatment without needing to leave their homes, which can be especially beneficial for those who have mobility issues or live in remote areas. The use of remote monitoring can allow healthcare providers to keep a closer eye on patients' health and take action more quickly if there are any concerning changes.

Another benefit of SCPH initiatives is the ability to leverage data and analytics to better understand the health needs of the population and target resources where they are needed most. For example, data on air and water quality, as well as data on the health of citizens, can be used to identify areas where there may be high levels of pollution or health issues, and take steps to address them. Data on the usage of healthcare services can be used to identify areas where there may be a shortage of doctors or other healthcare professionals, and take steps to address that shortage.

SCPH initiatives can also contribute to the overall well-being of citizens by promoting healthy lifestyles and reducing the risk of chronic diseases. For example, the use of technology to encourage physical activity, such as through the use of fitness tracking apps or gamified walking or cycling routes, can help to promote regular exercise. The use of technology to provide information on healthy eating and nutrition can help to educate citizens about the importance of a healthy diet.

GENERAL FRAMEWORK

The following framework is a general one and the actual implementation will depend on the specific requirements and constraints of the city.

1. Identify the specific public health concerns and objectives for the city.
2. Research and evaluate various technologies and solutions that can address those concerns and objectives.
3. Develop a plan for integrating telemedicine and remote monitoring technology into existing healthcare systems.
4. Establish a secure and compliant system for collecting and analyzing health data from citizens.
5. Partner with healthcare providers, insurance companies, and other stakeholders to ensure widespread adoption and use of the technology.
6. Implement training programs for healthcare professionals and citizens to effectively use the technology.
7. Develop policies and procedures to ensure the privacy and security of health data.
8. Implement a system for monitoring and evaluating the effectiveness of the technology in improving public health outcomes.

9. Create a communication plan to inform citizens about the availability and benefits of the technology.
10. Encourage participation and feedback from citizens to continuously improve the technology.
11. Develop a program to provide support and resources to disadvantaged communities to access the technology.
12. Identify and remove barriers to access, such as lack of internet connectivity and digital literacy.
13. Collaborate with other cities and organizations to share best practices and learn from their experiences.
14. Establish a governance structure to manage and oversee the technology.
15. Develop a budget and funding plan to sustain and scale the technology over time.
16. Identify and address any legal and regulatory issues related to the technology.
17. Create an emergency response plan in case of technical issues or data breaches.
18. Develop a plan to integrate the technology with other smart city initiatives, such as transportation and energy management.
19. Create a plan to measure and report on the impact of the technology on public health outcomes.
20. Establish a plan for continuous improvement and updates of the technology over time.
21. Engage with local government officials, community leaders, and other stakeholders to build support and buy-in for the program.
22. Develop a plan for educating citizens on how to maintain their health through technology.
23. Identify ways in which the technology can also be used to improve the health of the environment.
24. Develop a plan to ensure the technology is accessible to all citizens, including those with disabilities.
25. Create a plan to measure and report on the impact of the technology on reducing healthcare costs.

FREQUENTLY ASKED QUESTIONS

What is SMART CITY PUBLIC HEALTH (SCPH) and how does it improve community health? SCPH is a program that leverages technology and data to improve public health outcomes by identifying health risks and providing targeted interventions and resources to vulnerable populations.

Who is involved in the SCPH program? The SCPH program involves a variety of stakeholders, including healthcare providers, city officials, community organizations, and residents.

What kind of data is collected and analysed in SCPH initiatives? SCPH initiatives can collect and analyze data on a variety of health-related factors, such as air quality, water quality, disease outbreaks, and health disparities.

How is the data used to improve public health outcomes? Data collected through SCPH initiatives can be used to identify health risks and develop targeted interventions to address those risks. For example, if a community has a high rate of asthma, the SCPH program may work to improve air quality in that area.

How can the public get involved in SCPH initiatives? The public can get involved in SCPH initiatives by participating in health screenings, attending community events, and providing feedback on SCPH programs and initiatives.

What kind of technology is used in SCPH initiatives? SCPH initiatives can leverage a variety of technologies, such as mobile apps, wearable devices, and sensors, to collect and analyze health data and provide targeted interventions.

How does SCPH promote health equity and reduce health disparities? SCPH can promote health equity by focusing on the health needs of vulnerable populations and providing targeted interventions and resources to address health disparities.

How does SCPH help prepare for public health emergencies? SCPH can help prepare for public health emergencies by identifying health risks and vulnerabilities in the community and developing response plans to address those risks.

W**hat kind of partnerships are involved in SCPH initiatives?** SCPH initiatives can involve partnerships between healthcare providers, public health agencies, community organizations, and technology companies.

W**hat are some successful SCPH initiatives in other cities?** Successful SCPH initiatives in other cities have included programs that use technology to track disease outbreaks, mobile clinics that provide healthcare services to underserved communities, and community-based initiatives that promote healthy lifestyles and provide resources to address health disparities.

Conclusion: The implementation of SCPH initiatives can bring a range of benefits to citizens, including improved access to healthcare services, better targeting of resources, and a focus on promoting healthy lifestyles and reducing the risk of chronic diseases. By leveraging the power of technology, SCPH initiatives can help to improve the health and well-being of citizens, and contribute to the overall livability of the city.

SMART CITY PUBLIC SAFETY AND EMERGENCY MANAGEMENT (SCPSEM)

Using technology to enhance public safety and improve emergency response times.

Implementing a SCPSEM system can have several benefits for both citizens and city officials. One of the key benefits is the ability to enhance public safety by using technology to improve emergency response times. This can be achieved by using advanced communication systems, such as emergency notification systems and GPS tracking, to quickly and efficiently dispatch emergency services to the scene of an incident. Smart surveillance cameras and other sensors can be used to monitor public spaces and provide real-time data to emergency responders, allowing them to quickly assess and respond to incidents.

Another benefit of a SCPSEM system is the ability to improve incident management and coordination. This can be achieved by using advanced data analytics and visualization tools to quickly identify patterns and trends in incidents, and to provide real-time situational awareness to emergency responders. This can help to improve the efficiency of incident management and reduce response times.

Moreover, SCPSEM can also improve the safety of citizens by providing them with real-time information and alerts during emergency situations. This can include information on evacuation routes, emergency shelters, and other critical information. This can help to keep citizens safe and informed during emergencies, and can also reduce the likelihood of panic and confusion.

SCPSEM can also provide a platform for citizens to report incidents and provide feedback, which can help city officials to identify and address problem areas in the city. This can lead to a more responsive and efficient public safety system, and can also help to improve the overall quality of life for citizens.

GENERAL FRAMEWORK

The following framework is a general one and the actual implementation will depend on the specific requirements and constraints of the city.

1. Identify key public safety and emergency management challenges in the city.
2. Conduct a comprehensive risk assessment to identify potential hazards and vulnerabilities.
3. Develop a comprehensive communication plan for emergency response and recovery.
4. Invest in advanced technologies such as CCTV cameras, license plate recognition systems, and gunshot detection systems to enhance surveillance and improve response times.
5. Implement a city-wide emergency alert system to quickly notify residents of potential hazards and emergency situations.

6. Develop an incident command system to coordinate emergency response efforts across different agencies and organizations.
7. Establish partnerships with local hospitals and healthcare providers to improve emergency medical response capabilities.
8. Develop an emergency management training program for city employees and first responders.
9. Invest in advanced data analytics and visualization tools to improve decision-making during emergency situations.
10. Regularly conduct drills and exercises to test the city's emergency response capabilities and identify areas for improvement.
11. Implement a GIS based emergency management system to support incident management, resource tracking, and damage assessment.
12. Invest in IoT enabled smart devices and sensors to gather real-time information on environmental conditions, traffic, and other factors that could affect emergency response efforts.
13. Develop a community outreach program to educate residents on emergency preparedness and response.
14. Integrate existing systems and platforms to improve information sharing and collaboration among different agencies and organizations.
15. Invest in advanced technologies such as drones, robots and autonomous vehicles to support emergency response and recovery efforts.
16. Invest in a city-wide wireless network to support the communication and data transfer needs of emergency responders and other city services.
17. Develop a post-disaster recovery plan to guide the city's efforts to restore critical infrastructure and services.
18. Establish partnerships with private sector companies and organizations to enhance the city's emergency response capabilities.
19. Implement a system to track and analyze emergency calls to improve response times and identify patterns and trends.
20. Invest in advanced technologies such as facial recognition and biometrics to enhance security and improve emergency response efforts.
21. Develop a city-wide plan for emergency evacuation and sheltering.
22. Establish a city-wide emergency operations center to coordinate emergency response efforts.
23. Invest in advanced analytics and machine learning algorithms to improve the accuracy of predictions and forecasts in emergency situations.

24. Develop a program to provide emergency responders with real-time information on the location of critical infrastructure and other key assets.
25. Regularly review and update the city's emergency management plans and procedures to ensure they remain relevant and effective.

FREQUENTLY ASKED QUESTIONS

What is SMART CITY PUBLIC SAFETY AND EMERGENCY MANAGEMENT (SCPSEM) and how does it improve community safety? SCPSEM is a program that leverages technology and data to improve public safety outcomes by identifying and responding to emergency situations in a timely and efficient manner.

Who is involved in the SCPSEM program? The SCPSEM program involves a variety of stakeholders, including law enforcement, emergency responders, city officials, and residents.

What kind of data is collected and analysed in SCPSEM initiatives? SCPSEM initiatives can collect and analyze data on a variety of safety-related factors, such as crime rates, traffic accidents, natural disasters, and emergency response times.

How is the data used to improve public safety outcomes? Data collected through SCPSEM initiatives can be used to identify safety risks and develop targeted interventions to address those risks. For example, if a community has a high rate of traffic accidents, the SCPSEM program may work to improve road safety in that area.

How can the public get involved in SCPSEM initiatives? The public can get involved in SCPSEM initiatives by participating in community safety programs, reporting safety concerns to law enforcement, and providing feedback on SCPSEM programs and initiatives.

What kind of technology is used in SCPSEM initiatives? SCPSEM initiatives can leverage a variety of technologies, such as surveillance cameras, gunshot detection systems, and emergency alert systems, to collect and analyze safety data and respond to emergency situations.

How does SCPSEM promote safety equity and reduce safety disparities? SCPSEM can promote safety equity by focusing on the safety needs of vulnerable populations and providing targeted interventions and resources to address safety disparities.

How does SCPSEM help prepare for emergency situations? SCPSEM can help prepare for emergency situations by identifying safety risks and vulnerabilities in the community and developing response plans to address those risks.

What kind of partnerships are involved in SCPSEM initiatives? SCPSEM initiatives can involve partnerships between law enforcement agencies, emergency responders, city officials, community organizations, and technology companies.

What are some successful SCPSEM initiatives in other cities? Successful SCPSEM initiatives in other cities have included programs that use technology to detect and respond to gunfire, emergency alert systems that provide real-time information to residents during emergencies, and community-based initiatives that promote safety education and awareness.

Conclusion: *Implementing a SCPSEM system can have a wide range of benefits, including enhancing public safety, improving emergency response times, improving incident management and coordination, improving the safety of citizens, and providing a platform for citizens to report incidents and provide feedback.*

SMART CITY PUBLIC SERVICES (SCPS)

Using technology to improve the delivery of public services, such as online bill payments and service request portals.

The implementation of SCPS can bring numerous benefits to both citizens and city governments. One of the main benefits is the ability to improve the delivery of public services. By using technology such as online bill payments and service request portals, citizens can easily access and manage the services they need, without having to go to a physical location or wait in line. This not only saves time, but also makes it more convenient for citizens to access these services.

Another benefit of SCPS is the ability to increase government transparency and accountability. By making information about services and payments available online, citizens can easily see how their taxes are being spent and hold their government accountable for the services they are providing. By allowing citizens to submit service requests and track the status of their requests online, governments can show that they are actively working to address the needs and concerns of their citizens.

SCPS also has the potential to improve the efficiency and cost-effectiveness of public services. By using technology such as data analytics, governments can gain valuable insights into how services are being used, which areas are in most need of attention, and how to better allocate resources to meet those needs. This can help to reduce costs, improve service delivery and increase overall satisfaction of citizens with public services.

Moreover, SCPS can also help to improve communication between citizens and government. For example, by providing an online platform where citizens can access information about services, provide feedback and report problems, governments can increase their ability to respond to citizens' needs and concerns quickly and effectively.

GENERAL FRAMEWORK

The following framework is a general one and the actual implementation will depend on the specific requirements and constraints of the city.

1. Conduct a needs assessment to identify which public services can benefit from technology implementation.
2. Research and evaluate different technology solutions for improving the delivery of public services.
3. Develop a plan for integrating technology into existing systems and processes for public services.
4. Create a system for online bill payments and service request portals.
5. Develop a user-friendly interface for the public to access and use the new technology.
6. Train staff on the use and maintenance of the new technology.
7. Test the system for bugs and technical issues before deployment.
8. Implement the new technology and integrate it with existing systems and processes.
9. Monitor usage and gather feedback from the public to make any necessary adjustments.
10. Regularly update and maintain the system to ensure its effectiveness.
11. Use data analytics to track usage and performance of the new technology.
12. Monitor the budget and costs of implementing and maintaining the new technology.
13. Collaborate with other departments and agencies to ensure seamless integration of the technology.
14. Consider privacy and security measures to protect sensitive information.
15. Develop a communication plan to inform the public about the new technology and how to use it.
16. Create a system for tracking and resolving technical issues and complaints.
17. Develop a system for monitoring the performance and effectiveness of the new technology.
18. Assess the impact of the new technology on public services and make adjustments as necessary.
19. Regularly review and evaluate the system to ensure its continued effectiveness.
20. Consider scalability and future expansion of the system.
21. Investigate opportunities for integrating with other Smart City systems and services.
22. Develop a plan for disaster recovery and business continuity.
23. Create a system for ensuring accessibility for people with disabilities.
24. Create a system for monitoring energy consumption and cost savings.

25. Create a system for measuring and reporting on the performance and impact of the new technology on public services.

FREQUENTLY ASKED QUESTIONS

What is SMART CITY PUBLIC SERVICES (SCPS) and how does it improve the delivery of public services? SCPS is a program that leverages technology and data to improve the efficiency and effectiveness of public services, such as waste management, water management, and transportation.

How does SCPS use data to optimize public service delivery? SCPS uses data to identify patterns and trends in service delivery and to optimize resources to improve service delivery outcomes.

Who is involved in the SCPS program? The SCPS program involves a variety of stakeholders, including city officials, service providers, technology companies, and residents.

How can residents provide feedback on public services? Residents can provide feedback on public services through surveys, feedback forms, and by reporting service issues to the appropriate city officials.

What kind of technology is used in SCPS initiatives? SCPS initiatives can leverage a variety of technologies, such as smart sensors, data analytics platforms, and mobile apps, to collect and analyze service data and optimize service delivery.

How does SCPS promote sustainability and environmental responsibility? SCPS can promote sustainability by optimizing resources and reducing waste, such as reducing water consumption, optimizing waste management, and reducing emissions from transportation.

How does SCPS promote social equity and accessibility to public services? SCPS can promote social equity by improving access to public services for underserved communities and by providing targeted services to vulnerable populations.

How does SCPS help to reduce costs of public service delivery? SCPS helps to reduce costs by optimizing resources and improving service efficiency, reducing waste and duplication of efforts.

What kind of partnerships are involved in SCPS initiatives? SCPS initiatives can involve partnerships between city officials, service providers, technology companies, community organizations, and other stakeholders.

What are some successful SCPS initiatives in other cities? Successful SCPS initiatives in other cities have included programs that use smart sensors to optimize waste management, mobile apps that allow residents to report service issues in real-time, and data analytics platforms that help city officials to optimize service delivery resources.

Conclusion: The implementation of SCPS can greatly enhance the delivery of public services, increase government transparency and accountability, improve the efficiency and cost-effectiveness of public services and improve communication between citizens and government. By using technology such as online bill payments, service request portals, data analytics, and citizen feedback platforms, cities can create more efficient, cost-effective and responsive public services that meet the needs of citizens.

SMART CITY PUBLIC SPACES (SCPSP)

Using technology to enhance public spaces, such as smart parks and plazas, and create interactive experiences.

Implementing a SCPSP initiative can bring a wide range of benefits to citizens and the community as a whole. One major advantage is the ability to create more interactive and engaging public spaces. By incorporating technology such as Wi-Fi, digital displays, and interactive installations, these spaces can be transformed into vibrant community hubs that foster social interaction and engagement.

Another key benefit is the ability to optimize the use and management of public spaces. Smart technology can be used to monitor usage patterns, track maintenance needs, and adjust amenities to meet the needs of the community. This can help to ensure that public spaces are well-maintained, safe, and accessible for all citizens.

SCPSP can also promote sustainability and environmental stewardship. By incorporating green infrastructure such as rain gardens, green roofs, and urban forestry, these spaces can help to reduce the urban heat island effect, improve air quality, and promote biodiversity.

Using technology in public spaces can also improve accessibility and mobility. For example, by providing real-time information on transportation options and parking availability, it can make it easier for citizens to access public spaces and reduce congestion.

GENERAL FRAMEWORK

The following framework is a general one and the actual implementation will depend on the specific requirements and constraints of the city.

1. Identify public spaces in the city that can be enhanced with technology.
2. Conduct a survey to gather feedback from citizens on how they use and envision the use of these spaces.
3. Develop a plan for incorporating technology into the design of the public spaces.
4. Research and evaluate different technology options, such as smart lighting, Wi-Fi, and interactive kiosks.
5. Develop a budget for the technology implementation.
6. Create a project team to oversee the implementation of the technology.
7. Research and evaluate different technology providers and vendors.

8. Develop a contract with the chosen technology provider.
9. Design and develop a user interface for the technology.
10. Test the technology before full implementation.
11. Develop a training program for city staff to use and maintain the technology.
12. Develop a communication plan to inform citizens about the new technology and how to use it.
13. Install the technology and associated infrastructure in the public spaces.
14. Implement a monitoring and maintenance plan for the technology.
15. Develop a data management plan to collect and analyze data from the technology.
16. Integrate the technology with existing city systems, such as traffic management and public safety systems.
17. Develop a plan for future upgrades and expansion of the technology.
18. Collaborate with local businesses and community organizations to enhance and activate the public spaces.
19. Develop a plan for accessibility and inclusion for people with disabilities.
20. Develop a plan for emergency response and safety in the public spaces.
21. Develop a plan for sustainable energy usage for the technology.
22. Develop a plan for cybersecurity and data privacy for the technology.
23. Develop a plan for revenue generation from the technology.
24. Develop a plan for measuring the impact and success of the technology.
25. Continuously evaluate and improve the technology based on feedback from citizens and data analysis.

FREQUENTLY ASKED QUESTIONS

What is SMART CITY PUBLIC SPACES (SCPSP) and how does it improve public spaces? SCPSP is a program that uses technology to enhance the design and functionality of public spaces, making them more accessible and inviting for everyone.

Who is involved in the SCPSP program? The SCPSP program involves a variety of stakeholders, including city officials, architects, urban planners, community organizations, and residents.

How can residents provide input on SCPSP initiatives? Residents can provide input on SCPSP initiatives through public meetings, community workshops, and online feedback forms.

What kind of technology is used in SCPSP initiatives? SCPSP initiatives can leverage a variety of technologies, such as smart sensors, augmented reality, and interactive displays, to create engaging and interactive public spaces.

How does SCPSP promote community engagement and social interaction? SCPSP initiatives can promote community engagement by creating spaces that encourage social interaction and collaboration, such as community gardens, public art installations, and outdoor performance spaces.

How does SCPSP address issues of safety and security in public spaces? SCPSP initiatives can address safety and security issues by using smart lighting and surveillance systems, and by creating clear sightlines and open spaces that discourage criminal activity.

How does SCPSP promote sustainability and environmental responsibility? SCPSP can promote sustainability by incorporating green infrastructure, such as rain gardens and bioswales, and by promoting the use of public transportation and alternative modes of transportation.

What kind of partnerships are involved in SCPSP initiatives? SCPSP initiatives can involve partnerships between city officials, community organizations, technology companies, and other stakeholders.

What are some successful SCPSP initiatives in other cities? Successful SCPSP initiatives in other cities have included projects that use augmented reality to enhance public art installations, the creation of community gardens and parks, and the development of pedestrian-friendly public spaces.

How can SCPSP initiatives help to promote economic development in a city? SCPSP initiatives can help to promote economic development by creating attractive public spaces that attract visitors and businesses, and by promoting the use of public transportation, which can help to reduce traffic congestion and increase economic activity.

Conclusion: *SCPSP can help to create more livable, vibrant, and sustainable communities. By leveraging technology to enhance and optimize these spaces, cities can foster a sense of community, promote social interaction, and improve the quality of life for citizens.*

SMART CITY PUBLIC TRANSPORTATION (SCPT)

Using technology to improve the performance and accessibility of public transportation, such as real-time tracking and fare payment systems.

Implementing SCPT can bring a variety of benefits to the city and its citizens. One of the main advantages is the ability to improve the performance and reliability of public transportation systems. By using technology such as real-time tracking and fare payment systems, public transportation can become more efficient and convenient for riders. This can lead to increased ridership and a decrease in traffic congestion on the roads.

Another benefit is the ability to enhance accessibility for all citizens. SCPT can make it easier for individuals with disabilities, the elderly, and low-income individuals to access public transportation by providing real-time information and trip planning assistance. Technology such as automated fare collection can make it easier for riders to pay for their trips, reducing barriers to entry for those who may have difficulty with traditional fare methods.

SCPT can also help to reduce the environmental impact of transportation by decreasing the number of cars on the road. By providing reliable and efficient public transportation options, more people may choose to take public transportation instead of driving, leading to decreased greenhouse gas emissions and air pollution.

GENERAL FRAMEWORK

The following framework is a general one and the actual implementation will depend on the specific requirements and constraints of the city.

1. Identify the current state of public transportation infrastructure and services in the city.
2. Develop a comprehensive transportation plan that includes the integration of smart technologies.
3. Research and evaluate different smart transportation technology options, such as real-time tracking systems and fare payment systems.
4. Conduct a cost-benefit analysis to determine the financial feasibility of implementing the chosen technology.
5. Establish partnerships and collaborations with technology providers, transportation companies, and other stakeholders.
6. Develop a detailed implementation plan, including timelines, resources, and milestones.
7. Set up a project team to oversee the implementation and manage the different aspects of the project.
8. Implement real-time tracking systems for public transportation vehicles.
9. Implement fare payment systems that integrate with the real-time tracking systems.
10. Train public transportation staff on the use of the new technology.
11. Test and evaluate the new technology before full-scale deployment.
12. Roll out the new technology to all public transportation vehicles and infrastructure.
13. Monitor and evaluate the performance of the new technology.
14. Continuously gather feedback from users and make adjustments as necessary.
15. Develop and implement a plan for maintaining and updating the technology.
16. Develop and implement a data analytics system to track and analyze transportation usage patterns.

17. Develop and implement a system for real-time incident management and emergency response.
18. Develop and implement a system for providing real-time information to users, such as arrival times and service disruptions.
19. Develop and implement a system for integrating with other smart city systems and services, such as traffic management and parking management.
20. Develop and implement a system for integrating with other transportation options, such as bike-sharing and ride-sharing services.
21. Develop and implement a system for providing accessibility options for people with disabilities.
22. Develop and implement a communication and marketing plan to inform the public about the new technology and services.
23. Develop and implement a plan for monitoring and enforcing compliance with the new technology and services.
24. Develop and implement a plan for measuring and reporting on the success of the project.
25. Continuously evaluate and improve the smart public transportation system to ensure it meets the evolving needs of the city and its citizens.

FREQUENTLY ASKED QUESTIONS

How will smart city technology improve public transportation for commuters? Smart city technology can improve public transportation by providing real-time data on traffic, public transportation schedules, and the availability of parking spaces. This information can be used to optimize routes and schedules, reduce wait times and delays, and improve overall transportation efficiency.

Will smart city technology help reduce traffic congestion in cities? Yes, smart city technology can help reduce traffic congestion by providing real-time traffic data, optimizing traffic lights and routes, and encouraging the use of public transportation and alternative modes of transportation like bikes and scooters.

How will smart city technology improve safety on public transportation? Smart city technology can improve safety on public transportation by providing real-time data on potential safety issues, like accidents or crime, and allowing for quick response times. It can also provide better communication between riders and transportation officials in case of emergency.

Will smart city technology make public transportation more accessible to people with disabilities? Yes, smart city technology can make public transportation more accessible to people with disabilities by providing real-time data on the availability of accessible transportation, improving the functionality of public transportation apps, and allowing for better communication between transportation officials and people with disabilities.

How will smart city technology improve the overall rider experience on public transportation? Smart city technology can improve the overall rider experience on public transportation by providing real-time data on schedules, routes, and delays, improving the cleanliness and safety of public transportation vehicles and stations, and providing better communication between riders and transportation officials.

Will smart city technology encourage more people to use public transportation instead of driving? Yes, smart city technology can encourage more people to use public transportation instead of driving by providing real-time data on traffic, parking, and transportation schedules, making public transportation more accessible to people with disabilities, and encouraging the use of alternative modes of transportation like bikes and scooters.

How will smart city technology impact the cost of public transportation for riders? Smart city technology can impact the cost of public transportation for riders by optimizing routes and schedules, reducing delays and wait times, and improving overall transportation efficiency. This can potentially lead to lower costs for riders.

Will smart city technology allow for more environmentally-friendly public transportation options? Yes, smart city technology can allow for more environmentally-friendly public transportation options by encouraging the use of alternative modes of transportation like bikes and scooters, optimizing public transportation routes to reduce emissions, and providing real-time data on air quality and pollution levels.

How will smart city technology impact the jobs of public transportation workers? Smart city technology may impact the jobs of public transportation workers by automating certain tasks and improving overall efficiency. However, it can also create new job opportunities in fields like data analysis and transportation management.

Will smart city technology improve the overall reliability of public transportation? Yes, smart city technology can improve the overall reliability of public transportation by providing real-time data on traffic, schedules, and delays, optimizing routes and schedules, and allowing for quick response times to potential issues.

Conclusion: *Implementing SCPT can improve the performance and accessibility of public transportation, increase ridership, reduce traffic congestion and pollution, and enhance accessibility for all citizens.*

SMART CITY RENEWABLE ENERGY (SCRE)

Using technology to promote the use of renewable energy, such as solar and wind power, and reduce dependence on fossil fuels.

Implementing SCRE technology can have many benefits for a city and its citizens. One of the main benefits is the reduction of dependence on fossil fuels, which can help to reduce greenhouse gas emissions and improve air quality. The use of renewable energy sources can help to promote energy independence and reduce dependence on foreign energy sources.

Another key benefit of SCRE is the potential for cost savings. Renewable energy sources, such as solar and wind power, can provide a cost-effective alternative to traditional fossil fuel-based energy sources. This can be especially beneficial for cities that are looking to reduce their overall energy costs.

In addition to reducing dependence on fossil fuels and promoting cost savings, SCRE can also help to promote economic development. Renewable energy projects can create jobs in the construction, installation, and maintenance of renewable energy systems. This can help to stimulate economic growth and create new opportunities for businesses and entrepreneurs.

SCRE can also help to improve the overall quality of life for citizens. Clean, renewable energy sources can help to reduce pollution and improve public health, which can have a positive impact on the overall well-being of citizens.

GENERAL FRAMEWORK

The following framework is a general one and the actual implementation will depend on the specific requirements and constraints of the city.

1. Identify current energy usage and sources in the city.
2. Conduct a feasibility study to assess the potential for renewable energy generation in the city, including solar, wind, and hydro power.
3. Develop a plan for transitioning to renewable energy sources, including timelines and targets.
4. Implement solar power systems on public buildings, such as city halls and libraries.
5. Install wind turbines in appropriate locations within the city.
6. Develop policies and incentives to encourage the use of renewable energy in the private sector, such as building codes and tax breaks.
7. Invest in energy storage systems to ensure a reliable energy supply during peak demand.
8. Implement energy-efficient technologies in public buildings, such as LED lighting and building management systems.

9. Develop a smart grid system to manage energy distribution and consumption.
10. Partner with utility companies to provide incentives for residents and businesses to reduce energy consumption.
11. Develop educational programs and campaigns to raise awareness about renewable energy and energy conservation.
12. Establish a monitoring and evaluation system to track progress towards renewable energy goals.
13. Invest in research and development of new renewable energy technologies.
14. Create a Renewable Energy Resource Center to provide information and support to residents and businesses.
15. Invest in renewable energy projects such as solar parks, wind farms and hydroelectric power plants.
16. Implement Electric vehicle charging stations in public areas to promote the use of electric vehicles.
17. Develop a Renewable Energy Procurement Plan to encourage energy suppliers to provide green energy options.
18. Encourage private sector companies to invest in renewable energy projects.
19. Encourage the use of renewable energy in street lighting and traffic signalling systems.
20. Develop a micro-grid system to support the use of renewable energy in specific areas.
21. Encourage energy-efficient retrofitting of existing buildings.
22. Partner with local universities and research institutions to develop renewable energy technologies.
23. Develop a Renewable Energy Master Plan to guide the city's transition to renewable energy.
24. Create a task force to oversee the implementation of the renewable energy plan.
25. Regularly review and update the renewable energy plan to ensure it remains relevant and effective.

FREQUENTLY ASKED QUESTIONS

What is Smart City Renewable Energy (SCRE)? Smart City Renewable Energy (SCRE) is a program that leverages renewable energy sources and innovative technologies to promote sustainable energy practices in smart cities.

How does SCRE benefit the environment? SCRE helps reduce the carbon footprint of a city by replacing traditional, non-renewable sources of energy with clean, renewable sources like solar and wind power. This results in a reduction in harmful greenhouse gas emissions and helps combat climate change.

Who funds SCRE? SCRE projects can be funded by a variety of sources, including government grants, private investments, and partnerships with energy companies.

What types of renewable energy sources does SCRE use? SCRE focuses on several types of renewable energy, including solar, wind, geothermal, and hydropower.

How does SCRE ensure a reliable energy supply? SCRE incorporates energy storage systems and other technologies to ensure that renewable energy sources can provide a consistent and reliable source of power.

How can individuals get involved in SCRE? Individuals can get involved in SCRE by supporting renewable energy initiatives in their community, such as advocating for solar panels on public buildings, or investing in renewable energy projects.

How does SCRE benefit the local economy? SCRE can create jobs in the renewable energy industry and can also lower energy costs for residents and businesses, ultimately contributing to the local economy.

Can SCRE be implemented in existing cities or only in new smart city developments? SCRE can be implemented in both existing cities and new smart city developments. Existing cities can incorporate renewable energy sources into their infrastructure through retrofitting and upgrading existing systems.

Are there any challenges to implementing SCRE? One of the main challenges of implementing SCRE is the initial investment required to build and install renewable energy systems. However, over time, these investments can result in long-term cost savings and benefits to the environment.

What is the future of SCRE? The future of SCRE is promising, as more and more cities are adopting renewable energy practices in an effort to become more sustainable and combat climate change. With advancements in technology and the increasing demand for clean energy, SCRE is likely to play a significant role in the future of smart cities.

Conclusion: Implementing SCRE technology can have a wide range of benefits for a city and its citizens. From reducing dependence on fossil fuels and promoting cost savings, to promoting economic development and improving the overall quality of life, smart city renewable energy can help to create a more sustainable and livable city for all.

SMART CITY RESILIENCE (SCR)

Using technology to enhance the city's ability to withstand and recover from natural disasters, such as early warning systems and emergency response coordination.

Implementing a SCR strategy can bring a number of benefits for the city and its citizens. By using technology to enhance the city's ability to withstand and recover from natural disasters, cities can reduce the impact of these events on their communities and infrastructure. One of the main benefits is the ability to provide early warning systems and emergency response coordination. This can help to evacuate citizens before a disaster strikes, reducing the risk of injury or loss of life. SCR strategies can help cities to quickly and effectively respond to disasters, reducing the amount of damage caused and minimizing disruptions to essential services.

Another benefit of SCR strategies is the ability to improve the city's infrastructure and building codes to better withstand natural disasters. For example, by using data and analytics to identify vulnerable areas, cities can prioritize investments in infrastructure upgrades and retrofits that will increase the resilience of the city. This not only protects citizens and their property, but also helps to minimize the economic impact of disasters on the city.

SCR strategies can help cities to better manage and conserve resources, such as energy and water. By using technology to monitor and control these resources, cities can ensure that they are being used in the most efficient and sustainable way possible. This not only helps to conserve resources but also reduces the costs associated with their use.

GENERAL FRAMEWORK

The following framework is a general one and the actual implementation will depend on the specific requirements and constraints of the city.

1. Identify key areas of the city that are at risk of natural disasters, such as floods, earthquakes, and storms.
2. Gather data on past natural disasters and analyze patterns to identify potential vulnerabilities.
3. Develop a comprehensive resilience plan that addresses the identified vulnerabilities.
4. Implement early warning systems, such as weather monitoring and seismic sensors, to provide advance notice of potential natural disasters.
5. Create a communication plan to disseminate emergency information to residents and visitors.
6. Coordinate with local emergency management agencies to ensure that response plans are in place and that resources are available.
7. Use technology such as IoT sensors, drones, and GIS mapping to monitor and assess the impact of natural disasters in real-time.
8. Implement building codes and retrofits to increase the resilience of existing infrastructure.

9. Develop green infrastructure, such as rain gardens and permeable pavement, to reduce the impact of floods and storms.
10. Use data analytics to identify patterns and predict future natural disasters.
11. Develop public-private partnerships to provide funding and resources for resilience initiatives.
12. Establish a community-based approach to resilience and involve residents and community organizations in the planning and implementation process.
13. Implement a system for monitoring and reporting on the effectiveness of resilience measures.
14. Develop a post-disaster recovery plan that outlines the steps that will be taken to restore critical services and infrastructure.
15. Implement a system for monitoring energy consumption and reducing energy waste to increase energy efficiency.
16. Develop a network of charging stations for electric vehicles to reduce dependence on fossil fuels.
17. Develop a plan for integrating renewable energy sources into the city's energy mix.
18. Develop a system for monitoring and reporting on the city's carbon footprint.
19. Implement a program to educate residents and visitors about the importance of sustainability and resilience.
20. Develop a system for monitoring and managing waste and recycling.
21. Implement a program to promote sustainable transportation, such as bike-sharing and electric vehicles.
22. Develop a plan to promote sustainable land use and reduce urban sprawl.
23. Develop a system for monitoring and managing water resources.
24. Implement a program to promote sustainable agriculture and local food production.
25. Develop a plan to promote sustainable tourism and recreation.

FREQUENTLY ASKED QUESTIONS

What is Smart City Resilience (SCRE) and why is it important?
Smart City Resilience (SCRE) is the ability of a city to prepare, adapt, and respond to disruptive events or shocks while maintaining essential functions and recovering quickly. It is important because

disruptions can come in many forms, including natural disasters, cyber-attacks, and pandemics, and cities need to be prepared to ensure the safety and well-being of their citizens.

How does SCRE differ from traditional emergency management? SCRE is a more comprehensive and proactive approach to emergency management that focuses on building long-term capacity to handle disruptions, rather than simply responding to them when they occur. It involves a range of measures, including risk assessment, planning, training, and investment in infrastructure and technology.

Who is responsible for implementing SCRE initiatives? Responsibility for implementing SCRE initiatives falls to city governments, which must work collaboratively with a range of stakeholders, including businesses, community organizations, and citizens. It is important that all parties are involved in the planning and implementation process to ensure a coordinated and effective response.

What role does technology play in SCRE? Technology plays a crucial role in SCRE by enabling the collection and analysis of data, which can inform risk assessment and help identify areas of vulnerability. It can also facilitate communication and response during disruptive events, as well as support the development of new infrastructure and systems.

How can citizens get involved in SCRE efforts? Citizens can get involved in SCRE efforts by participating in community planning and engagement initiatives, such as public forums and surveys. They can also take steps to prepare themselves and their families for potential disruptions, such as creating emergency kits and developing family communication plans.

What are some examples of SCRE initiatives in action? Examples of SCRE initiatives in action include the use of predictive analytics to forecast weather events and inform emergency response planning, the deployment of smart sensors to monitor critical infrastructure, and the development of community-led programs to build local capacity and promote social cohesion.

What are the challenges associated with implementing SCRE initiatives? The challenges associated with implementing SCRE initiatives include the need for significant investment in infrastructure and technology, the complexity of coordinating diverse stakeholders, and the potential for resistance to change among citizens and other groups.

How can SCRE initiatives be funded? SCRE initiatives can be funded through a range of mechanisms, including public-private partnerships, grants and loans from federal and state governments, and revenue generated through taxes and fees.

What are some of the benefits of SCRE? Benefits of SCRE include increased safety and security for citizens, improved response times and reduced recovery periods following disruptions, and enhanced economic and social sustainability.

How can SCRE be integrated with other smart city initiatives? SCRE can be integrated with other smart city initiatives through a coordinated and holistic approach to planning and implementation. For example, technologies used for energy efficiency can also be used to monitor and manage critical infrastructure, while citizen engagement programs can help build social resilience and promote sustainability.

Conclusion: SCR strategies can help cities to better prepare for, respond to, and recover from natural disasters. By using technology to enhance the city's ability to withstand and recover from these events, cities can reduce the impact of natural disasters on their communities, minimize disruptions to essential services, and protect the city's infrastructure and resources.

SMART CITY RETAIL (SCRT)

Using technology to enhance the shopping experience and optimize inventory management, such as digital storefronts and RFID tags.

Implementing SCRT technology can bring a variety of benefits to both retailers and consumers. One of the main benefits is the ability to enhance the shopping experience through digital storefronts and other interactive technologies. For example, retailers can use digital kiosks, augmented reality displays, and mobile apps to provide customers with more information about products, personalized recommendations, and other interactive features. Retailers can use technology such as RFID tags to optimize inventory management, reducing the risk of stockouts and overstocking.

SCRT technology can also help retailers to better understand their customers' preferences and buying habits through data analytics. This data can be used to inform marketing strategies, improve product selection, and provide more personalized customer service. Retailers can use technology such as beacon technology and geofencing to send targeted offers and promotions to customers' mobile devices based on their location.

The use of SCRT technology can also bring benefits to consumers. One of the main benefits is the ability to access more information about products and retailers through digital storefronts and mobile apps. This can help consumers to make more informed purchasing decisions. Consumers can use technology such as mobile payment systems to make purchases quickly and easily, without the need to carry cash or credit cards.

GENERAL FRAMEWORK

The following framework is a general one and the actual implementation will depend on the specific requirements and constraints of the city.

1. Identify key areas for improvement in the city's retail sector.
2. Conduct research on available technology solutions for enhancing the shopping experience and optimizing inventory management.
3. Develop a plan for implementing technology solutions, such as digital storefronts and RFID tags.
4. Identify potential partners and vendors for the implementation of technology solutions.
5. Create a budget and secure funding for the implementation of technology solutions.
6. Develop and implement a training program for retail employees on the use of new technology.
7. Implement digital storefronts in key retail areas.
8. Roll out RFID tagging system for inventory management.
9. Implement data analytics to monitor and optimize the use of technology solutions.
10. Develop an evaluation process to measure the success of the technology implementation.
11. Implement a customer feedback system to gather input on the technology solutions.
12. Continuously monitor and update technology solutions to improve the shopping experience.
13. Invest in digital marketing strategies to promote the use of new technology.
14. Establish a communication plan to inform customers about the implementation of new technology.
15. Invest in cybersecurity measures to protect customer data and prevent data breaches.
16. Integrate technology solutions with existing retail management systems.
17. Collaborate with other cities to share best practices and lessons learned.
18. Develop a plan for maintaining and updating technology solutions.
19. Develop a plan for addressing any technical issues that may arise.
20. Identify potential barriers to the successful implementation of technology solutions and develop strategies to address them.
21. Develop a plan to measure the environmental impact of technology solutions.
22. Implement a sustainability plan to ensure that technology solutions are energy-efficient.
23. Develop a plan to support small businesses in adopting technology solutions.

24. Create a feedback loop to gather input from small businesses on the effectiveness of technology solutions.
25. Develop a plan to address any potential workforce displacement resulting from the implementation of technology solutions.

FREQUENTLY ASKED QUESTIONS

What is Smart City Retail (SCRT) and how does it work? Smart City Retail (SCRT) is an innovative approach to retail that uses technology to improve the shopping experience for customers. It integrates different digital tools such as mobile apps, online shopping, and social media to engage customers in new ways, enhance the in-store experience, and make shopping more convenient.

How can Smart City Retail (SCRT) benefit retailers? Smart City Retail (SCRT) can help retailers improve their operational efficiency, increase sales, and build brand loyalty. It allows retailers to better understand their customers' needs and preferences, personalize the shopping experience, and provide better customer service.

How does Smart City Retail (SCRT) improve the shopping experience for customers? Smart City Retail (SCRT) provides customers with more options and convenience, allowing them to shop when and where they want. It enables personalized shopping experiences, such as personalized product recommendations, which can help customers make informed decisions and find products that meet their needs.

What kind of technology is used in Smart City Retail (SCRT)? Smart City Retail (SCRT) uses a range of technologies, such as the Internet of Things (IoT), mobile apps, cloud computing, and artificial intelligence (AI). These technologies are used to provide real-time data and insights to retailers, as well as to enhance the customer experience.

How does Smart City Retail (SCRT) address privacy concerns? Smart City Retail (SCRT) takes privacy concerns seriously and adheres to strict data protection regulations. Retailers must obtain customers' consent before collecting any personal data, and they must ensure that data is stored securely and used only for legitimate purposes.

Can small businesses benefit from Smart City Retail (SCRT)? Yes, small businesses can benefit from Smart City Retail (SCRT) just as much as large businesses. SCRT provides small businesses with access to tools and technologies that can help them compete with larger retailers.

How does Smart City Retail (SCRT) impact local communities? Smart City Retail (SCRT) can have a positive impact on local communities by creating more job opportunities and stimulating economic growth. It also provides customers with a wider range of shopping options, which can help to support local businesses.

How can retailers get started with Smart City Retail (SCRT)? Retailers can get started with Smart City Retail (SCRT) by first understanding their customers' needs and preferences. They can then explore different technologies and tools that can help them meet these needs, such as mobile apps, social media, and online shopping platforms.

Is Smart City Retail (SCRT) cost-effective for retailers? Yes, Smart City Retail (SCRT) can be cost-effective for retailers. It can help to improve operational efficiency, reduce costs, and increase sales, leading to a positive return on investment.

How can Smart City Retail (SCRT) continue to evolve in the future? Smart City Retail (SCRT) is constantly evolving and adapting to new technologies and trends. In the future, we can expect to see more advanced AI and machine learning tools, as well as greater use of virtual and augmented reality in the retail space. Additionally, retailers will continue to focus on providing more personalized and convenient shopping experiences for their customers.

Conclusion: Implementing SCRT technology can help retailers to optimize their operations and provide a more engaging shopping experience for consumers. It can help retailers to better understand their *customers and offer them more personalized services, while it can also make the shopping experience more convenient for consumers.*

SMART CITY SECURITY (SCS)

Using technology to enhance public safety and security, such as surveillance cameras and facial recognition systems.

Implementing a SCS system can bring a number of benefits to the city and its residents. One of the main benefits is increased public safety and security. By using technology such as surveillance cameras and facial recognition systems, city officials can monitor and track potential criminal activity in real-time, allowing for faster response times and increased effectiveness in preventing and solving crimes.

Another benefit of a SCS system is the ability to detect and respond to potential threats more quickly. With the use of advanced technology, city officials can be alerted to potential security risks and respond to them in a timely and efficient manner. This can include everything from natural disasters and terrorist attacks to cyber threats and other forms of criminal activity.

In addition to public safety and security, a SCS system can also help to improve the overall quality of life for residents. For example, the use of surveillance cameras and facial recognition systems can help to deter crime and make residents feel safer in their own community. This can lead to a more positive and productive environment for everyone.

SCS also can help in reducing the burden on the local police force as well as reduce the cost of security by utilizing advanced technology to monitor the city 24/7, which can help to free up resources and allow police and other city officials to focus on more pressing issues.

GENERAL FRAMEWORK

The following framework is a general one and the actual implementation will depend on the specific requirements and constraints of the city.

1. Conduct a thorough analysis of current public safety and security measures in the city to identify areas for improvement.
2. Develop a comprehensive smart city security plan that outlines specific goals, objectives, and strategies.
3. Identify and evaluate different technologies and systems that can be used to enhance public safety and security.
4. Develop a budget and secure funding for the implementation of the smart city security plan.
5. Identify key stakeholders and partners, including government agencies, law enforcement, and private sector companies.
6. Develop a communication plan to keep residents and businesses informed about the smart city security plan and its implementation.
7. Implement surveillance cameras in key areas throughout the city, such as public parks, transportation hubs, and high-crime areas.
8. Implement facial recognition systems for security and identification purposes.
9. Develop a system for monitoring and analyzing data from surveillance cameras and other security systems in real-time.
10. Develop a system for emergency response coordination and communication, including a dedicated emergency response team and emergency notification systems.
11. Establish partnerships with local law enforcement agencies to improve the sharing of information and coordination of efforts.
12. Train law enforcement and other public safety personnel on the use of new technologies and systems.
13. Establish a system for public reporting of security concerns and incidents.
14. Develop a system for incident response management and investigation.
15. Conduct regular audits and evaluations of the smart city security plan to identify areas for improvement.
16. Regularly update and maintain all security systems and technologies.
17. Develop a system for data privacy protection and data security.
18. Develop a continuity plan for maintaining security during a crisis or emergency.

19. Develop a system for incident reporting and analysis for data-driven decision making.
20. Develop a system for monitoring the movement of people and vehicles in the city using license plate recognition and GPS tracking technology.
21. Develop an alert system for suspicious activities and suspicious individuals.
22. Develop a system for monitoring the city's perimeter using smart cameras and sensors.
23. Develop a system for managing access to restricted areas and buildings using smart card or biometric technologies.
24. Develop a system for incident management and crisis communication.
25. Implement a comprehensive public education campaign to raise awareness about the smart city security plan and its benefits.

FREQUENTLY ASKED QUESTIONS

What is Smart City Security? Smart City Security (SCS) refers to the use of advanced technologies such as video surveillance, sensors, and data analysis to enhance the safety and security of a city's residents and infrastructure.

How does SCS work? SCS works by using a network of connected devices, sensors, and cameras to collect data on a city's environment. This data is analysed in real-time to detect potential security threats, and appropriate action can be taken to prevent incidents or respond quickly to them.

What are some of the benefits of SCS? SCS can provide several benefits, such as early detection and prevention of crime, faster response times to incidents, reduced costs of security operations, and enhanced public safety and well-being.

What types of technology are used in SCS? Some of the technologies used in SCS include surveillance cameras, facial recognition software, license plate recognition, gunshot detection systems, and predictive analytics.

Is SCS invasive to people's privacy? The use of SCS can raise privacy concerns, as the technology can be used to monitor the movements and activities of individuals. However, proper guidelines and regulations can be put in place to balance security needs with individual privacy rights.

Can SCS be used to prevent terrorist attacks? Yes, SCS can be an effective tool in preventing terrorist attacks by identifying and tracking suspicious activities and individuals.

Is SCS expensive to implement? The cost of implementing SCS can vary depending on the size and complexity of the city's infrastructure. However, the benefits of enhanced security and reduced crime can outweigh the costs in the long run.

Who is responsible for managing SCS? The management of SCS can be the responsibility of various stakeholders, including the city government, law enforcement agencies, and private security firms.

What are some examples of successful SCS implementations? Some examples of successful SCS implementations include the use of CCTV cameras in London, the ShotSpotter gunshot detection system in Chicago, and the use of predictive analytics to reduce crime in Los Angeles.

How can citizens get involved in SCS? Citizens can get involved in SCS by reporting suspicious activities to law enforcement, volunteering for neighbourhood watch programs, and participating in community events aimed at promoting public safety.

Conclusion: Implementing a SCS system can bring a wide range of benefits to the city and its residents, including increased public safety and security, improved response times to potential threats, and a more positive and productive environment for everyone.

SMART CITY SERVICES (SCSR)

Using technology to improve the delivery of city services, such as online permit applications and service request portals.

The implementation of SCSR has the potential to bring about significant benefits for both citizens and city governments. One of the main advantages of using technology to improve the delivery of city services is that it can make the process more efficient and convenient for citizens. For example, by implementing online permit applications and service request portals, citizens can easily access and submit the necessary forms and documents from the comfort of their own homes. This eliminates the need to physically visit city hall or other government offices, which can save citizens time and money.

Another key benefit of SCSR is that it can help city governments to better manage and track their services. By implementing technology such as online portals and digital records, city governments can easily access and analyze data on service requests, permit applications, and other transactions. This can provide valuable insights into the needs and preferences of citizens, which can help city governments to improve the delivery of their services. Digital records can help to prevent errors and reduce the risk of fraud, which can further improve the quality and integrity of city services.

SCSR also has the potential to increase transparency and accountability in the delivery of city services. By making information on service requests, permit applications, and other transactions available online, citizens can easily access and review the status of their requests and see how city services are being provided. This can help to increase public trust in city governments and improve communication between citizens and city officials.

GENERAL FRAMEWORK

The following framework is a general one and the actual implementation will depend on the specific requirements and constraints of the city.

1. Identify and prioritize the city services that would benefit most from being made available online.
2. Develop and implement an online service request portal for residents and businesses to easily access city services.
3. Develop and implement an online permit application system for building and development projects.
4. Establish a secure and reliable system for online payments for city services and permits.
5. Develop a comprehensive database of city services and permit information to be made available online.
6. Implement a system for tracking and monitoring service requests and permit applications.
7. Develop a system for communicating updates and status of service requests and permit applications to residents and businesses.
8. Establish a system for collecting and analyzing data on the usage and effectiveness of the online service and permit systems.
9. Develop and implement a system for authenticating users accessing the online service and permit systems.
10. Develop and implement a system for managing and updating the content on the online service and permit systems.
11. Train city staff on the use and maintenance of the online service and permit systems.
12. Develop a plan for promoting the online service and permit systems to residents and businesses.
13. Develop a system for providing technical support for residents and businesses using the online service and permit systems.
14. Establish a system for monitoring and addressing any issues or errors with the online service and permit systems.
15. Develop and implement a system for ensuring compliance with any relevant laws and regulations regarding online city services and permits.
16. Develop a system for integrating the online service and permit systems with other city systems and databases.
17. Develop a system for collecting and analyzing feedback from residents and businesses on the online service and permit systems.

18. Develop a system for making the online service and permit systems accessible to people with disabilities.
19. Develop and implement a system for ensuring the security and confidentiality of personal information collected through the online service and permit systems.
20. Develop a system for managing and archiving any physical documents that are still required for city services and permits.
21. Develop a system for ensuring continuity of service in case of technical difficulties or system failures.
22. Develop a system for integrating the online service and permit systems with any relevant regional or national systems.
23. Develop a system for monitoring and addressing any potential fraud or misuse of the online service and permit systems.
24. Develop a plan for regularly reviewing and updating the online service and permit systems to ensure they remain effective and efficient.
25. Develop a system for measuring the overall impact of the online service and permit systems on the delivery of city services and the satisfaction of residents and businesses.

FREQUENTLY ASKED QUESTIONS

What are the services offered under the Smart City Services project? The Smart City Services (SCSR) project offers a wide range of services, including waste management, water management, energy management, traffic management, public safety, and healthcare.

How will the SCSR project benefit the residents of the city? The SCSR project aims to improve the quality of life of the residents by providing them with efficient and reliable services that are easily accessible and available 24/7.

Will the SCSR project be accessible through mobile apps? Yes, the SCSR project will be accessible through mobile apps that allow residents to access and avail the services provided by the project.

What are the key features of the SCSR project? The key features of the SCSR project include real-time monitoring and management, data analytics, automation, and citizen engagement.

How will the SCSR project help in reducing the city's carbon footprint? The SCSR project will help in reducing the city's carbon footprint by promoting energy-efficient practices and encouraging the use of renewable energy sources.

How will the SCSR project ensure the safety and security of the residents? The SCSR project will have a robust public safety system that includes surveillance, emergency response, and disaster management services.

Will the SCSR project be implemented in all areas of the city? Yes, the SCSR project aims to cover all areas of the city and provide services to all residents regardless of their location.

How will the SCSR project be funded? The SCSR project will be funded through a combination of public and private investment, grants, and loans.

How will the SCSR project ensure the privacy of the residents' data? The SCSR project will have a data privacy and security framework in place that will ensure the protection of the residents' data.

How can residents provide feedback on the services provided by the SCSR project? Residents can provide feedback on the services provided by the SCSR project through various channels, including mobile apps, social media, and dedicated feedback mechanisms.

Conclusion: The implementation of SCSR has the potential to bring about significant benefits for both citizens and city governments. By making city services more efficient and convenient, improving data management and transparency, and increasing accountability, SCSR can help to improve the overall quality and integrity of city services, and enhance citizen's trust in the government.

SMART CITY STANDARDIZATION (SCSZ)

Using technology to ensure interoperability and standardization of Smart City systems and services, such as common data models and protocols.

Implementing SCSZ using technology can bring numerous benefits to a city. By ensuring interoperability and standardization of smart city systems and services, cities can improve the efficiency and effectiveness of their operations. This can be achieved by using common data models and protocols, which allow different systems and services to communicate and work together seamlessly.

One of the key benefits of SCSZ is that it enables cities to make better use of their data. By standardizing data models and protocols, cities can ensure that data is collected and stored in a consistent and structured manner. This allows for more accurate and reliable analysis of the data, which can be used to inform decision-making and improve the delivery of city services. Standardization enables cities to share data between different systems and services, which can lead to the development of new and innovative solutions to city challenges.

Another benefit of SCSZ is that it can help to lower the costs of implementing and maintaining smart city systems and services. By using common data models and protocols, cities can reduce the need for costly and time-consuming custom integrations between different systems. This can also make it easier for cities to replace or upgrade systems and services in the future, as they will be able to use solutions that are already compatible with their existing infrastructure.

SCSZ also helps in providing better service to citizens by providing a seamless and integrated experience. The citizens can access the various city services through a single platform, which is easily accessible and user-friendly. This can help improve citizen satisfaction and engagement, as well as increase trust in government services.

GENERAL FRAMEWORK

The following framework is a general one and the actual implementation will depend on the specific requirements and constraints of the city.

1. Identify the current systems and services in use by the city and assess their compatibility with each other.
2. Research and select a set of industry-standard protocols and data models to be used as the foundation for interoperability.
3. Develop a plan for migrating existing systems and services to the chosen protocols and data models.
4. Create guidelines for future systems and services to be developed by the city to ensure they are compatible with the chosen protocols and data models.
5. Implement a monitoring and testing system to ensure ongoing compliance with the chosen protocols and data models.
6. Establish a process for addressing and resolving interoperability issues as they arise.
7. Create a database of all systems and services in use by the city and their corresponding protocols and data models.
8. Develop a training program for city staff to ensure they are familiar with the chosen protocols and data models.
9. Create a communication plan to inform stakeholders and the public about the standardization efforts.
10. Establish partnerships with other cities and organizations to share information and best practices related to standardization.
11. Develop a system for tracking and reporting on the progress of the standardization efforts.
12. Create a budget for the implementation of standardization efforts and allocate resources accordingly.
13. Establish a governance structure to oversee and manage the standardization efforts.
14. Identify and engage key stakeholders and partners to be involved in the standardization efforts.
15. Conduct a thorough risk assessment and develop a plan to mitigate identified risks.
16. Create a process for regularly reviewing and updating the chosen protocols and data models to ensure they remain current.
17. Develop a system for capturing and analyzing data on the use and effectiveness of the chosen protocols and data models.

18. Establish a process for addressing and resolving any issues related to data privacy and security.
19. Develop a plan for scaling the standardization efforts to other areas of the city as needed.
20. Create a system for measuring and reporting on the cost savings and efficiency gains resulting from standardization.
21. Establish a process for addressing and resolving any challenges or barriers encountered during the standardization process.
22. Develop a plan for maintaining and supporting the standardization efforts over time.
23. Create a system for sharing the city's standardization efforts and results with other cities and organizations.
24. Identify potential future standardization efforts and develop a plan for implementing them.
25. Establish a process for evaluating the overall effectiveness of the standardization efforts and making necessary adjustments.

FREQUENTLY ASKED QUESTIONS

What is Smart City Standardization and why is it important? Smart City Standardization (SCSZ) refers to the development and adoption of common technical and organizational standards for smart city technologies and services. It is important because it promotes interoperability, scalability, and cost-effectiveness in smart city deployments, and allows cities to share best practices and learn from each other.

Who is responsible for setting SCSZ? There is no single organization responsible for setting SCSZ. Rather, standards are developed by a range of organizations including international standards bodies, industry consortia, and city networks, among others.

How can SCSZ benefit citizens? SCSZ can benefit citizens by ensuring that services and technologies are accessible, reliable, and secure. By establishing common technical and organizational standards, citizens can expect a seamless and consistent user experience across different smart city services and applications.

What are some of the key SCSZ that are currently being developed or implemented? Some of the key SCSZ that are currently being developed or implemented include those related to data management and privacy, interoperability, security, and sustainability.

How do SCSZ ensure data privacy and security? SCSZ related to data management and privacy require that all data collected by smart city technologies is protected and only used for its intended purpose. Standards related to security ensure that smart city systems are designed to prevent and respond to cybersecurity threats.

How can cities ensure that they are following SCSZ? Cities can ensure that they are following SCSZ by participating in relevant standardization organizations, adhering to best practices, and requiring that vendors and partners comply with established standards.

How do SCSZ promote collaboration and knowledge sharing? SCSZ promote collaboration and knowledge sharing by establishing a common language and framework for smart city technologies and services. This allows cities to learn from each other and share best practices, ultimately leading to more effective and efficient smart city deployments.

Can SCSZ help cities save money? Yes, SCSZ can help cities save money by promoting interoperability and scalability, which reduces the need for custom development and allows cities to more easily adopt off-the-shelf solutions.

Is SCSZ legally binding? SCSZ is typically voluntary, meaning that cities and vendors are not required to comply with them. However, adherence to established standards is often viewed as a best practice and can help ensure that smart city technologies and services are accessible, reliable, and secure.

How can citizens get involved in the development of SCSZ? Citizens can get involved in the development of SCSZ by providing input and feedback to city officials and standardization organizations, and by participating in public forums and consultations related to smart city development.

Conclusion: *Implementing SCSZ using technology can bring many benefits to cities, including improved efficiency and effectiveness of operations, better use of data, lower costs, and improved citizen engagement and satisfaction. It is an essential step in building a truly smart city.*

SMART CITY STORMWATER MANAGEMENT (SCSWM)

Using technology to manage stormwater and reduce flooding, such as smart drainage systems and green infrastructure.

SCSWM is a technology-based approach to managing stormwater and reducing the risk of flooding in urban areas. By using smart drainage systems and green infrastructure, cities can improve their ability to handle heavy rainfall and protect their citizens and property from the negative effects of flooding.

One of the key benefits of SCSWM is that it can help to reduce the amount of pollutants that enter waterways. Traditional stormwater management systems, such as concrete drainage channels, can become clogged with debris and pollutants, which can contaminate local rivers and streams. Smart systems, on the other hand, are designed to filter out pollutants and prevent them from entering waterways.

SCSWM can also help to improve the overall health of urban ecosystems. Green infrastructure, such as rain gardens and green roofs, can help to reduce the urban heat island effect and improve air quality. These systems also provide habitats for wildlife and can help to improve the overall aesthetic of a city.

Another benefit of SCSWM is that it can be integrated with other smart city technologies, such as sensors and real-time monitoring systems. This allows cities to respond quickly to changing weather conditions and take proactive measures to protect citizens and property. Smart systems can be connected to weather forecasting systems, allowing cities to anticipate heavy rainfall and take action before flooding occurs.

GENERAL FRAMEWORK

The following framework is a general one and the actual implementation will depend on the specific requirements and constraints of the city.

1. Identify areas of the city that are at high risk of flooding and prioritize them for stormwater management efforts.
2. Conduct a comprehensive assessment of the current stormwater infrastructure, including drainage systems and green infrastructure.
3. Develop a plan for upgrading and expanding the existing infrastructure to better manage stormwater and reduce the risk of flooding.
4. Invest in smart technologies, such as sensors and real-time monitoring systems, to improve the efficiency and effectiveness of the stormwater management system.
5. Implement green infrastructure solutions, such as rain gardens and green roofs, to capture and retain stormwater on site.
6. Develop a system for early warning and emergency response in the event of heavy rainfall or flooding.
7. Implement a public education and outreach program to increase awareness of the importance of stormwater management and the actions citizens can take to reduce the risk of flooding.
8. Partner with other organizations and agencies, such as the National Weather Service, to access real-time weather data and improve forecasting and warning capabilities.
9. Develop a comprehensive data management system to collect and analyze data on weather patterns, stormwater flow, and infrastructure performance.
10. Use data analytics to identify patterns and trends in stormwater management, and use this information to inform future decision-making.
11. Develop a maintenance plan for the stormwater management infrastructure to ensure it is functioning properly and efficiently.

12. Establish a system for monitoring and reporting on the performance of the stormwater management system, including metrics such as volume of stormwater captured, flooding incidents, and maintenance needs.
13. Develop a funding strategy for the ongoing maintenance and improvement of the stormwater management system.
14. Continuously monitor and evaluate the performance of the stormwater management system and make adjustments as needed.
15. Develop a plan to integrate smart city stormwater management system with other smart city systems, such as transportation and energy management.
16. Develop a plan for sharing data and information with other stakeholders, such as neighbouring cities and utility companies.
17. Establish a governance structure to oversee the implementation and operation of the smart city stormwater management system.
18. Develop a plan for engaging with the community and soliciting input on the stormwater management system.
19. Develop a plan for training city staff and other stakeholders on the operation and maintenance of the stormwater management system.
20. Develop a plan for testing and piloting new technologies and approaches before implementing them on a larger scale.
21. Develop a plan for addressing any potential legal or regulatory challenges that may arise in the implementation of the stormwater management system.
22. Develop a plan for addressing any potential privacy and security concerns related to the collection and use of data in the stormwater management system.
23. Develop a plan for addressing any potential social equity concerns related to the implementation of the stormwater management system.
24. Develop a plan for addressing any potential environmental concerns related to the implementation of the stormwater management system.
25. Develop a plan for evaluating the long-term sustainability and scalability of the stormwater management system.

FREQUENTLY ASKED QUESTIONS

What is Smart City Stormwater Management and why is it important? Smart City Stormwater Management (SCSWM) is a system designed to manage the flow of stormwater through a city using sensors and data analysis. It's important because it can help prevent flooding and water pollution, which are both major concerns in urban areas.

How does SCSWM work? SCSWM uses sensors to monitor rainfall, water levels, and flow rates in stormwater systems. The data collected is then analysed to predict potential flooding and pollution events, and to optimize the operation of the stormwater system.

Who is responsible for implementing SCSWM? The responsibility for implementing SCSWM falls on the city government, specifically the department in charge of water management.

Can SCSWM be integrated with other Smart City systems? Yes, SCSWM can be integrated with other Smart City systems such as Smart Traffic Management and Smart Building Management to create a comprehensive and efficient urban management system.

How does SCSWM benefit the community? SCSWM benefits the community by reducing the risk of flooding, protecting water quality, and improving the overall health of the city's waterways.

Is SCSWM expensive to implement? The cost of implementing SCSWM will vary depending on the size of the city and the complexity of the system. However, the long-term benefits of the system can outweigh the initial costs.

How does SCSWM impact the environment? SCSWM helps to reduce the amount of pollutants that are discharged into waterways, which can help improve the overall health of the environment.

How does SCSWM impact traffic flow? SCSWM can impact traffic flow by redirecting traffic during flooding events or by helping to prevent flooding altogether.

Can SCSWM be customized for different cities? Yes, SCSWM can be customized for different cities based on their unique needs and characteristics.

How can I get involved in supporting the implementation of SCSWM in my city? You can get involved by contacting your local government representatives and expressing your support for the implementation of SCSWM in your community. You can also join local advocacy groups that are focused on improving urban water management.

Conclusion: *SCSWM is a technology-based approach to managing stormwater and reducing the risk of flooding in urban areas. By using smart drainage systems and green infrastructure, cities can improve their ability to handle heavy rainfall and protect their citizens and property from the negative effects of flooding. SCSWM can help to improve the overall health of urban ecosystems, reduce pollutants that enter waterways, and can be integrated with other smart city technologies, such as sensors and real-time monitoring systems.*

SMART CITY STREET FURNITURE (SCSF)

Using technology to make street furniture more functional and interactive, such as smart benches and trash cans.

Implementing SCSF can bring several benefits to a city. By integrating technology into street furniture, cities can make them more functional and interactive. For example, smart benches can provide charging stations for devices, and have built-in WiFi connectivity. Smart trash cans can alert maintenance crews when they need to be emptied and also have recycling options.

Smart street furniture can also improve the overall aesthetic of a city. Through the integration of digital displays, cities can showcase public art, provide information to citizens, and even create interactive experiences. This can help to improve the overall livability of a city and enhance the public's perception of their city.

Another benefit of SCSF is the ability to collect data and analyze it. For example, smart benches can collect data on usage patterns and popular locations. This data can then be used to optimize the placement of street furniture, making them more accessible to citizens. Cities can use data collected from smart trash cans to identify areas that generate more waste and take necessary steps to reduce it.

GENERAL FRAMEWORK

The following framework is a general one and the actual implementation will depend on the specific requirements and constraints of the city.

1. Identify areas where street furniture is needed, such as high-traffic areas, parks, and public plazas.
2. Research and evaluate different types of smart street furniture, including smart benches, trash cans, and lighting systems.
3. Determine the specific features and functionality needed for the chosen smart street furniture, such as charging ports, Wi-Fi connectivity, and environmental sensors.
4. Develop a plan for the installation and maintenance of the smart street furniture, including the necessary infrastructure and equipment.
5. Coordinate with local government and utility companies to ensure proper permits and permissions are obtained.
6. Identify potential vendors and suppliers of the smart street furniture and evaluate their capabilities and pricing.
7. Develop a budget and funding plan for the project, including potential grants and partnerships.
8. Prepare detailed specifications for the smart street furniture and issue a Request for Proposal (RFP) to potential vendors.

9. Review and evaluate responses to the RFP and select a vendor to provide the smart street furniture.
10. Coordinate with the vendor to finalize the design and functionality of the smart street furniture.
11. Schedule and plan for the installation of the smart street furniture, including any necessary street closures or detours.
12. Coordinate with any necessary parties, such as local utility companies, to ensure a smooth installation process.
13. Test and commission the smart street furniture to ensure it is functioning properly.
14. Develop a plan for ongoing maintenance and support of the smart street furniture.
15. Train city staff on how to properly maintain and troubleshoot the smart street furniture.
16. Develop a plan for the collection and analysis of data from the smart street furniture, such as usage statistics and environmental data.
17. Integrate the smart street furniture with any necessary city systems, such as lighting and waste management systems.
18. Develop a plan for the use of the data collected from the smart street furniture, such as identifying patterns and trends in usage.
19. Develop a user interface for the smart street furniture, such as a mobile app or website, to allow users to access information and interact with the furniture.
20. Promote the smart street furniture to the public and educate them on its features and benefits.
21. Monitor the usage and effectiveness of the smart street furniture and make any necessary adjustments.
22. Continuously evaluate and improve the smart street furniture based on feedback and usage data.
23. Work with city partners to explore new and innovative ways to use smart street furniture to improve the city.
24. Develop a plan to ensure the security and privacy of data collected from the smart street furniture.
25. Continuously evaluate the smart street furniture to ensure they are meeting the goals of the project.

FREQUENTLY ASKED QUESTIONS

What is Smart City Street Furniture and how does it work? Smart City Street Furniture (SCSF) refers to any urban outdoor furniture equipped with smart technology, such as benches with built-in charging stations, bus shelters with digital displays, and waste bins with sensors to monitor waste levels. The technology is designed to improve the functionality and sustainability of public spaces, while also providing data and connectivity for smart city management.

How does SCSF help improve public spaces? SCSF provides numerous benefits, such as improved accessibility, increased safety, and enhanced convenience for citizens. For example, benches with charging stations provide a place for people to relax and recharge their devices, while digital displays at bus shelters provide real-time information about bus routes and schedules.

What are some examples of SCSF? SCSF can include a wide range of items, such as benches, bus shelters, waste bins, bike racks, and even planters. Some examples of SCSF include solar-powered benches that charge devices, waste bins with sensors that alert city workers when they need to be emptied, and bike racks that offer digital locks and charging stations.

How is data collected and used from SCSF? SCSF uses sensors and other technology to collect data on usage patterns, environmental conditions, and other factors. This data is then analysed by city managers and used to make informed decisions about urban planning and development.

Are there any privacy concerns with SCSF? Privacy concerns are always a consideration when it comes to smart city technology, and SCSF is no exception. However, most SCSF is designed to collect anonymous data, such as usage patterns and environmental conditions, rather than personal information about individuals.

Who is responsible for maintaining SCSF? In most cases, the city or municipality is responsible for maintaining SCSF, just as they are responsible for maintaining other public infrastructure. However, some SCSF may be privately owned and maintained, depending on the specific arrangement.

How are SCSF projects funded? SCSF projects can be funded in a variety of ways, such as through public-private partnerships, government grants, and corporate sponsorships. The specific funding source depends on the project and the city in question.

What are some of the challenges associated with implementing SCSF? Implementing SCSF can be challenging due to factors such as infrastructure limitations, funding constraints, and public acceptance. It can also be difficult to strike a balance between incorporating the latest technology and maintaining the aesthetic appeal of public spaces.

How does SCSF contribute to sustainability? SCSF can help promote sustainability in a number of ways, such as through the use of solar power, recycled materials, and energy-efficient technology. By reducing waste and improving energy efficiency, SCSF can help cities achieve their sustainability goals.

How is public feedback incorporated into the design of SCSF? Public feedback is an important consideration in the design of SCSF, and many designers and city planners seek input from citizens during the design process. This feedback can help ensure that SCSF is functional, aesthetically pleasing, and meets the needs of the community.

Conclusion: *Implementing SCSF can help to improve the overall livability and functionality of a city, while also providing valuable data for future city planning and development.*

SMART CITY STREETS (SCST)

Using technology to enhance the performance and safety of streets, such as smart traffic signals and road sensors.

Implementing SCST technology to enhance the performance and safety of streets can bring a wide range of benefits. One of the main benefits is the ability to optimize traffic flow through the use of smart traffic signals. These signals can adjust in real-time based on traffic conditions, reducing congestion and improving overall traffic flow. Road sensors can be used to detect and respond to hazards or incidents, such as accidents or road closures. This can greatly improve emergency response times and overall safety on the streets.

Another key benefit of implementing SCST technology on streets is the ability to gather and analyze data on traffic patterns. This data can be used to identify areas of congestion and inform infrastructure improvements to increase capacity and improve overall mobility. This data can be used to inform transportation planning and policy decisions, such as the development of new bike lanes or the implementation of carpooling programs.

SCST technology can also be used to enhance the functionality of street furniture, such as smart benches and trash cans. These can include features such as built-in charging stations for electronic devices, Wi-Fi connectivity, and real-time information on bus and train schedules.

GENERAL FRAMEWORK

The following framework is a general one and the actual implementation will depend on the specific requirements and constraints of the city.

1. Conduct a thorough assessment of current street infrastructure and identify areas for improvement.

2. Develop a comprehensive plan for implementing smart street technology, including specific goals and objectives.
3. Identify and evaluate potential technology solutions, such as smart traffic signals, road sensors, and connected vehicle systems.
4. Develop a budget and secure funding for the implementation of smart street technology.
5. Establish partnerships and collaborations with relevant stakeholders, including city departments, technology vendors, and community organizations.
6. Develop a communication plan to inform and engage the public about the implementation of smart street technology.
7. Implement smart traffic signals and road sensors on selected streets.
8. Evaluate the performance and effectiveness of the technology on an ongoing basis.
9. Continuously monitor and collect data from the smart street technology.
10. Use the data to optimize traffic flow and improve safety.
11. Implement a system for maintaining and upgrading the smart street technology.
12. Develop plans for expanding the use of smart street technology to other streets in the city.
13. Use smart street technology to improve accessibility for all users, including pedestrians, bicyclists, and people with disabilities.
14. Use smart street technology to enhance the livability and sustainability of streetscapes.
15. Use smart street technology to support city-wide transportation planning and management.
16. Use smart street technology to support emergency management and incident response.
17. Coordinate with other smart city initiatives, such as smart parking and public transportation.
18. Develop a cybersecurity plan to protect the data and systems associated with smart street technology.
19. Develop a plan for data governance and management, including privacy and data sharing.
20. Establish a governance structure to oversee the planning, implementation, and operation of smart street technology.
21. Develop a plan for integrating smart street technology with other smart city systems and services.
22. Develop a plan for standardizing data and interfaces across different smart street technology systems.

23. Develop a plan for ensuring the scalability and future-proofing of smart street technology.
24. Develop a plan for ensuring the accessibility and inclusivity of smart street technology.
25. Continuously review and update the plan based on feedback, evaluation, and new developments.

FREQUENTLY ASKED QUESTIONS

What is the purpose of Smart City Streets (SCST) initiative? The main purpose of the Smart City Streets (SCST) initiative is to improve the safety, accessibility, and sustainability of urban streets by using innovative technologies and data-driven solutions.

How will the SCST initiative benefit the general public? The SCST initiative will benefit the general public by providing safer and more accessible streets, reducing traffic congestion, improving air quality, and promoting sustainable transportation modes.

What types of technologies will be used in the SCST initiative? The SCST initiative will use a variety of technologies such as smart traffic signals, sensors, cameras, and mobile applications to collect and analyze data, monitor traffic patterns, and optimize transportation services.

How will the SCST initiative impact the environment? The SCST initiative is expected to have a positive impact on the environment by reducing vehicle emissions, promoting sustainable transportation modes, and improving overall air quality.

What are some of the challenges of implementing the SCST initiative? Some of the challenges of implementing the SCST initiative include high costs, lack of interoperability between different technologies, and concerns over data privacy and security.

How will the SCST initiative impact emergency services? The SCST initiative is expected to have a positive impact on emergency services by improving response times and reducing congestion in high-traffic areas.

What is the role of the public in the SCST initiative? The public has an important role in the SCST initiative as they are the main beneficiaries of the improved street infrastructure. Public input and feedback are also important for the successful implementation of the initiative.

How will the SCST initiative impact the transportation industry? The SCST initiative is expected to have a significant impact on the transportation industry by promoting the use of new technologies, creating new business models, and improving the overall efficiency of transportation services.

Will the SCST initiative lead to job creation? Yes, the SCST initiative is expected to create new jobs in areas such as data analysis, technology development, and street maintenance.

What is the timeline for the implementation of the SCST initiative? The timeline for the implementation of the SCST initiative varies depending on the specific project and location. However, many cities are already implementing different components of the initiative and have long-term plans for its full implementation.

Conclusion: *Implementing SCST technology on streets can greatly improve the performance and safety of streets, enhance the functionality of street furniture, and make transportation more efficient and sustainable.*

SMART CITY SUSTAINABILITY (SCSU)

Using technology to promote sustainable development and reduce the city's environmental footprint, such as carbon footprint tracking and sustainable procurement.

Implementing SCSU initiatives can bring a number of benefits to a city and its residents. One of the key advantages is the ability to track and reduce the city's carbon footprint. By using technology such as sensors, data analytics, and IoT devices, cities can monitor and measure their energy consumption, waste management, and transportation patterns. This data can then be used to identify areas where improvements can be made to reduce the city's environmental impact.

Another benefit of SCSU is the ability to promote sustainable development and reduce dependence on fossil fuels. By using technology such as renewable energy systems, energy storage, and smart grid systems, cities can increase their use of clean, renewable energy sources such as solar and wind power. This can help to reduce greenhouse gas emissions, improve air quality, and promote long-term energy independence.

SCSU initiatives can also improve the quality of life for residents. By promoting sustainable transportation options, such as bike-sharing and electric vehicles, cities can reduce traffic congestion, improve air quality and reduce dependence on fossil fuels. By implementing green infrastructure such as parks, gardens and rainwater harvesting systems, cities can improve residents' access to green spaces and promote healthier living.

SCSU initiatives can also help cities to save money. By using technology such as building automation systems, cities can reduce their energy consumption and costs. By using sustainable procurement strategies, cities can reduce their costs associated with the purchase of goods and services.

GENERAL FRAMEWORK

The following framework is a general one and the actual implementation will depend on the specific requirements and constraints of the city.

1. Identify specific sustainability goals for the city, such as reducing carbon emissions or increasing energy efficiency.
2. Conduct a comprehensive inventory of the city's current energy use, emissions, and waste streams.
3. Develop a plan for reducing the city's environmental footprint, including specific targets and timelines.
4. Implement energy-efficient technologies and practices in city buildings and operations.
5. Promote the use of renewable energy sources, such as solar and wind power.
6. Implement sustainable transportation options, such as bike-sharing programs and electric vehicle charging stations.
7. Encourage green building practices, such as LEED certification and energy-efficient design.
8. Develop a recycling and waste management program to reduce the amount of waste sent to landfills.
9. Implement a green procurement policy to encourage the purchase of environmentally-friendly products and services.
10. Develop a monitoring and reporting system to track the city's progress towards sustainability goals.
11. Implement a city-wide carbon footprint tracking system.
12. Develop a public education and outreach campaign to increase awareness and participation in sustainability efforts.
13. Develop a robust public transportation system to reduce car dependence.
14. Develop a system to track and manage the city's water resources.
15. Implement green spaces, parks, and recreation areas to increase community engagement
16. Develop a system for tracking and managing air quality.
17. Implement a green finance system to fund sustainability projects.
18. Encourage the use of electric vehicles through incentives and regulations.
19. Develop a system for tracking and managing the city's biodiversity.
20. Develop a system for tracking and managing the city's sustainable food systems.
21. Implement a system for tracking and managing the city's sustainable waste management systems.
22. Develop a system for tracking and managing the city's sustainable transportation systems.
23. Develop a system for tracking and managing the city's sustainable energy systems.

24. Develop a system for tracking and managing the city's sustainable land use systems.
25. Develop a system for tracking and managing the city's sustainable natural resources systems.

FREQUENTLY ASKED QUESTIONS

What is Smart City Sustainability? Smart City Sustainability (SCSU) refers to the use of technology and data to promote sustainable development in urban areas. This includes initiatives aimed at reducing energy consumption, waste, and greenhouse gas emissions, as well as improving air and water quality.

What are the benefits of SCSU? SCSU can help reduce the environmental impact of cities while also promoting economic growth and improving quality of life for residents. It can also help cities save money on energy and other costs.

What kinds of technologies are used in SCSU? Technologies used in SCSU can include smart lighting, green infrastructure, energy management systems, and intelligent transportation systems, among others.

How can individuals get involved in SCSU initiatives? Individuals can get involved in SCSU by supporting local initiatives, reducing their own energy consumption and waste, and advocating for sustainable policies and infrastructure.

How does SCSU impact climate change? SCSU initiatives can help reduce greenhouse gas emissions, which contribute to climate change. By reducing energy consumption, promoting green infrastructure, and encouraging sustainable transportation, cities can play a significant role in addressing climate change.

Are there any risks associated with SCSU? While there are some potential risks associated with the use of technology in cities, such as privacy concerns, these risks can be mitigated through proper data management and security protocols.

How can SCSU help address environmental justice issues? By promoting sustainable development in all areas of the city, including low-income and marginalized communities, SCSU initiatives can help address environmental justice issues and ensure that everyone benefits from a healthier and more sustainable environment.

What role does data play in SCSU? Data is a key component of SCSU, as it can be used to monitor and optimize energy use, traffic flow, and other factors that impact sustainability. However, it is important to ensure that data is managed responsibly and securely.

What kinds of policy changes are needed to support SCSU? Policy changes that support SCSU can include incentives for sustainable infrastructure and practices, zoning regulations that promote mixed-use development and public transportation, and funding for research and development of sustainable technologies.

How can SCSU help prepare cities for future challenges? SCSU initiatives can help cities become more resilient to future challenges, such as climate change, by promoting sustainable practices that can adapt to changing conditions. It can also help cities become more efficient and cost-effective in the long run.

Conclusion: *Implementing SCSU initiatives can help cities to promote sustainable development, improve the quality of life for residents, and save money in the long run.*

SMART CITY SUSTAINABLE ENERGY (SCSE)

Implementing renewable energy sources such as solar and wind power to decrease dependence on fossil fuels and reduce carbon emissions.

Implementing Smart City Sustainable Energy (SCSE), can bring a multitude of benefits to a city. One of the most significant is the ability to reduce dependence on fossil fuels and decrease carbon emissions. This can have a significant impact on the environment, helping to combat climate change and improve air quality.

Another benefit of SCSE is the potential for cost savings. Renewable energy sources such as solar and wind power can often be less expensive than traditional fossil fuels in the long run, especially as technology and infrastructure improve. This can lead to cost savings for both the city and its residents.

Implementing SCSE can also help to promote energy independence and resilience. By generating energy locally, cities can become less reliant on external sources and be better prepared to handle disruptions in energy supply.

Another benefit of SCSE is that it can also lead to job creation and economic development. The installation and maintenance of renewable energy systems can create jobs, and the use of local renewable energy sources can also attract businesses and investors.

SCSE can also help to improve the quality of life for residents. Clean energy can improve air and water quality, and also decrease noise pollution.

GENERAL FRAMEWORK

The following framework is a general one and the actual implementation will depend on the specific requirements and constraints of the city.

1. Conduct a comprehensive energy audit to identify areas of energy consumption and areas where renewable energy sources can be implemented.
2. Develop a plan for implementing renewable energy sources, including the types of energy to be used and the timeline for implementation.

3. Identify the best locations for renewable energy systems, such as solar panels or wind turbines.
4. Obtain necessary permits and approvals for the installation of renewable energy systems.
5. Develop a financing plan for the implementation of renewable energy systems.
6. Select a contractor to design and install the renewable energy systems.
7. Develop a monitoring and maintenance plan for the renewable energy systems.
8. Coordinate with utility companies to ensure the integration of renewable energy into the grid.
9. Develop educational programs and outreach efforts to educate the community about the benefits of renewable energy.
10. Implement energy-efficient measures in buildings and infrastructure to reduce overall energy consumption.
11. Invest in research and development of new renewable energy technologies.
12. Develop policies and incentives to encourage the use of renewable energy by businesses and residents.
13. Collaborate with other cities to share best practices and resources for implementing renewable energy.
14. Develop a plan for the decommissioning and disposal of renewable energy systems at the end of their lifespan.
15. Implement a system for tracking and reporting on the city's renewable energy usage and carbon emissions.
16. Collaborate with local universities and research institutions to promote renewable energy education and research.
17. Develop a plan for emergency management and backup power in case of power outages.
18. Develop a plan for the integration of electric vehicle charging infrastructure.
19. Identify and mitigate potential negative impacts of renewable energy systems on the environment and local communities.
20. Develop a plan for the integration of energy storage systems to maximize the use of renewable energy.
21. Develop a plan for the integration of demand response programs to optimize energy usage during peak times.
22. Develop a plan for the integration of microgrids to improve energy resilience and reduce dependence on the traditional grid.

23. Collaborate with local and regional organizations to advocate for renewable energy policies at the state and national level.
24. Develop a plan for the integration of building energy management systems to optimize energy usage in buildings.
25. Develop a plan for the integration of energy-efficient lighting and appliances in public spaces.

FREQUENTLY ASKED QUESTIONS

What is Smart City Sustainable Energy and how does it work? Smart City Sustainable Energy (SCSE) is a project that focuses on the use of renewable energy to power cities. This is achieved by using various technologies such as solar panels, wind turbines, and geothermal energy. The energy generated is then stored in batteries for later use.

How can SCSE help to reduce carbon emissions in cities? SCSE helps to reduce carbon emissions by promoting the use of renewable energy. This reduces the need for fossil fuels which are major contributors to greenhouse gas emissions.

How does SCSE ensure that energy is distributed evenly across a city? SCSE ensures that energy is distributed evenly across a city by using a smart grid system. This system allows for the efficient distribution of energy and ensures that areas that require more energy receive it.

What are the benefits of using renewable energy in a Smart City? The benefits of using renewable energy in a Smart City include reduced carbon emissions, improved air quality, lower energy costs, increased energy security, and improved quality of life for residents.

How can individuals get involved in the SCSE project? Individuals can get involved in the SCSE project by supporting the use of renewable energy in their homes and businesses. This includes installing solar panels or using wind turbines to generate energy.

What challenges does SCSE face in implementing renewable energy in a city? The challenges that SCSE faces in implementing renewable energy in a city include high initial costs, lack of infrastructure, and resistance from traditional energy providers.

How can the public be assured of the safety of renewable energy sources? The public can be assured of the safety of renewable energy sources through rigorous testing and certification by regulatory bodies. Additionally, renewable energy sources are generally safer than traditional energy sources.

How can SCSE promote energy efficiency in buildings and homes? SCSE can promote energy efficiency in buildings and homes by encouraging the use of energy-efficient appliances and promoting the use of smart home systems. This reduces the amount of energy needed to power buildings and homes.

How does SCSE ensure the reliability of energy supply in a city? SCSE ensures the reliability of energy supply in a city by using a combination of renewable energy sources and traditional energy sources. This ensures that energy is available even when renewable sources are not producing energy.

How does SCSE ensure that renewable energy sources do not impact the environment negatively? SCSE ensures that renewable energy sources do not impact the environment negatively by carefully selecting sites for renewable energy projects and using technologies that minimize the impact on the environment. Additionally, renewable energy sources produce much less pollution than traditional energy sources.

Conclusion: *Implementing SCSE can bring many benefits to a city, including reducing dependence on fossil fuels, decreasing carbon emissions, saving costs, promoting energy independence and resilience, creating jobs and economic development, and improving the quality of life for residents.*

SMART CITY SUSTAINABLE TRANSPORTATION (SCSTR)

Using technology to promote sustainable transportation options, such as bike-sharing, electric vehicles and carpooling.

Implementing SCSTR technologies can bring a wide range of benefits to a city and its citizens. One of the main benefits is the reduction of carbon emissions and air pollution. By promoting sustainable transportation options such as bike-sharing, electric vehicles and carpooling, cities can decrease their dependence on fossil fuels and decrease the amount of emissions from transportation. This can lead to improved air quality and public health.

Promoting sustainable transportation options can also reduce traffic congestion and improve mobility within the city. This can lead to more efficient use of public transportation and reduced travel times for citizens.

Another benefit of implementing SCSTR is the potential for cost savings for both individuals and the city as a whole. For example, bike-sharing systems and electric vehicles can provide a low-cost alternative to traditional forms of transportation such as cars. This can lead to reduced transportation costs for citizens and decreased demand for expensive infrastructure such as roads and parking facilities. Sustainable transportation options can also increase the economic development of a city by attracting new businesses and tourists.

Promoting sustainable transportation options can also help to create more livable and equitable cities. By providing citizens with more transportation options, cities can help to reduce social and economic disparities and improve access to jobs, services and amenities.

GENERAL FRAMEWORK

The following framework is a general one and the actual implementation will depend on the specific requirements and constraints of the city.

1. Conduct a thorough needs assessment to determine the transportation challenges in the city.
2. Identify the most promising sustainable transportation options (e.g. bike-sharing, electric vehicles, carpooling) for the city.
3. Develop a comprehensive plan to implement the selected transportation options, including specific objectives, timelines, and budgets.
4. Establish partnerships with relevant stakeholders, including local government, businesses, and community organizations.
5. Design and implement an information and education campaign to raise awareness about the benefits of sustainable transportation options.
6. Invest in technology infrastructure to support the implementation of sustainable transportation options, such as a bike-sharing system, electric vehicle charging stations, and carpooling platforms.
7. Encourage active participation of the community, by engaging them in decision-making and feedback processes.
8. Provide incentives and subsidies to encourage the use of sustainable transportation options, such as discounted or free access to bike-sharing and carpooling services.
9. Develop policies and regulations that promote the use of sustainable transportation options and discourage the use of single-occupancy vehicles.
10. Establish a comprehensive monitoring and evaluation system to track the progress and impact of the sustainable transportation program.
11. Provide ongoing support and technical assistance to ensure the effective and efficient implementation of sustainable transportation options.
12. Implement traffic management strategies to prioritize sustainable transportation options, such as dedicated bike lanes and traffic-calming measures.
13. Develop a plan for the maintenance and expansion of sustainable transportation infrastructure, including bike-sharing stations, electric vehicle charging stations, and carpooling platforms.
14. Encourage the use of electric vehicles by offering tax credits, subsidies, and other incentives.
15. Promote carpooling by offering financial incentives, dedicated carpool lanes, and parking incentives for car-poolers.
16. Develop a plan for the integration of sustainable transportation options with the existing public transportation system, such as bike-

sharing and electric vehicle charging stations at public transportation hubs.
17. Encourage the private sector to adopt sustainable transportation options by offering incentives, subsidies, and tax credits.
18. Implement a robust data management system to track and analyze transportation trends and patterns.
19. Collaborate with academic institutions to research and develop new sustainable transportation technologies and solutions.
20. Develop a crisis management plan to address unexpected challenges and disruptions in sustainable transportation services.
21. Ensure that sustainable transportation options are accessible to all segments of the population, including low-income communities and individuals with disabilities.
22. Provide training and support to transportation providers, such as bike-share companies and carpooling services, to ensure the highest standards of safety and service.
23. Foster a culture of sustainability and environmental responsibility through community engagement and education initiatives.
24. Regularly review and evaluate the sustainable transportation program to identify areas for improvement and make necessary adjustments.
25. Continuously engage with stakeholders to ensure that the sustainable transportation program remains relevant and responsive to changing needs and conditions.

FREQUENTLY ASKED QUESTIONS

What is Smart City Sustainable Transportation (SCSTR)? SCSTR refers to the use of advanced technology to make transportation more efficient and environmentally friendly, with a focus on reducing carbon emissions and promoting sustainable mobility.

How does SCSTR benefit the general public? SCSTR can improve air quality, reduce traffic congestion, and provide residents with more accessible, affordable and reliable transportation options, such as electric buses and bike-sharing programs.

What is the role of the government in promoting SCSTR? The government can invest in infrastructure for sustainable transportation, such as bike lanes and charging stations for electric vehicles, and create policies that encourage the use of public transportation.

What role do private companies play in SCSTR? Private companies can develop and provide sustainable transportation technologies, such as electric or hybrid vehicles, and offer innovative services such as car-sharing or ride-hailing apps.

How can individuals contribute to SCSTR? Individuals can choose sustainable transportation options, such as walking, biking or taking public transportation, and can also advocate for more sustainable transportation policies and infrastructure.

What are some of the challenges facing the implementation of SCSTR? Challenges include high costs, lack of infrastructure, limited public awareness and resistance to change from traditional transportation systems.

What are some examples of SCSTR initiatives? Examples include electric vehicle charging stations, bike-sharing programs, dedicated bus lanes, and pedestrian zones in urban centers.

What are the benefits of electric vehicles in SCSTR? Electric vehicles are a sustainable transportation option that reduces carbon emissions and lowers the cost of transportation for individuals and the government.

How can SCSTR help reduce traffic congestion? By providing alternative transportation options, such as public transportation, bike-sharing or carpooling, SCSTR can reduce the number of cars on the road and alleviate traffic congestion.

How can SCSTR help promote social equity? SCSTR can provide more affordable and accessible transportation options for underserved communities, such as low-income households or people with disabilities, improving mobility and reducing social inequalities.

Conclusion: *Implementing SCSTR technologies can bring significant benefits to a city and its citizens, including improved air quality and public health, reduced traffic congestion and travel times, cost savings, economic development, and more livable and equitable cities.*

SMART CITY TOURISM (SCTO)

Using technology to enhance the tourism experience and promote sustainable tourism practices, such as interactive city guides and digital concierge services.

Implementing SCTO can bring a number of benefits to a city. By using technology to enhance the tourism experience, cities can attract more visitors and increase revenue from tourism. Interactive city guides, for example, can provide tourists with information about the city's history, landmarks, and events. Digital concierge services can help tourists plan their visit and make reservations for hotels, restaurants, and activities.

Sustainable tourism practices can also be promoted through technology. For example, a city could use an app to encourage tourists to use public transportation or walk or bike instead of driving. This can help reduce traffic congestion and air pollution. The app could also provide information about sustainable tourism activities and accommodations.

Another benefit of SCTO is that it can help create a more inclusive and accessible city for tourists. By providing information in multiple languages and offering accessibility features, such as audio descriptions and large print, cities can make it easier for tourists with disabilities to access tourist information and enjoy the city.

GENERAL FRAMEWORK

The following framework is a general one and the actual implementation will depend on the specific requirements and constraints of the city.

1. Conduct a comprehensive market research to understand the tourism landscape in the city and identify areas for improvement.
2. Establish partnerships with relevant stakeholders, including local government, tourism boards, and hospitality organizations.
3. Develop a comprehensive plan for the implementation of smart city tourism initiatives, including specific objectives, timelines, and budgets.
4. Invest in technology infrastructure to support smart city tourism, such as interactive city guides, digital concierge services, and online booking platforms.
5. Design and implement an information and education campaign to raise awareness about smart city tourism initiatives and the benefits they bring.
6. Provide training and support to tourism and hospitality providers to ensure they are equipped to provide top-notch services.
7. Encourage the private sector to adopt smart city tourism initiatives by offering incentives, subsidies, and tax credits.
8. Develop a comprehensive data management system to track and analyze tourism trends and patterns.
9. Collaborate with academic institutions to research and develop new smart city tourism technologies and solutions.
10. Develop a crisis management plan to address unexpected challenges and disruptions in tourism services.
11. Ensure that smart city tourism initiatives are accessible to all segments of the population, including low-income communities and individuals with disabilities.
12. Foster a culture of sustainability and environmental responsibility through community engagement and education initiatives.
13. Regularly review and evaluate the smart city tourism program to identify areas for improvement and make necessary adjustments.
14. Continuously engage with stakeholders to ensure that the smart city tourism program remains relevant and responsive to changing needs and conditions.
15. Implement policies and regulations that promote sustainable tourism practices and discourage practices that are harmful to the environment and local communities.
16. Provide incentives and subsidies to encourage the use of sustainable tourism options, such as eco-friendly accommodations and low-emission transportation.
17. Encourage the use of digital payment systems to reduce waste and promote sustainability.

18. Develop a plan for the integration of smart city tourism initiatives with the existing public transportation system, such as digital concierge services at public transportation hubs.
19. Provide ongoing support and technical assistance to ensure the effective and efficient implementation of smart city tourism initiatives.
20. Develop a plan for the maintenance and expansion of smart city tourism infrastructure, such as interactive city guides and digital concierge services.
21. Implement traffic management strategies to prioritize sustainable tourism options and reduce congestion in popular tourist areas.
22. Develop a plan to promote responsible tourism practices and mitigate the negative impacts of tourism on local communities and the environment.
23. Encourage the use of sustainable tourism options, such as eco-friendly accommodations and low-emission transportation.
24. Foster a culture of innovation and experimentation by encouraging the development of new and creative solutions to enhance the tourism experience.
25. Leverage data and analytics to continuously improve the tourism experience and promote sustainable practices.

FREQUENTLY ASKED QUESTIONS

What is the role of tourism in the development of a smart city? Tourism plays a significant role in the development of a smart city, as it contributes to the local economy, job creation, and helps to enhance the overall quality of life for residents.

How can a smart city improve the tourist experience? A smart city can improve the tourist experience by leveraging technology to enhance accessibility, provide real-time information on local attractions, and offer tailored services and experiences.

What are some examples of smart city initiatives that support tourism? Smart city initiatives that support tourism include intelligent transportation systems, smart parking, public Wi-Fi, smart wayfinding, and mobile apps that provide real-time information and personalized recommendations.

How can a smart city balance the needs of tourists with those of local residents? A smart city can balance the needs of tourists with those of local residents by implementing policies and infrastructure that are sustainable, equitable, and enhance the overall quality of life for both tourists and residents.

What impact can smart tourism have on the environment? Smart tourism can have a positive impact on the environment by reducing carbon emissions through smart transportation systems, promoting sustainable tourism practices, and leveraging renewable energy sources.

What types of data are collected in a smart city to support tourism? Data collected in a smart city to support tourism includes information on tourist behaviour, preferences, and demographics, as well as real-time data on traffic, weather, and local events.

How can a smart city promote local tourism? A smart city can promote local tourism by highlighting local attractions, supporting local businesses, and offering personalized recommendations based on user data.

How can a smart city ensure the safety and security of tourists? A smart city can ensure the safety and security of tourists by implementing surveillance systems, smart lighting, emergency response systems, and providing real-time alerts and notifications.

What is the impact of smart tourism on the local economy? Smart tourism can have a positive impact on the local economy by creating jobs, stimulating local businesses, and attracting investment to the region.

How can tourists benefit from the data collected in a smart city? Tourists can benefit from the data collected in a smart city by receiving personalized recommendations, real-time information on local attractions, and access to tailored services and experiences that enhance the overall tourist experience.

Conclusion: Implementing SCTO can help cities attract more visitors, promote sustainable tourism practices, and create a more inclusive and accessible city for tourists. This can lead to increased revenue and a better quality of life for residents.

SMART CITY TOURIST ACCESSIBILITY (SCTA)

Using technology to improve accessibility for tourists with disabilities, such as audio descriptions and assistive technology.

Implementing SCTA can have a wide range of benefits for both tourists with disabilities and the city as a whole. One major benefit is increased accessibility for tourists with disabilities, which can lead to a more inclusive and welcoming environment for all visitors. This can be achieved through the use of technology such as audio descriptions and assistive technology, which can help individuals with visual or auditory impairments navigate the city and access information.

Another benefit is the potential for increased tourism and economic growth. By making the city more accessible to a wider range of visitors, the city may attract more tourists, which can lead to increased revenue for local businesses and the city itself. Providing accessible technology can also enhance the overall tourism experience, which can lead to positive word-of-mouth and encourage repeat visits.

Implementing SCTA can also benefit the city's reputation as a leader in innovation and social responsibility. By making the city more accessible and inclusive, the city can showcase its commitment to creating a sustainable and equitable environment for all citizens and visitors.

GENERAL FRAMEWORK

The following framework is a general one and the actual implementation will depend on the specific requirements and constraints of the city.

1. Conduct a comprehensive assessment of the accessibility needs of tourists with disabilities in the city.
2. Establish partnerships with relevant stakeholders, including local government, tourism boards, and disability organizations.
3. Develop a comprehensive plan for the implementation of smart city tourist accessibility initiatives, including specific objectives, timelines, and budgets.
4. Invest in technology infrastructure to support smart city tourist accessibility, such as audio descriptions, assistive technology, and digital accessibility solutions.
5. Design and implement an information and education campaign to raise awareness about smart city tourist accessibility initiatives and the benefits they bring.
6. Provide training and support to tourism and hospitality providers to ensure they are equipped to provide accessible services.
7. Encourage the private sector to adopt smart city tourist accessibility initiatives by offering incentives, subsidies, and tax credits.
8. Develop a comprehensive data management system to track and analyze accessibility trends and patterns.
9. Collaborate with academic institutions to research and develop new smart city tourist accessibility technologies and solutions.
10. Develop a crisis management plan to address unexpected challenges and disruptions in accessibility services.
11. Ensure that smart city tourist accessibility initiatives are accessible to all segments of the population, including low-income communities and individuals with disabilities.
12. Foster a culture of inclusion and accessibility through community engagement and education initiatives.
13. Regularly review and evaluate the smart city tourist accessibility program to identify areas for improvement and make necessary adjustments.
14. Continuously engage with stakeholders to ensure that the smart city tourist accessibility program remains relevant and responsive to changing needs and conditions.

15. Implement policies and regulations that promote accessibility for tourists with disabilities and discourage practices that are harmful to accessibility.
16. Provide incentives and subsidies to encourage the use of accessible tourism options, such as audio descriptions and assistive technology.
17. Encourage the use of digital payment systems to reduce waste and promote accessibility.
18. Develop a plan for the integration of smart city tourist accessibility initiatives with the existing public transportation system, such as audio descriptions and assistive technology at public transportation hubs.
19. Provide ongoing support and technical assistance to ensure the effective and efficient implementation of smart city tourist accessibility initiatives.
20. Develop a plan for the maintenance and expansion of smart city tourist accessibility infrastructure, such as audio descriptions and assistive technology.
21. Implement traffic management strategies to prioritize accessible tourism options and reduce congestion in popular tourist areas.
22. Develop a plan to promote responsible tourism practices that take into account the needs of tourists with disabilities.
23. Encourage the use of accessible tourism options, such as audio descriptions and assistive technology.
24. Foster a culture of innovation and experimentation by encouraging the development of new and creative solutions to improve accessibility for tourists with disabilities.
25. Leverage data and analytics to continuously improve accessibility for tourists with disabilities and promote inclusive practices.

FREQUENTLY ASKED QUESTIONS

What is Smart City Tourism Accessibility and how does it work? Smart City Tourism Accessibility (SCTA) refers to the use of technology to make tourist destinations more accessible to people with disabilities. This can include the use of smart devices and apps to provide information about accessible facilities, and the use of assistive technology to help people with disabilities navigate their surroundings.

How can SCTA benefit people with disabilities? SCTA can help people with disabilities to enjoy more independent and stress-free travel experiences. By providing information about accessible facilities and assistive technologies, people with disabilities can plan and enjoy their trips with greater confidence.

What are some examples of SCTA technologies? Examples of SCTA technologies include wheelchair accessibility maps, voice-activated assistants for people with visual impairments, and sign language interpretation services.

How can SCTA help to boost tourism in a city or region? SCTA can make a city or region more attractive to people with disabilities, who may be more likely to choose destinations that are easier to navigate and more accommodating. This can help to increase tourism and bring economic benefits to the region.

What role do businesses and local governments play in promoting SCTA? Businesses and local governments have a key role to play in promoting SCTA by ensuring that their facilities are accessible, providing information about accessibility, and investing in assistive technology and other accessibility measures.

How can SCTA be integrated with other smart city initiatives? SCTA can be integrated with other smart city initiatives by using the same infrastructure and technology to provide information and services to people with disabilities. For example, smart traffic systems can be used to improve transportation for people with disabilities.

How can SCTA help to promote social inclusion? SCTA can help to promote social inclusion by making tourist destinations more accessible to people with disabilities, who may otherwise be excluded from travel and tourism experiences.

How can SCTA help to improve the quality of life for people with disabilities? SCTA can help to improve the quality of life for people with disabilities by providing greater access to leisure and travel experiences, and by promoting greater independence and social participation.

What are some of the challenges facing the implementation of SCTA? Challenges facing the implementation of SCTA include the cost of technology and infrastructure, the need for coordination between different stakeholders, and the need to address the specific accessibility needs of different types of disabilities.

How can people with disabilities get involved in the development of SCTA? People with disabilities can get involved in the development of SCTA by providing feedback and input on the design of accessible facilities and technology, and by advocating for greater accessibility in their communities.

Conclusion: *Implementing SCTA can have a positive impact on the city's economic growth, reputation, and inclusivity. It can also provide a more enjoyable and accessible experience for tourists with disabilities, which can ultimately lead to increased tourism and economic growth for the city.*

SMART CITY TOURIST ACCOMMODATION MANAGEMENT (SCTAM)

Using technology to manage tourist accommodation, such as online booking and real-time availability.

Implementing smart city technology to manage tourist accommodation can bring a wide range of benefits to both tourists and the city. One major benefit is the ability to easily book and manage accommodation online. This can make the process more convenient for tourists, allowing them to quickly find and book a suitable place to stay. Real-time availability updates can help tourists to find available accommodation in a shorter time frame.

Another benefit is the ability to improve the overall efficiency of the accommodation management process. SCTAM can provide real-time data on occupancy rates, which can help managers to optimize pricing and occupancy strategies. It can also help to streamline the check-in and check-out process, making it more efficient for both tourists and staff.

SCTAM can also help to improve the overall sustainability of tourist accommodation. For example, it can be used to track and monitor energy consumption, and identify ways to reduce waste and increase recycling. This can help to reduce the environmental impact of tourist accommodation and promote more sustainable practices.

GENERAL FRAMEWORK

The following framework is a general one and the actual implementation will depend on the specific requirements and constraints of the city.

1. Define the scope and objectives of the SCTAM project.
2. Conduct a market analysis to determine the demand for online booking of tourist accommodations.
3. Select a suitable technology platform for the SCTAM project.
4. Develop the front-end user interface and back-end functionality of the platform.
5. Integrate the platform with existing systems, such as hotel management systems, payment gateways, and GPS systems.
6. Set up and configure the platform, including customizing its features and functionalities.
7. Develop a robust data management system for storing and managing guest information, booking details, and real-time availability.
8. Establish security measures, such as encryption and access controls, to ensure data privacy and security.
9. Train staff members on how to use the platform and how to provide customer support.
10. Launch the platform and promote it to potential users, such as tourists and hotels.
11. Monitor and analyze user feedback to identify areas for improvement.

12. Continuously update and maintain the platform to ensure it is up-to-date and functioning optimally.
13. Implement analytics tools to track the platform's performance, user engagement, and conversion rates.
14. Create a marketing plan to increase awareness of the platform and to drive bookings.
15. Offer incentives and rewards to users who use the platform, such as loyalty programs and special discounts.
16. Partner with hotels and other tourist accommodations to increase the platform's offerings and reach.
17. Conduct user testing and research to gather insights into the platform's usability and effectiveness.
18. Integrate with third-party services and platforms, such as travel agencies and travel apps, to expand the platform's reach.
19. Foster collaboration and partnerships with other stakeholders, such as government agencies and tourism boards.
20. Develop and implement strategies to mitigate risks and challenges, such as data breaches and platform downtime.
21. Establish an effective customer service and support system to assist users with their needs and inquiries.
22. Conduct regular audits and security assessments to ensure the platform's compliance with regulations and standards.
23. Monitor market trends and technologies and make appropriate changes to the platform to stay competitive.
24. Continuously evaluate and measure the platform's performance and impact on the tourist industry.
25. Continuously seek feedback from users and stakeholders to improve the platform's offerings and user experience.

FREQUENTLY ASKED QUESTIONS

What is Smart City Tourism Accommodation Management (SCTAM)? SCTAM is an innovative solution that uses smart technology to enhance the management of accommodations in a smart city. It improves the experience of tourists and residents alike by providing real-time information on available accommodations, reducing wait times, and increasing efficiency.

How does SCTAM work? SCTAM uses a combination of sensors, data analytics, and digital platforms to monitor and manage the availability and quality of accommodations. It enables tourists and residents to search for and book accommodations using mobile apps and other digital tools.

What benefits does SCTAM offer? SCTAM offers numerous benefits, including improved convenience, increased efficiency, better customer service, and reduced environmental impact. It enhances the overall tourism experience and helps smart cities to become more sustainable and livable.

How does SCTAM address the issue of overcrowding in tourist destinations? SCTAM helps to manage the flow of tourists by providing real-time information on the availability of accommodations. This helps to spread out the tourism flow and reduce overcrowding in popular destinations.

How does SCTAM ensure the safety and security of tourists? SCTAM uses digital platforms to monitor and manage the safety and security of tourists. It provides real-time alerts on potential safety risks and enables immediate response to emergencies.

Can SCTAM help to reduce the environmental impact of tourism? Yes, SCTAM can help to reduce the environmental impact of tourism by enabling better management of resources and reducing waste. It also promotes sustainable tourism practices and supports the transition to a more sustainable future.

What role do local businesses play in SCTAM? Local businesses are key partners in SCTAM, providing accommodations and other services to tourists. SCTAM enables local businesses to better manage their resources and improve their customer service, enhancing their competitiveness in the tourism industry.

How does SCTAM promote inclusivity and accessibility in tourism? SCTAM provides real-time information on the accessibility of accommodations and other tourist facilities, enabling people with disabilities to more easily plan their trips. It also promotes inclusivity by making tourism more accessible to people from diverse backgrounds.

How can SCTAM support the growth of the tourism industry in a smart city? SCTAM can help to support the growth of the tourism industry by improving the overall tourism experience, increasing efficiency, and enhancing the competitiveness of local businesses. It can also attract new tourists and promote the city as a desirable tourist destination.

What is the future of SCTAM in smart cities? SCTAM is a key component of the smart city of the future, enabling better management of accommodations and other tourism resources. As smart cities continue to evolve, SCTAM will play an increasingly important role in promoting sustainability, inclusivity, and competitiveness in the tourism industry.

Conclusion: Implementing SCTAM to manage tourist accommodation can bring a wide range of benefits that can improve the experience for tourists and make the process more efficient and sustainable for the city.

SMART CITY TOURIST ACTIVITY MANAGEMENT (SCTAC)

Using technology to manage and promote tourist activities, such as digital booking and real-time availability.

SCTAC is a solution that leverages technology to manage and promote tourist activities. The use of technology in this context helps to streamline the process of booking tourist activities and provides real-time availability information. In this way, SCTAC offers several key benefits to both tourists and the organizations that manage tourist activities.

One of the major benefits of SCTAC is increased efficiency. With digital booking, tourists can quickly and easily reserve activities, reducing the time and effort required to make a reservation. This results in a better user experience and a more seamless process for tourists. Moreover, real-time availability information makes it easier for tourists to find and book the activities they are interested in, reducing the frustration of unavailability or long wait times.

SCTAC also provides improved data management and analysis capabilities. By automating the booking process, organizations can gather data about tourist activities, such as the number of bookings, the types of activities booked, and the time of day when bookings occur. This data can be used to optimize activity offerings and pricing, resulting in better decision-making and improved overall performance.

Another key benefit of SCTAC is increased visibility and accessibility. By promoting tourist activities through the platform, organizations can reach a wider audience and make it easier for tourists to discover and book activities. This can result in increased sales and a more successful tourist activity business. Digital booking and real-time availability information can be accessed from anywhere, at any time, providing increased flexibility for both tourists and organizations.

SCTAC can lead to cost savings for organizations. By automating manual processes, organizations can reduce the cost of manual labour and increase the speed and accuracy of their operations. The ability to track and analyze data can help organizations to make more informed decisions, reducing the costs associated with ineffective decision-making.

GENERAL FRAMEWORK

The following framework is a general one and the actual implementation will depend on the specific requirements and constraints of the city.

1. Conduct market research to understand tourist needs and preferences for activities.

2. Identify key stakeholders in the tourist activity industry.
3. Assess current processes for managing and promoting tourist activities.
4. Identify technology solutions that can support the management and promotion of tourist activities.
5. Determine the specific features and functionality required for the technology solution.
6. Select a technology partner to implement the solution.
7. Define project scope and timeline.
8. Develop a detailed project plan, including budget and resource allocation.
9. Conduct a thorough data audit to identify and categorize data required for the solution.
10. Define data management and security protocols.
11. Set up digital infrastructure and tools to support the solution.
12. Integrate the technology solution with existing systems and processes, as needed.
13. Develop a digital marketing strategy to promote the tourist activities and the technology solution.
14. Establish a training program for stakeholders, including tourist activity providers and end-users.
15. Launch the technology solution and digital marketing campaign.
16. Monitor and measure the effectiveness of the solution, including metrics such as user engagement and conversion rates.
17. Continuously improve the solution through data-driven decision making and regular updates and upgrades.
18. Foster collaboration and partnerships with key stakeholders in the tourist activity industry.
19. Encourage feedback from users and stakeholders to inform continuous improvement.
20. Implement measures to ensure data privacy and security.
21. Evaluate the impact of the solution on the tourist activity industry and the wider community.
22. Develop and implement strategies to overcome challenges and mitigate risks.
23. Foster a culture of innovation and continuous improvement within the organization.
24. Regularly review and update the technology solution to stay up-to-date with market trends and changing user needs.
25. Evaluate and refine the digital marketing strategy to ensure optimal performance and return on investment.

FREQUENTLY ASKED QUESTIONS

What is SCTAC and how does it impact tourism in a Smart City? SCTAC is the Smart City Tourism Activity Management system that helps manage and regulate tourism-related activities in a Smart City. It ensures that tourists have access to information and facilities while also preventing any adverse impact on the local community.

Who benefits from the SCTAC system in a Smart City? SCTAC benefits both tourists and the local community by providing access to information about tourism activities and ensuring that they are managed sustainably and in a way that is beneficial for everyone.

How does SCTAC contribute to sustainable tourism in a Smart City? SCTAC helps to manage tourism activities in a way that minimizes negative impact on the environment and the local community. It ensures that tourism activities are sustainable and contribute to the long-term economic, social and environmental wellbeing of the community.

How does SCTAC improve the overall tourism experience in a Smart City? SCTAC helps to improve the tourism experience by providing tourists with access to information about local attractions, events and activities. It also ensures that these activities are well-managed and that tourists have access to high-quality services.

How does SCTAC promote responsible tourism in a Smart City? SCTAC promotes responsible tourism by ensuring that tourism activities are sustainable and that they have a positive impact on the environment and the local community. It also encourages tourists to act responsibly and respect the local culture and customs.

What role does technology play in SCTAC? Technology plays a key role in SCTAC by providing a platform for tourists to access information about tourism activities and services. It also helps to manage and regulate these activities in a way that is sustainable and beneficial for the community.

How does SCTAC ensure that tourism activities are inclusive for everyone? SCTAC ensures that tourism activities are inclusive by providing information and facilities that are accessible for people with disabilities, the elderly and other groups. It also ensures that tourism activities are designed in a way that is respectful of cultural and social differences.

How does SCTAC manage the balance between tourism and the needs of the local community? SCTAC manages the balance between tourism and the needs of the local community by ensuring that tourism activities are sustainable and respectful of the local environment and culture. It also provides opportunities for the local community to benefit from tourism through the development of local businesses and services.

What is the process for implementing SCTAC in a Smart City? The process for implementing SCTAC in a Smart City involves a range of stakeholders including tourism operators, local businesses, community groups and government agencies. It involves developing a shared vision for sustainable tourism and then implementing a range of initiatives and services that support this vision.

What are some of the challenges involved in implementing SCTAC in a Smart City? Some of the challenges involved in implementing SCTAC in a Smart City include engaging with a range of stakeholders and managing conflicting interests. There may also be challenges around the availability of resources and technology, and the need to ensure that SCTAC is integrated with other Smart City systems and initiatives.

Conclusion: The use of SCTAC technology in the management and promotion of tourist activities has numerous benefits for both tourists and organizations. From increased efficiency and better data management to increased visibility and accessibility, SCTAC provides a range of benefits that can improve the tourist experience and drive business success.

SMART CITY TOURIST ATTRACTIONS MANAGEMENT (SCTOA)

Using technology to manage tourist attractions, such as digital ticketing and real-time visitor information.

Implementing SCTOA can bring a wide range of benefits for both the city and its visitors. One key benefit is the ability to improve the overall visitor experience through digital ticketing and real-time visitor information. For example, with digital ticketing, visitors can easily purchase and access tickets to attractions through their mobile devices, eliminating the need to wait in long lines. Real-time visitor information can be made available through digital kiosks or mobile apps, providing visitors with up-to-date information on wait times, special exhibits, and other important details. This can help visitors plan their time more effectively and make the most of their visit.

Another benefit of using technology for tourist attraction management is the ability to increase operational efficiency and reduce costs. Digital ticketing and real-time visitor information systems can help reduce the need for manual labour and increase automation, resulting in cost savings for the city and attraction operators. Technology can also help optimize the use of resources, such as reducing the amount of paper used for ticketing and improving visitor flow through attractions.

SCTOA Technology can also help promote sustainable tourism practices by collecting data on visitor patterns, which can help city planners and attraction operators to make more informed decisions about how to best manage their resources and reduce negative impacts on the environment. Also, it can help to promote the attraction to a wider range of people, providing a more inclusive experience for everyone.

GENERAL FRAMEWORK

The following framework is a general one and the actual implementation will depend on the specific requirements and constraints of the city.

1. Conduct market research to understand tourist needs and preferences for attractions.
2. Identify key stakeholders in the tourist attraction industry.
3. Assess current processes for managing and promoting tourist attractions.
4. Identify technology solutions that can support the management of tourist attractions.
5. Determine the specific features and functionality required for the technology solution.
6. Select a technology partner to implement the solution.
7. Define project scope and timeline.
8. Develop a detailed project plan, including budget and resource allocation.
9. Conduct a thorough data audit to identify and categorize data required for the solution.
10. Define data management and security protocols.
11. Set up digital infrastructure and tools to support the solution.
12. Integrate the technology solution with existing systems and processes, as needed.
13. Develop a digital marketing strategy to promote the tourist attractions and the technology solution.
14. Establish a training program for stakeholders, including attraction managers and staff.
15. Launch the technology solution and digital marketing campaign.
16. Monitor and measure the effectiveness of the solution, including metrics such as visitor volume and ticket sales.
17. Continuously improve the solution through data-driven decision making and regular updates and upgrades.
18. Foster collaboration and partnerships with key stakeholders in the tourist attraction industry.
19. Encourage feedback from visitors and stakeholders to inform continuous improvement.
20. Implement measures to ensure data privacy and security.
21. Evaluate the impact of the solution on the tourist attraction industry and the wider community.

22. Develop and implement strategies to overcome challenges and mitigate risks.
23. Foster a culture of innovation and continuous improvement within the organization.
24. Regularly review and update the technology solution to stay up-to-date with market trends and changing user needs.
25. Evaluate and refine the digital marketing strategy to ensure optimal performance and return on investment.

FREQUENTLY ASKED QUESTIONS

What is the main goal of the SCTOA project? The main goal of the SCTOA project is to use technology to enhance the management and accessibility of tourist attractions in the smart city.

How will SCTOA improve the tourist experience in the smart city? SCTOA will improve the tourist experience in the smart city by providing real-time information and insights about attractions, allowing for better planning and exploration of the city's offerings.

Will SCTOA affect the cost of visiting tourist attractions in the smart city? No, SCTOA will not directly affect the cost of visiting tourist attractions in the smart city. However, it may allow for more efficient use of resources, potentially leading to cost savings for both visitors and attraction managers.

How will SCTOA benefit attraction managers in the smart city? SCTOA will benefit attraction managers in the smart city by providing data-driven insights into visitor behaviour and preferences, allowing for more effective management of attractions and resources.

Is there a privacy risk associated with using SCTOA as a tourist? SCTOA has been designed with privacy in mind, and personal data will only be used for the purpose of enhancing the tourist experience. Any data collected will be handled according to relevant privacy laws and regulations.

How will SCTOA integrate with existing tourist infrastructure in the smart city? SCTOA is designed to be easily integrated with existing tourist infrastructure in the smart city, and can work alongside existing systems to enhance their capabilities.

Will SCTOA be available in multiple languages for non-English speaking tourists? Yes, SCTOA will be available in multiple languages to ensure that all tourists can benefit from its features.

How can tourists access SCTOA? Tourists can access SCTOA through a mobile app or website, which will be made available for free.

Will SCTOA provide discounts or special offers for visiting certain tourist attractions in the smart city? While SCTOA will not provide discounts or special offers directly, it may allow for more efficient management of resources, potentially leading to cost savings that can be passed on to visitors.

How will SCTOA be funded and maintained? SCTOA will be funded and maintained through a combination of public and private funding, and will be overseen by a dedicated team responsible for its ongoing development and maintenance.

Conclusion: *Implementing SCTOA can help enhance the visitor experience, increase operational efficiency, and promote sustainable tourism practices. It can make the city more attractive to visitors and improve the city's economy.*

SMART CITY TOURIST CROWD MANAGEMENT (SCTCM)

Using technology to manage and optimize the flow of tourists, such as real-time crowd monitoring and dynamic pricing.

Implementing SCTCM can bring a number of benefits to both the city and its visitors. One of the main benefits is the ability to better manage and optimize the flow of tourists in popular areas. This can be achieved through the use of real-time crowd monitoring technology which allows city officials to track the number of visitors in a particular area and adjust staffing or pricing accordingly. This can help to ensure that popular areas remain safe and enjoyable for visitors, while also reducing the risk of overcrowding and congestion.

Another benefit of SCTCM is the ability to implement dynamic pricing. This can be used to encourage visitors to visit less popular areas or attractions during peak periods, helping to distribute tourists more evenly throughout the city. This technology can be used to provide real-time information to visitors about crowd levels at different attractions, allowing them to make more informed decisions about where to go and when to visit.

SCTCM also has the potential to bring economic benefits to the city. By managing crowds more effectively, the city can attract more visitors and increase revenue from tourism. By reducing the risk of overcrowding and congestion, the city can create a more positive experience for visitors, encouraging them to return in the future.

GENERAL FRAMEWORK

The following framework is a general one and the actual implementation will depend on the specific requirements and constraints of the city.

1. Conduct market research to understand tourist needs and preferences for crowd management.
2. Identify key stakeholders in the tourist industry and relevant government agencies.
3. Assess current processes for managing the flow of tourists.
4. Identify technology solutions that can support the management of tourist crowds.
5. Determine the specific features and functionality required for the technology solution.
6. Select a technology partner to implement the solution.
7. Define project scope and timeline.

8. Develop a detailed project plan, including budget and resource allocation.
9. Conduct a thorough data audit to identify and categorize data required for the solution.
10. Define data management and security protocols.
11. Set up digital infrastructure and tools to support the solution.
12. Integrate the technology solution with existing systems and processes, as needed.
13. Develop a digital marketing strategy to promote the tourist crowd management solution.
14. Establish a training program for stakeholders, including tourist industry providers and government agencies.
15. Launch the technology solution and digital marketing campaign.
16. Monitor and measure the effectiveness of the solution, including metrics such as visitor volume and crowd density.
17. Continuously improve the solution through data-driven decision making and regular updates and upgrades.
18. Foster collaboration and partnerships with key stakeholders in the tourist industry and relevant government agencies.
19. Encourage feedback from tourists and stakeholders to inform continuous improvement.
20. Implement measures to ensure data privacy and security.
21. Evaluate the impact of the solution on the tourist industry and the wider community.
22. Develop and implement strategies to overcome challenges and mitigate risks.
23. Foster a culture of innovation and continuous improvement within the organization.
24. Regularly review and update the technology solution to stay up-to-date with market trends and changing user needs.
25. Evaluate and refine the digital marketing strategy to ensure optimal performance and return on investment.

FREQUENTLY ASKED QUESTIONS

What is Smart City Tourism Crowd Management (SCTCM)? SCTCM is a system that uses technology to manage crowds in tourist areas by analyzing data and providing real-time information to visitors and city officials.

How does SCTCM work? SCTCM works by using various sensors and cameras to collect data on the number of people in a specific area. This data is then analysed to provide real-time information on the current crowd levels, allowing officials to take measures to manage the crowds.

How can SCTCM benefit tourists? SCTCM can benefit tourists by providing them with real-time information on crowd levels in tourist areas, allowing them to plan their visits accordingly and avoid crowded areas.

What are some examples of SCTCM in action? Some examples of SCTCM in action include the use of digital displays to provide real-time information on crowd levels in tourist areas and the use of mobile apps to provide visitors with real-time information on the crowd levels in different parts of the city.

How can SCTCM help city officials manage tourist areas? SCTCM can help city officials manage tourist areas by providing them with real-time information on crowd levels and allowing them to take measures to manage the crowds and ensure the safety of visitors.

How does SCTCM address issues related to safety in tourist areas? SCTCM addresses safety issues in tourist areas by providing officials with real-time information on crowd levels, allowing them to take measures to prevent overcrowding and ensure the safety of visitors.

What are some of the challenges of implementing SCTCM? Some of the challenges of implementing SCTCM include the need for significant investment in infrastructure and technology, concerns around data privacy, and the need for effective collaboration between city officials and other stakeholders.

How can tourists access real-time information on crowd levels in tourist areas? Tourists can access real-time information on crowd levels in tourist areas through digital displays, mobile apps, and other information kiosks that are available throughout the city.

How can SCTCM be used to improve the overall tourism experience? SCTCM can be used to improve the overall tourism experience by allowing visitors to plan their visits more effectively, avoid overcrowded areas, and have a more enjoyable and safe visit to the city.

What are some of the future developments in SCTCM? Some of the future developments in SCTCM include the use of artificial intelligence and machine learning algorithms to analyze data more effectively, the development of new sensors and cameras that can collect more accurate data, and the integration of SCTCM with other smart city systems to improve overall efficiency and effectiveness.

Conclusion: Implementing SCTCM is a great way to enhance the experience of visitors to the city while also improving the overall functionality and efficiency of the city's tourism industry. By using technology to manage and optimize the flow of tourists, cities can create a more enjoyable and sustainable experience for all.

SMART CITY TOURIST DATA ANALYSIS (SCTDA)

Using technology to analyze tourist data, such as visitor demographics and spending patterns.

Implementing SCTDA can bring a range of benefits for both city officials and tourists alike. By using technology to analyze data on visitor demographics and spending patterns, city officials can gain valuable insights into how tourists are engaging with their city. This information can be used to optimize the tourism experience and promote sustainable tourism practices. For example, if data analysis reveals that a particular attraction is particularly popular among a certain demographic, city officials can focus their marketing efforts on that demographic. Similarly, if data analysis reveals that tourists are spending the majority of their money on shopping, city officials can work to attract more high-end retailers to the city.

By using SCTDA technology to analyze data on visitor demographics and spending patterns, city officials can also identify areas where improvements are needed. For example, if data analysis reveals that a particular area of the city is not attracting many tourists, city officials can work to develop new attractions or improve existing ones in that area, plus using SCTDA technology to analyze data on visitor demographics and spending patterns, city officials can also gain insight into how tourists are traveling to and around the city. This information can be used to optimize transportation options and reduce congestion, making the city more attractive to tourists.

In addition to the benefits for city officials, SCTDA can also help tourists. By providing real-time visitor information, tourists can make more informed decisions about where to go and what to do, leading to a more enjoyable and efficient tourism experience. By providing dynamic pricing based on real-time visitor information, tourists can also save money by visiting attractions and booking accommodations during off-peak times.

GENERAL FRAMEWORK

The following framework is a general one and the actual implementation will depend on the specific requirements and constraints of the city.

1. Conduct market research to understand the tourist industry and its key stakeholders.
2. Identify existing data sources that can be leveraged for tourist data analysis.
3. Determine the specific data points and KPIs required for the analysis.
4. Develop a data management strategy, including data governance and security protocols.
5. Choose technology tools and platforms to support the data analysis process.
6. Define the data analysis process, including data cleaning, transformation and modelling.
7. Establish a data integration plan to ensure data quality and accuracy.
8. Develop a data visualization plan to present insights and results to stakeholders.

9. Clean and process existing data sets to prepare them for analysis.
10. Analyze the data to uncover insights and identify patterns and trends.
11. Develop predictive models to support decision making.
12. Implement data visualization dashboards to present insights and results.
13. Train stakeholders on the data analysis process and the use of the data visualization tools.
14. Continuously monitor and update the data to ensure accuracy and completeness.
15. Collaborate with stakeholders to integrate data analysis insights into decision making processes.
16. Regularly evaluate and refine the data analysis process to ensure efficiency and effectiveness.
17. Ensure data privacy and security through strict data governance and security protocols.
18. Evaluate the impact of the data analysis on the tourist industry and the wider community.
19. Develop and implement strategies to overcome challenges and mitigate risks.
20. Foster a culture of data-driven decision making within the organization.
21. Regularly review and update the technology tools and platforms to stay up-to-date with market trends and changing user needs.
22. Foster collaboration and partnerships with key stakeholders in the tourist industry and relevant government agencies.
23. Encourage feedback from stakeholders to inform continuous improvement.
24. Evaluate the return on investment from the data analysis initiative.
25. Continuously monitor and measure the effectiveness of the data analysis and refine as necessary.

FREQUENTLY ASKED QUESTIONS

What is Smart City Tourism Data Analysis and how does it work?
Smart City Tourism Data Analysis (SCTDA) is a system that collects and analyses data from tourism activities in a city. The system uses various data sources, such as social media, foot traffic, and transportation patterns, to generate insights and predictions for tourism stakeholders.

How can SCTDA help cities attract more tourists? SCTDA can help cities better understand tourist behaviours and preferences, allowing them to tailor their offerings to meet demand. It can also help identify areas that need improvement, such as traffic congestion or safety concerns.

Is SCTDA only useful for large cities? No, SCTDA can be used in any city, regardless of its size. In fact, smaller cities may benefit more from the insights generated by the system, as they may have fewer resources to conduct their own data analysis.

How does SCTDA protect the privacy of tourists? SCTDA only collects data from publicly available sources and does not identify individual tourists. Any personal information that is collected is anonymized and aggregated to protect privacy.

How accurate is the data collected by SCTDA? The accuracy of the data depends on the quality of the sources used. SCTDA uses multiple sources to generate insights, which helps improve accuracy. However, no system can be 100% accurate, and it's important to validate insights with other data sources.

Who has access to the data collected by SCTDA? Access to the data is restricted to authorized stakeholders, such as city officials, tourism organizations, and businesses. The data is used to inform decision-making and improve tourism outcomes.

How much does it cost to implement SCTDA? The cost of implementing SCTDA can vary depending on the size of the city and the complexity of the system. However, many cities have found that the benefits of the system outweigh the costs.

How can SCTDA be used to improve sustainability in tourism? SCTDA can help identify areas of high foot traffic or transportation congestion, allowing cities to implement measures to reduce emissions and improve sustainability. It can also help identify areas that are being negatively impacted by tourism, allowing for targeted interventions.

How is SCTDA different from traditional tourism data collection methods? SCTDA uses real-time data from multiple sources to generate insights, whereas traditional methods often rely on surveys or visitor logs. The system allows for more timely and accurate data analysis, which can help improve decision-making.

How can cities use the insights generated by SCTDA to attract specific types of tourists? SCTDA can help identify the interests and behaviours of specific tourist groups, allowing cities to tailor their offerings to meet demand. For example, if the system identifies a high volume of social media posts about outdoor activities, a city may promote its outdoor attractions more heavily to attract those tourists.

Conclusion: *Implementing SCTDA can bring a range of benefits for both city officials and tourists. It can help city officials to optimize the tourism experience and promote sustainable tourism practices, while also providing valuable information for tourists to make more informed decisions and save money.*

SMART CITY TOURIST DESTINATION MARKETING (SCTDM)

Using technology to promote tourist destinations, such as social media marketing and digital advertising.

Implementing SCTDM can bring a wide range of benefits to both tourists and the city. By using technology to promote tourist destinations, cities can attract more visitors and increase tourism revenue. Social media marketing and digital advertising are powerful tools that allow cities to reach a wider audience and target specific demographics.

This can result in more efficient and effective marketing campaigns, which in turn can lead to increased visitor numbers and higher revenue for the city. By using data analytics, cities can gain valuable insights into what types of visitors they are attracting and what types of activities and attractions they are most interested in.

This information can be used to tailor future marketing campaigns and make them even more effective. Moreover, by using data analysis, cities can track the performance of their marketing campaigns in real-time, which allows them to make adjustments and optimize their efforts as needed.

GENERAL FRAMEWORK

The following framework is a general one and the actual implementation will depend on the specific requirements and constraints of the city.

1. Conduct market research to understand the target tourist segments and their preferences.
2. Define the key messages and unique selling points of the tourist destination.
3. Develop a comprehensive destination marketing plan.
4. Choose technology tools and platforms to support the destination marketing efforts.
5. Establish a social media presence and develop a content strategy.
6. Develop a website and optimize it for search engines.
7. Implement digital advertising campaigns on relevant platforms.
8. Utilize data and analytics to monitor and optimize the performance of the marketing efforts.
9. Leverage influencer marketing to reach target segments.
10. Utilize email marketing to engage with prospects and customers.
11. Create and distribute destination-related content such as videos and articles.
12. Utilize virtual and augmented reality technologies to showcase the destination.
13. Partner with relevant tourism organizations and government agencies to promote the destination.
14. Organize and participate in events, trade shows and webinars to promote the destination.
15. Develop and maintain relationships with the media to secure coverage and exposure.
16. Offer special promotions and incentives to attract visitors.

17. Utilize customer feedback to continuously improve the marketing efforts.
18. Monitor and evaluate the performance of the marketing efforts and adjust as necessary.
19. Foster a culture of innovation and creativity within the organization.
20. Continuously monitor and analyze the tourist industry trends and adjust the marketing plan accordingly.
21. Foster a strong sense of community and collaboration among stakeholders.
22. Ensure brand consistency and alignment across all marketing channels.
23. Continuously evaluate and refine the destination marketing plan.
24. Regularly update and enhance the website and social media presence.
25. Continuously measure and evaluate the impact of the marketing efforts and make necessary adjustments.

FREQUENTLY ASKED QUESTIONS

What is Smart City Tourism Destination Marketing (SCTDM)? Smart City Tourism Destination Marketing (SCTDM) is a marketing strategy that utilizes smart technologies to promote a city's tourism offerings. This includes utilizing various digital technologies such as mobile apps, virtual reality, and social media to enhance the visitor experience.

How does SCTDM differ from traditional destination marketing? Traditional destination marketing primarily focuses on promoting a city's physical attributes, such as its attractions, culture, and heritage. SCTDM, on the other hand, takes a more holistic approach, incorporating digital technologies to enhance the visitor experience and improve accessibility.

What are some examples of smart technologies used in SCTDM? Examples of smart technologies used in SCTDM include mobile apps that provide real-time information on attractions and events, virtual reality experiences that showcase a city's offerings, and social media campaigns that target specific demographics.

How can SCTDM benefit a city's tourism industry? SCTDM can benefit a city's tourism industry by increasing visitor numbers, improving the visitor experience, and enhancing a city's reputation as a modern and innovative destination. It can also help generate new revenue streams by encouraging visitors to spend more time and money in the city.

How can the general public get involved with SCTDM? The general public can get involved with SCTDM by participating in social media campaigns, providing feedback on mobile apps and virtual reality experiences, and sharing their experiences with others through word of mouth and online reviews.

Who is responsible for implementing SCTDM? SCTDM is typically implemented by a collaboration between government agencies, tourism boards, and private sector stakeholders such as hotels, restaurants, and attractions.

What are some challenges that may arise when implementing SCTDM? Some challenges that may arise when implementing SCTDM include the need for significant investment in digital technologies, data privacy concerns, and the potential for technology failures that may negatively impact the visitor experience.

How can SCTDM help promote sustainable tourism? SCTDM can help promote sustainable tourism by encouraging visitors to use public transportation, reducing waste through the use of digital technologies, and promoting eco-friendly attractions and accommodations.

How can SCTDM help promote cultural heritage preservation? SCTDM can help promote cultural heritage preservation by utilizing virtual reality experiences to showcase historical sites and cultural landmarks, and by promoting responsible tourism practices that respect local traditions and customs.

How can SCTDM adapt to the post-COVID-19 world? SCTDM can adapt to the post-COVID-19 world by incorporating contactless technologies, such as virtual tours and digital check-ins, promoting outdoor attractions and activities, and emphasizing health and safety measures to reassure visitors.

Conclusion: *Implementing SCTDM can help cities attract more visitors and increase tourism revenue, while also providing valuable insights that can be used to improve future marketing efforts and optimize the visitor experience.*

SMART CITY TOURIST EMERGENCY SERVICES (SCTES)

Using technology to provide emergency services for tourists, such as emergency response coordination and medical assistance.

The implementation of SCTES offers numerous benefits to both tourists and the city in which they are visiting. By leveraging technology to provide emergency services, the city is able to provide a safer and more secure experience for tourists. This not only helps to protect tourists from potential harm, but also helps to ensure that they are able to get the help they need in a timely and efficient manner.

One of the key benefits of SCTES is that it enables the coordination of emergency response efforts in real-time. This can greatly improve response times and help to ensure that tourists receive the assistance they need as quickly as possible. In the event of a medical emergency, for example, SCTES can help to quickly dispatch the appropriate medical personnel to the scene. This can greatly reduce the risk of serious injury or illness and help to improve the overall health outcomes for tourists.

Another benefit of SCTES is that it can help to provide tourists with greater peace of mind while they are traveling. Knowing that they have access to reliable and effective emergency services can help to reduce anxiety and improve overall confidence in the safety of the city. This can also help to foster greater trust in the city, which can translate into increased tourism and economic benefits for the city over time.

GENERAL FRAMEWORK

The following framework is a general one and the actual implementation will depend on the specific requirements and constraints of the city.

1. Define the scope of SCTES and its objectives.
2. Assess current emergency services infrastructure and identify areas for improvement.
3. Conduct a feasibility study to evaluate the technical and financial aspects of implementing SCTES.
4. Develop a plan for integrating technology into the existing emergency services infrastructure.
5. Select a technology solution that fits the city's needs and budget.
6. Implement real-time emergency response coordination and communication systems.
7. Develop protocols for medical assistance and emergency medical services.
8. Establish partnerships with local hospitals and medical facilities to provide medical services to tourists.
9. Develop a system for collecting and analyzing data on emergency incidents and medical assistance.
10. Train emergency services personnel on the use of new technology and protocols.
11. Implement a public awareness campaign to inform tourists about the availability of SCTES.
12. Install tourist information kiosks and provide mobile app support for emergency services.
13. Develop a comprehensive data management system for emergency services.
14. Ensure that data privacy and security measures are in place.
15. Monitor and evaluate the performance of SCTES on a regular basis.
16. Conduct regular training and certification programs for emergency services personnel.
17. Regularly update the technology infrastructure to keep pace with advances in technology.

18. Foster partnerships with local businesses and community organizations to support SCTES.
19. Develop a contingency plan to address any potential issues or challenges.
20. Evaluate and improve the overall user experience for tourists.
21. Collaborate with other smart city initiatives to ensure compatibility and integration of systems.
22. Implement a system for evaluating the satisfaction of tourists with SCTES.
23. Establish a feedback mechanism to continuously improve SCTES.
24. Regularly review and update protocols to ensure they are up-to-date and effective.
25. Monitor the overall impact of SCTES on the city and its tourists, and make adjustments as needed.

FREQUENTLY ASKED QUESTIONS

What is Smart City Tourism Emergency Services (SCTES)? Smart City Tourism Emergency Services (SCTES) is an emergency response system that utilizes smart technologies to provide quick and effective emergency services to tourists in a city. This includes utilizing digital technologies such as mobile apps and sensors to detect and respond to emergency situations.

How does SCTES differ from traditional emergency services? Traditional emergency services are designed to respond to emergency situations in general, while SCTES is specifically designed to respond to emergency situations involving tourists. SCTES incorporates digital technologies to improve response times and provide tailored emergency services to tourists.

What are some examples of smart technologies used in SCTES? Examples of smart technologies used in SCTES include mobile apps that provide real-time information on emergency services, sensors that detect emergency situations, and virtual reality simulations that train emergency responders on responding to emergency situations involving tourists.

How can SCTES benefit tourists visiting a city? SCTES can benefit tourists by providing quick and effective emergency services in case of emergency situations. This can improve the overall safety and security of the tourist destination and increase visitor confidence.

How can the general public get involved with SCTES? The general public can get involved with SCTES by downloading and using the mobile app, reporting emergency situations, and participating in training programs for emergency responders.

Who is responsible for implementing SCTES? SCTES is typically implemented by a collaboration between government agencies, emergency services providers, and private sector stakeholders such as hotels, restaurants, and attractions.

What are some challenges that may arise when implementing SCTES? Some challenges that may arise when implementing SCTES include the need for significant investment in digital technologies, ensuring data privacy and security, and the potential for technology failures that may negatively impact emergency response times.

How can SCTES help promote sustainable tourism? SCTES can help promote sustainable tourism by improving the safety and security of tourist destinations, reducing the impact of emergency situations on the environment, and promoting responsible tourism practices that minimize the risk of emergency situations.

How can SCTES help promote cultural heritage preservation? SCTES can help promote cultural heritage preservation by ensuring the safety and security of cultural heritage sites and landmarks, promoting responsible tourism practices that respect local traditions and customs, and training emergency responders on responding to emergency situations involving cultural heritage sites.

How can SCTES adapt to the post-COVID-19 world? SCTES can adapt to the post-COVID-19 world by incorporating contactless technologies, such as remote triage and telemedicine, promoting health and safety measures to tourists, and training emergency responders on responding to emergency situations involving infectious diseases.

Conclusion: *The implementation of SCTES offers numerous benefits to both tourists and the city in which they are visiting. By leveraging technology to provide emergency services, the city is able to provide a safer and more secure experience for tourists, while also helping to improve response times, promote health outcomes, and provide peace of mind. This, in turn, can help to foster greater trust in the city and contribute to increased tourism and economic benefits for the city.*

SMART CITY TOURIST EVENT MANAGEMENT (SCTEM)

Using technology to manage and promote tourist events, such as digital ticketing and event scheduling.

Implementing technology for SCTEM can bring a range of benefits for both city officials and visitors. One major benefit is the ability to improve the overall coordination and organization of events. With digital ticketing and scheduling tools, it becomes easier to keep track of event attendance, manage ticket sales, and plan for resources such as staff and equipment. This can lead to more efficient use of resources and a better overall experience for visitors.

Another benefit is the ability to promote events more effectively. Digital marketing and advertising tools can help reach a wider audience and target specific demographics, leading to higher attendance and more successful events. Real-time updates and notifications can be used to keep visitors informed and engaged throughout the event.

SCTEM can help to improve safety and security during events. By using real-time monitoring and emergency response coordination, officials can quickly respond to any issues that may arise. This can help to ensure the safety of visitors and minimize the impact of any disruptions.

GENERAL FRAMEWORK

The following framework is a general one and the actual implementation will depend on the specific requirements and constraints of the city.

1. Define the scope and objectives of SCTEM.
2. Assess the current event management infrastructure and identify areas for improvement.
3. Conduct a feasibility study to evaluate the technical and financial aspects of implementing SCTEM.
4. Develop a plan for integrating technology into the existing event management infrastructure.
5. Select a technology solution that fits the city's needs and budget.
6. Implement a system for digital ticketing and event scheduling.
7. Develop a comprehensive database of tourist events in the city.
8. Establish partnerships with local businesses and organizations to promote tourist events.
9. Develop a system for collecting and analyzing data on tourist events.
10. Train event management personnel on the use of new technology and protocols.
11. Implement a public awareness campaign to inform tourists about SCTEM.
12. Develop a mobile app for tourists to access information about events and purchase tickets.
13. Ensure that data privacy and security measures are in place.
14. Monitor and evaluate the performance of SCTEM on a regular basis.
15. Conduct regular training and certification programs for event management personnel.
16. Regularly update the technology infrastructure to keep pace with advances in technology.
17. Foster partnerships with local businesses and community organizations to support SCTEM.
18. Develop a contingency plan to address any potential issues or challenges.
19. Evaluate and improve the overall user experience for tourists.
20. Collaborate with other smart city initiatives to ensure compatibility and integration of systems.
21. Implement a system for evaluating the satisfaction of tourists with SCTEM.
22. Establish a feedback mechanism to continuously improve SCTEM.

23. Regularly review and update protocols to ensure they are up-to-date and effective.
24. Monitor the overall impact of SCTEM on the city and its tourists, and make adjustments as needed.
25. Continuously strive to improve and enhance the quality and variety of tourist events in the city.

FREQUENTLY ASKED QUESTIONS

What is Smart City Tourism Event Management (SCTEM)? Smart City Tourism Event Management (SCTEM) is a system that utilizes smart technologies to manage events in a city. This includes utilizing digital technologies such as mobile apps and sensors to streamline event planning, promotion, and execution.

How can SCTEM benefit event organizers? SCTEM can benefit event organizers by providing real-time data on event attendance, feedback, and engagement. This can help organizers make informed decisions and improve the overall event experience for attendees.

How can SCTEM benefit attendees of tourism events? SCTEM can benefit attendees by providing real-time information on event schedules, locations, and activities. This can help attendees make informed decisions and have a more enjoyable and immersive event experience.

What are some examples of smart technologies used in SCTEM? Examples of smart technologies used in SCTEM include mobile apps that provide real-time information on events, sensors that detect and respond to changes in event attendance, and virtual reality simulations that train event organizers on managing events in real-time.

How can the general public get involved with SCTEM? The general public can get involved with SCTEM by downloading and using the mobile app, providing feedback on events attended, and participating in community forums to suggest improvements for future events.

Who is responsible for implementing SCTEM? SCTEM is typically implemented by a collaboration between government agencies, event management companies, and private sector stakeholders such as hotels, restaurants, and attractions.

What are some challenges that may arise when implementing SCTEM? Some challenges that may arise when implementing SCTEM include the need for significant investment in digital technologies, ensuring data privacy and security, and the potential for technology failures that may negatively impact event experiences.

How can SCTEM help promote sustainable tourism? SCTEM can help promote sustainable tourism by providing event organizers with real-time data on resource usage and waste generation, and promoting responsible event management practices that minimize environmental impact.

How can SCTEM help promote cultural heritage preservation? SCTEM can help promote cultural heritage preservation by providing event organizers with real-time information on cultural heritage sites and landmarks, promoting responsible event management practices that respect local traditions and customs, and encouraging the inclusion of local cultural activities in event programming.

How can SCTEM adapt to the post-COVID-19 world? SCTEM can adapt to the post-COVID-19 world by incorporating contactless technologies, such as virtual and hybrid events, promoting health and safety measures to attendees, and providing real-time data on event attendance to help enforce social distancing measures.

Conclusion: *Implementing technology for SCTEM can lead to more efficient, effective, and safe events, ultimately bringing a better experience for tourists and helping to promote the city as a destination.*

SMART CITY TOURIST EXPERIENCE MANAGEMENT (SCTXM)

Using technology to enhance the tourist experience, such as virtual and augmented reality, and gamification.

Implementing SCTXM can greatly enhance the overall experience for tourists visiting a city. By using technology such as virtual and augmented reality, tourists can have a more immersive and interactive experience while exploring the city's attractions.

This can be particularly useful for historical and cultural sites, allowing tourists to gain a deeper understanding and appreciation of the city's history and heritage. The use of gamification can make the experience more engaging and fun, encouraging tourists to explore more of the city and discover hidden gems they may have otherwise missed.

This technology can also provide valuable data on how tourists interact with the city and which experiences are most popular, allowing for more effective and targeted marketing and planning for future events. Real-time monitoring and analysis of the tourist's experiences can also help identify and address any pain points or issues, ensuring a smooth and enjoyable experience for all visitors.

GENERAL FRAMEWORK

The following framework is a general one and the actual implementation will depend on the specific requirements and constraints of the city.

1. Define the goals and objectives of SCTXM.
2. Identify key tourist areas and attractions.
3. Conduct a feasibility study to assess the potential for using technology.
4. Determine the appropriate technology to use, such as virtual reality, augmented reality, or gamification.
5. Develop a detailed project plan with timelines and budgets.

6. Acquire necessary resources, such as technology, hardware, and software.
7. Identify and engage key stakeholders, including local government, tourist boards, and technology partners.
8. Build partnerships with technology providers and other relevant organizations.
9. Develop a marketing and communication strategy to promote the SCTXM initiative.
10. Train staff and stakeholders in the use of the technology.
11. Implement the technology in key tourist areas and attractions.
12. Evaluate the effectiveness of the technology and identify areas for improvement.
13. Gather feedback from tourists and use it to improve the experience.
14. Continuously update and maintain the technology.
15. Foster a culture of innovation and encourage ongoing experimentation and development.
16. Encourage collaboration and sharing of best practices among stakeholders.
17. Invest in research and development to stay ahead of the curve in technology and innovation.
18. Utilize data and analytics to make informed decisions about the SCTXM initiative.
19. Foster a culture of sustainability and ensure that the technology is used in a responsible and ethical manner.
20. Ensure that the technology is accessible to all visitors, including those with disabilities.
21. Develop partnerships with tourism industry associations and organizations to promote SCTXM.
22. Develop a robust security and privacy framework to protect visitor data.
23. Ensure that the technology is easy to use and does not detract from the overall tourist experience.
24. Foster a culture of continuous improvement and seek to continuously enhance the tourist experience.
25. Evaluate the overall impact of SCTXM on the tourist industry and the local economy.

FREQUENTLY ASKED QUESTIONS

What is Smart City Tourism Experience Management (SCTXM)? Smart City Tourism Experience Management (SCTXM) is a system that uses smart technologies to improve the overall tourism experience in a city. It includes the use of digital technologies to enhance the interaction between tourists and their environment.

How does SCTXM work to improve the tourism experience? SCTXM works to improve the tourism experience by providing personalized and real-time information to tourists, such as recommendations for attractions, events, and restaurants. This enhances the overall tourism experience and ensures that tourists have a memorable visit.

What are some examples of smart technologies used in SCTXM? Examples of smart technologies used in SCTXM include mobile apps that provide personalized recommendations, sensors that collect data on tourist behaviour, and augmented reality applications that enhance the visitor experience.

Who is responsible for implementing SCTXM? The implementation of SCTXM is typically the responsibility of government agencies, tourism boards, and private sector stakeholders such as hotels, restaurants, and attractions.

How can tourists access the SCTXM system? Tourists can access the SCTXM system through mobile apps, interactive kiosks, and digital signage located throughout the city.

How can SCTXM help promote sustainable tourism? SCTXM can help promote sustainable tourism by providing tourists with real-time information on sustainable tourism practices and promoting responsible tourism behaviour. This can help minimize the impact of tourism on the environment.

What are some challenges that may arise when implementing SCTXM? Challenges that may arise when implementing SCTXM include the need for significant investment in digital technologies, ensuring data privacy and security, and the potential for technology failures that may negatively impact the tourist experience.

How can SCTXM help promote cultural heritage preservation? SCTXM can help promote cultural heritage preservation by providing tourists with real-time information on cultural heritage sites and landmarks, promoting responsible tourism behaviour that respects local traditions and customs, and encouraging the inclusion of local cultural activities in the tourism experience.

How can SCTXM help to create a seamless tourism experience? SCTXM can help to create a seamless tourism experience by providing tourists with personalized recommendations, real-time information, and interactive experiences. This ensures that tourists have a hassle-free and enjoyable visit.

How can SCTXM adapt to the post-COVID-19 world? SCTXM can adapt to the post-COVID-19 world by incorporating contactless technologies, promoting health and safety measures to tourists, and providing real-time information on crowd density and social distancing measures to help enforce safety protocols.

Conclusion: *Using SCTXM technology to enhance the tourist experience can greatly improve the overall satisfaction and enjoyment of visitors, leading to increased tourism and economic benefits for the city.*

SMART CITY TOURIST FEEDBACK AND REVIEWS (SCTFR)

Using technology to collect and analyze tourist feedback and reviews, such as online surveys and review platforms.

Implementing SCTFR using technology can bring a variety of benefits to both the city and its visitors. One major benefit is the ability to gather real-time feedback from tourists on their experiences. This can help city officials and businesses identify areas of improvement and make necessary adjustments to enhance the overall tourist experience.

The use of online surveys and review platforms can provide valuable insights into the demographics and spending patterns of tourists, allowing for more targeted and effective destination marketing. This can lead to increased tourism and revenue for the city.

Moreover, collecting and analyzing tourist feedback and reviews can also help to improve the overall reputation of the city as a tourist destination. Positive reviews and high ratings can attract more visitors, while addressing negative feedback can lead to improvements that can enhance the overall reputation of the city. The analysis of feedback and reviews can also be used to identify best practices and trends that can be replicated in other areas of the city.

Integrating feedback and reviews with other smart city systems, such as digital ticketing, can provide a holistic view of the tourist experience and help to optimize the flow of visitors.

GENERAL FRAMEWORK

The following framework is a general one and the actual implementation will depend on the specific requirements and constraints of the city.

1. Define the goals and objectives of SCTFR.
2. Identify key tourist areas and attractions to collect feedback from.
3. Develop a plan for collecting feedback and reviews, including online surveys and review platforms.
4. Acquire necessary resources, such as technology, hardware, and software.
5. Identify and engage key stakeholders, including local government, tourist boards, and technology partners.
6. Build partnerships with technology providers and other relevant organizations.
7. Develop a marketing and communication strategy to promote SCTFR initiative.

8. Train staff and stakeholders in the use of the technology.
9. Implement the technology for collecting feedback and reviews in key tourist areas and attractions.
10. Evaluate the effectiveness of the technology and identify areas for improvement.
11. Gather and analyze feedback and reviews to identify trends and areas for improvement.
12. Use feedback and reviews to inform decision-making and improve the tourist experience.
13. Continuously update and maintain the technology.
14. Foster a culture of innovation and encourage ongoing experimentation and development.
15. Encourage collaboration and sharing of best practices among stakeholders.
16. Invest in research and development to stay ahead of the curve in technology and innovation.
17. Utilize data and analytics to make informed decisions about the SCTFR initiative.
18. Foster a culture of sustainability and ensure that the technology is used in a responsible and ethical manner.
19. Ensure that the technology is accessible to all visitors, including those with disabilities.
20. Develop partnerships with tourism industry associations and organizations to promote SCTFR.
21. Develop a robust security and privacy framework to protect visitor data.
22. Ensure that the technology is easy to use and does not detract from the overall tourist experience.
23. Foster a culture of continuous improvement and seek to continuously enhance the tourist experience.
24. Evaluate the overall impact of SCTFR on the tourist industry and the local economy.
25. Encourage tourists to provide feedback and reviews, and incentivize them to do so.

FREQUENTLY ASKED QUESTIONS

What is Smart City Tourist Feedback and Reviews (SCTFR)?
Smart City Tourist Feedback and Reviews (SCTFR) is a system that

collects and analyses feedback from tourists to improve the overall tourism experience in a city.

How does SCTFR work? SCTFR collects feedback from tourists through various channels such as mobile apps, social media, and digital kiosks. The data is then analysed to identify areas of improvement and to provide insights into the tourist experience.

Who is responsible for implementing SCTFR? SCTFR is typically implemented by government agencies, tourism boards, and private sector stakeholders such as hotels, restaurants, and attractions.

How can tourists provide feedback through SCTFR? Tourists can provide feedback through various channels such as mobile apps, social media, and digital kiosks. They can rate their experiences, provide comments, and share photos and videos.

How is the feedback collected through SCTFR used to improve the tourist experience? The feedback collected through SCTFR is analysed to identify areas of improvement, and to provide insights into the tourist experience. This information can then be used to make data-driven decisions that can enhance the overall tourist experience.

How can SCTFR ensure the privacy and security of tourists' data? SCTFR can ensure the privacy and security of tourists' data by implementing robust data protection measures, such as data encryption, access controls, and regular security audits.

How can SCTFR be used to improve sustainability in tourism? SCTFR can be used to improve sustainability in tourism by collecting feedback on sustainable tourism practices and using this information to promote responsible tourism behaviour. This can help minimize the impact of tourism on the environment.

How can SCTFR help to promote local businesses? SCTFR can help to promote local businesses by collecting feedback on their services and promoting those that receive positive reviews. This can help to drive more business to local establishments and create a more vibrant local economy.

How can SCTFR help to promote cultural diversity in tourism? SCTFR can help to promote cultural diversity in tourism by collecting feedback on cultural experiences and promoting those that receive positive reviews. This can encourage tourists to explore local cultures and traditions and promote cross-cultural understanding.

How can SCTFR adapt to changes in tourism trends and preferences? SCTFR can adapt to changes in tourism trends and preferences by using real-time data analysis to identify emerging trends and changing tourist preferences. This can help tourism stakeholders to make informed decisions that keep up with the changing demands of the market.

Conclusion: *Implementing SCTFR using technology can help to improve the overall tourist experience, increase tourism and revenue, and enhance the reputation of the city as a tourist destination.*

SMART CITY TOURIST INFORMATION SERVICES (SCTIS)

Using technology to provide real-time tourist information, such as digital concierge services and interactive city guides.

Implementing SCTIS can bring a wide range of benefits for both tourists and the city. One of the main benefits is the ability to provide real-time information to tourists. Digital concierge services and interactive city guides can offer tourists up-to-date information on things like weather, transportation, and local events, which can help them plan their trip more effectively and make the most of their time in the city.

Another benefit of SCTIS is the ability to personalize the tourist experience. By collecting and analyzing data on tourists' preferences and behaviour, cities can tailor the information and recommendations they provide to individual visitors, making their experience more enjoyable and efficient. This can also help to improve the overall satisfaction and likelihood of return visits of tourists.

Smart tourist information services can also help to promote sustainable tourism practices. By providing tourists with information on sustainable transportation options and eco-friendly activities and accommodations, cities can encourage them to make more environmentally-conscious choices during their visit.

SCTIS can also bring benefits for the city itself. By having real-time information on tourist numbers and behaviours, cities can better plan and manage resources like transportation, accommodation, and public spaces. This can help to optimize the flow of tourists, reducing congestion and improving overall visitor satisfaction.

GENERAL FRAMEWORK

The following framework is a general one and the actual implementation will depend on the specific requirements and constraints of the city.

1. Define the goals and objectives of SCTIS.
2. Identify key tourist areas and attractions to provide information services.
3. Develop a plan for providing real-time tourist information, such as digital concierge services and interactive city guides.
4. Acquire necessary resources, such as technology, hardware, and software.
5. Identify and engage key stakeholders, including local government, tourist boards, and technology partners.
6. Build partnerships with technology providers and other relevant organizations.
7. Develop a marketing and communication strategy to promote SCTIS initiative.
8. Train staff and stakeholders in the use of the technology.
9. Implement the technology for providing real-time tourist information in key tourist areas and attractions.

10. Evaluate the effectiveness of the technology and identify areas for improvement.
11. Continuously update and maintain the technology.
12. Foster a culture of innovation and encourage ongoing experimentation and development.
13. Encourage collaboration and sharing of best practices among stakeholders.
14. Invest in research and development to stay ahead of the curve in technology and innovation.
15. Utilize data and analytics to make informed decisions about the SCTIS initiative.
16. Foster a culture of sustainability and ensure that the technology is used in a responsible and ethical manner.
17. Ensure that the technology is accessible to all visitors, including those with disabilities.
18. Develop partnerships with tourism industry associations and organizations to promote SCTIS.
19. Develop a robust security and privacy framework to protect visitor data.
20. Ensure that the technology is easy to use and does not detract from the overall tourist experience.
21. Foster a culture of continuous improvement and seek to continuously enhance the tourist experience.
22. Evaluate the overall impact of SCTIS on the tourist industry and the local economy.
23. Provide clear and comprehensive information, including maps, schedules, and recommendations.
24. Offer a range of information services, including live chat and phone support.
25. Ensure that the information services are available in multiple languages to accommodate visitors from different countries.

FREQUENTLY ASKED QUESTIONS

What are Smart City Tourist Information Services (SCTIS)? Smart City Tourist Information Services (SCTIS) are digital platforms that provide information to tourists about the attractions, events, services, and amenities in a city.

How can I access SCTIS as a tourist? SCTIS can be accessed through various channels such as mobile apps, websites, and digital kiosks, which are typically located at major tourist hubs in the city.

What kind of information can I find on SCTIS? SCTIS provides information on a wide range of topics such as tourist attractions, events, accommodations, restaurants, transportation, and safety information.

How accurate is the information provided by SCTIS? SCTIS strives to provide accurate and up-to-date information to tourists. However, it is important to note that some information may be subject to change, and tourists should confirm the information before making plans.

Can I use SCTIS to plan my itinerary for my trip? Yes, SCTIS provides a wealth of information that can be used to plan a detailed itinerary for your trip.

How can SCTIS help me save time and money during my trip? SCTIS can help you save time and money by providing information on the most efficient routes to tourist attractions, and by recommending affordable and high-quality accommodations and restaurants.

Who is responsible for maintaining SCTIS? SCTIS is typically maintained by government agencies, tourism boards, and private sector stakeholders such as hotels, restaurants, and attractions.

How can SCTIS help to promote sustainable tourism? SCTIS can help to promote sustainable tourism by providing information on eco-friendly activities and accommodations, and by promoting responsible tourism behaviour.

Can I provide feedback on SCTIS? Yes, SCTIS typically provides a feedback mechanism that allows tourists to rate the quality of the information provided, and to provide suggestions for improvement.

How can SCTIS be improved to better serve tourists? SCTIS can be improved by continuously updating and improving the quality of the information provided, by incorporating emerging technologies such as AI and machine learning, and by providing personalized recommendations based on the tourist's preferences and behaviour.

Conclusion: *SCTIS can provide real-time information to tourists, personalize their experience, promote sustainable tourism, and help the city manage resources more effectively.*

SMART CITY TOURIST LANGUAGE ASSISTANCE (SCTLA)

Using technology to provide language assistance for tourists, such as translation and interpretation services.

Implementing SCTLA can bring a number of benefits to the city and its visitors. One of the main benefits is the ability to provide real-time translation and interpretation services for tourists, which can improve their overall experience and make it easier for them to navigate the city. This can include providing translation services for signs, menus, and other information, as well as offering interpretation services for phone calls and in-person interactions.

Another benefit of SCTLA is the ability to attract more international visitors to the city. Many tourists may be hesitant to visit a city if they are not fluent in the local language, so providing language assistance can make the city more accessible and appealing to a wider range of visitors.

In addition to improving the experience of individual tourists, SCTLA can also benefit local businesses. For example, providing translation services for menus and other information can make it easier for tourists to understand what is being offered and make informed decisions about where to eat and shop.

SCTLA can also improve safety for tourists by providing them with the ability to communicate with emergency services in their native language, which can be especially important in emergency situations.

GENERAL FRAMEWORK

The following framework is a general one and the actual implementation will depend on the specific requirements and constraints of the city.

1. Define the goals and objectives of SCTLA.
2. Identify key tourist areas and attractions where language assistance is needed.
3. Develop a plan for providing language assistance services, such as translation and interpretation services.
4. Acquire necessary resources, such as technology, hardware, and software.
5. Identify and engage key stakeholders, including local government, tourist boards, and language service providers.
6. Build partnerships with language service providers and other relevant organizations.
7. Develop a marketing and communication strategy to promote SCTLA initiative.
8. Train staff and stakeholders in the use of the technology.
9. Implement the technology for providing language assistance services in key tourist areas and attractions.
10. Evaluate the effectiveness of the technology and identify areas for improvement.
11. Continuously update and maintain the technology.
12. Foster a culture of innovation and encourage ongoing experimentation and development.
13. Encourage collaboration and sharing of best practices among stakeholders.
14. Invest in research and development to stay ahead of the curve in technology and innovation.
15. Utilize data and analytics to make informed decisions about the SCTLA initiative.
16. Foster a culture of sustainability and ensure that the technology is used in a responsible and ethical manner.

17. Ensure that the technology is accessible to all visitors, including those with disabilities.
18. Develop partnerships with language industry associations and organizations to promote SCTLA.
19. Develop a robust security and privacy framework to protect visitor data.
20. Ensure that the technology is easy to use and does not detract from the overall tourist experience.
21. Foster a culture of continuous improvement and seek to continuously enhance the tourist experience.
22. Evaluate the overall impact of SCTLA on the tourist industry and the local economy.
23. Provide language assistance services in multiple languages to accommodate visitors from different countries.
24. Offer a range of language assistance services, including live chat and phone support.
25. Ensure that language assistance services are available 24/7 to accommodate visitors' needs.

FREQUENTLY ASKED QUESTIONS

What is Smart City Tourist Language Assistance (SCTLA)? SCTLA is a service that provides real-time translation and interpretation services to tourists in their native language.

How can I access SCTLA as a tourist? SCTLA can be accessed through various channels such as mobile apps, websites, and digital kiosks, which are typically located at major tourist hubs in the city.

Which languages are supported by SCTLA? SCTLA supports a wide range of languages, depending on the city and the service provider. Typically, the service supports the most commonly spoken languages in the city.

Is SCTLA available 24/7? The availability of SCTLA depends on the service provider. Some providers may offer 24/7 support, while others may have limited hours of operation.

Can SCTLA help me communicate with locals who don't speak my language? Yes, SCTLA can help you communicate with locals who don't speak your language by providing real-time translation and interpretation services.

How accurate is the translation and interpretation provided by SCTLA? SCTLA strives to provide accurate and high-quality translation and interpretation services. However, it is important to note that some nuances and cultural differences may be lost in translation.

How much does SCTLA cost? The cost of SCTLA depends on the service provider. Some providers may offer the service for free, while others may charge a fee.

Who is responsible for providing SCTLA? SCTLA is typically provided by government agencies, tourism boards, and private sector stakeholders such as hotels, restaurants, and attractions.

How can SCTLA help to improve the tourist experience? SCTLA can help to improve the tourist experience by breaking down language barriers, and by allowing tourists to communicate effectively with locals and navigate the city more easily.

Can I provide feedback on the quality of SCTLA? Yes, SCTLA typically provides a feedback mechanism that allows tourists to rate the quality of the translation and interpretation provided, and to provide suggestions for improvement.

Conclusion: *Implementing SCTLA can help to improve the experience of tourists visiting the city, increase the number of international visitors, and support local businesses, while also improving safety.*

SMART CITY TOURIST NAVIGATION (SCTN)

Using technology to provide tourists with real-time navigation and route planning, such as GPS and map services.

Implementing SCTN technology can bring a wide range of benefits for both tourists and the city itself. One major benefit is the ability to provide tourists with real-time navigation and route planning, which can greatly enhance the overall tourism experience. This can include GPS and map services, which can help tourists easily find their way around the city and navigate to popular destinations, as well as providing information on public transportation and other modes of transportation.

Another benefit of SCTN technology is the ability to optimize the flow of tourists in the city. This can be done through real-time crowd monitoring and dynamic pricing, which can help manage the number of tourists visiting popular destinations, and prevent overcrowding and long wait times.

SCTN technology can also be used to collect and analyze data on tourist behaviour and spending patterns. This data can be used to inform city planning and decision-making, as well as provide valuable insights to local businesses and tourist industry stakeholders.

By providing tourists with real-time information and navigation services, SCTN technology can also improve the safety and security of tourists. This includes providing emergency services such as emergency response coordination and medical assistance.

GENERAL FRAMEWORK

The following framework is a general one and the actual implementation will depend on the specific requirements and constraints of the city.

1. Define the goals and objectives of SCTN.
2. Identify key tourist areas and attractions that require navigation and route planning services.
3. Conduct a comprehensive review of existing navigation and map services.
4. Identify technology solutions that can be used for SCTN, such as GPS and mapping software.

5. Develop a detailed plan for the implementation of SCTN.
6. Acquire necessary resources, including hardware and software.
7. Train staff and stakeholders in the use of the technology.
8. Develop a marketing and communication strategy to promote SCTN.
9. Implement the technology for real-time navigation and route planning services in key tourist areas and attractions.
10. Evaluate the effectiveness of the technology and identify areas for improvement.
11. Continuously update and maintain the technology.
12. Foster a culture of innovation and encourage ongoing experimentation and development.
13. Encourage collaboration and sharing of best practices among stakeholders.
14. Invest in research and development to stay ahead of the curve in technology and innovation.
15. Utilize data and analytics to make informed decisions about the SCTN initiative.
16. Foster a culture of sustainability and ensure that the technology is used in a responsible and ethical manner.
17. Ensure that the technology is accessible to all visitors, including those with disabilities.
18. Develop partnerships with technology companies and organizations to promote SCTN.
19. Develop a robust security and privacy framework to protect visitor data.
20. Ensure that the technology is easy to use and does not detract from the overall tourist experience.
21. Foster a culture of continuous improvement and seek to continuously enhance the tourist experience.
22. Evaluate the overall impact of SCTN on the tourist industry and the local economy.
23. Provide navigation services in multiple languages to accommodate visitors from different countries.
24. Offer real-time traffic updates and detour planning to improve the navigation experience.
25. Ensure that navigation services are available 24/7 to accommodate visitors' needs.

FREQUENTLY ASKED QUESTIONS

What is Smart City Tourist Navigation (SCTN)? SCTN is a service that helps tourists navigate the city by providing real-time information about transportation, traffic, and local attractions.

How can I access SCTN as a tourist? SCTN can be accessed through various channels such as mobile apps, websites, and digital kiosks, which are typically located at major tourist hubs in the city.

Can SCTN provide information about public transportation? Yes, SCTN can provide real-time information about public transportation, such as bus and train schedules, delays, and routes.

Can SCTN provide walking directions? Yes, SCTN can provide walking directions to help tourists navigate the city on foot.

How accurate is the information provided by SCTN? SCTN strives to provide accurate and up-to-date information, but it is important to note that some information may be subject to change, particularly in the case of traffic and public transportation.

Can SCTN provide information about local attractions and events? Yes, SCTN can provide information about local attractions and events, such as museums, restaurants, and festivals.

How much does SCTN cost? The cost of SCTN depends on the service provider. Some providers may offer the service for free, while others may charge a fee.

Who is responsible for providing SCTN? SCTN is typically provided by government agencies, tourism boards, and private sector stakeholders such as hotels, restaurants, and attractions.

How can SCTN help to improve the tourist experience? SCTN can help to improve the tourist experience by making it easier for tourists to navigate the city and find local attractions, restaurants, and events.

Can I provide feedback on the quality of SCTN? Yes, SCTN typically provides a feedback mechanism that allows tourists to rate the quality of the service and to provide suggestions for improvement.

Conclusion: *Implementing SCTN technology can greatly enhance the overall tourism experience, as well as provide a range of benefits for both tourists and the city. It can improve the flow of tourists, increase safety and security, and provide valuable data for informed decision-making.*

SMART CITY TOURIST PARKING MANAGEMENT (SCTPM)

Using technology to manage and optimize parking availability for tourists, such as real-time parking guidance and dynamic pricing.

Implementing SCTPM can bring a number of benefits to both tourists and city officials. By using technology to manage and optimize parking availability, cities can help tourists easily find parking spots and reduce the time spent searching for a place to park. This can improve the overall tourist experience and make the city more attractive as a destination.

Real-time parking guidance can be provided through a number of technologies, such as GPS and map services, to help tourists find available parking spots quickly and easily. This can also help to reduce congestion on the roads and make the city more efficient. Dynamic pricing can be used to adjust the cost of parking based on demand, which can help to ensure that parking spots are always available for tourists when they need them.

SCTPM can also help city officials to better understand the usage patterns of parking spots and make data-driven decisions about how to manage them. By collecting data on parking usage, cities can identify areas where more parking is needed and make adjustments to increase availability. This can also be used to identify areas where parking is underutilized and repurpose those spaces for other uses.

GENERAL FRAMEWORK

The following framework is a general one and the actual implementation will depend on the specific requirements and constraints of the city.

1. Define the goals and objectives of SCTPM.
2. Identify key tourist areas and attractions that require parking management services.
3. Conduct a comprehensive review of existing parking management systems.
4. Identify technology solutions that can be used for SCTPM, such as real-time parking guidance and dynamic pricing.
5. Develop a detailed plan for the implementation of SCTPM.
6. Acquire necessary resources, including hardware and software.
7. Train staff and stakeholders in the use of the technology.
8. Develop a marketing and communication strategy to promote SCTPM.
9. Implement the technology for real-time parking guidance and dynamic pricing in key tourist areas and attractions.
10. Evaluate the effectiveness of the technology and identify areas for improvement.
11. Continuously update and maintain the technology.
12. Foster a culture of innovation and encourage ongoing experimentation and development.
13. Encourage collaboration and sharing of best practices among stakeholders.
14. Invest in research and development to stay ahead of the curve in technology and innovation.
15. Utilize data and analytics to make informed decisions about the SCTPM initiative.
16. Foster a culture of sustainability and ensure that the technology is used in a responsible and ethical manner.
17. Ensure that the technology is accessible to all visitors, including those with disabilities.
18. Develop partnerships with technology companies and organizations to promote SCTPM.
19. Develop a robust security and privacy framework to protect visitor data.
20. Ensure that the technology is easy to use and does not detract from the overall tourist experience.

21. Foster a culture of continuous improvement and seek to continuously enhance the tourist experience.
22. Evaluate the overall impact of SCTPM on the tourist industry and the local economy.
23. Provide parking management services in multiple languages to accommodate visitors from different countries.
24. Offer real-time parking availability updates and parking reservation options to improve the parking experience.
25. Ensure that parking management services are available 24/7 to accommodate visitors' needs.

FREQUENTLY ASKED QUESTIONS

What is Smart City Tourist Parking Management (SCTPM)? SCTPM is a service that manages parking for tourists, helping them find available parking spots and providing real-time information about parking availability and pricing.

How can I access SCTPM as a tourist? SCTPM can be accessed through various channels such as mobile apps, websites, and digital kiosks, which are typically located at major tourist hubs in the city.

How does SCTPM help me find parking spots? SCTPM uses real-time data to show available parking spots in the city, helping tourists save time and avoid frustration.

Can SCTPM help me find the cheapest parking options? Yes, SCTPM can provide information about parking prices and help tourists find the most affordable parking options.

Can I reserve a parking spot in advance through SCTPM? Yes, some SCTPM providers allow tourists to reserve parking spots in advance through mobile apps or websites.

How does SCTPM benefit the city and its residents? SCTPM helps to reduce traffic congestion and carbon emissions by minimizing the time spent searching for parking, while also providing revenue for the city and local businesses.

How secure is the payment process for parking through SCTPM? SCTPM providers typically use secure payment gateways to ensure that transactions are safe and secure.

Who is responsible for providing SCTPM? SCTPM is typically provided by government agencies, parking companies, and private sector stakeholders such as hotels, restaurants, and attractions.

How can SCTPM help to improve the tourist experience? SCTPM can help to improve the tourist experience by making it easier for tourists to find parking and by reducing stress and frustration associated with parking in a new city.

How can I provide feedback on the quality of SCTPM? SCTPM typically provides a feedback mechanism that allows tourists to rate the quality of the service and to provide suggestions for improvement.

Conclusion: *SCTPM can greatly improve the experience of tourists visiting the city, make the city more efficient and reduce congestion on the roads. It can also help city officials to better understand how parking is used in the city and make data-driven decisions about how to manage it effectively.*

SMART CITY TOURIST PERSONALIZATION (SCTP)

Using technology to provide personalized tourist experiences, such as recommendations and customized itineraries.

Implementing technology for SCTP can have a wide range of benefits for both tourists and the city itself. One of the main benefits is the ability to provide personalized recommendations and customized itineraries to tourists based on their preferences and interests. This can enhance their overall experience and satisfaction, as they are able to discover and explore the city in a way that is tailored to them. By collecting and analyzing data on tourists' preferences and behaviour, cities can gain valuable insights into what visitors are looking for and use this information to improve the overall tourism offering.

Another benefit of SCTP is the ability to improve the efficiency and convenience of tourist services. For example, using technology to provide real-time information and recommendations can help tourists navigate the city more easily and make the most of their time. It can also help them find and book accommodations, activities, and transportation that meet their specific needs. This can lead to a more enjoyable and stress-free experience for tourists, which can increase their likelihood of returning to the city in the future.

Moreover, SCTP can also lead to a more sustainable tourism industry. By providing tourists with personalized recommendations, they may be more likely to explore lesser-known areas of the city, which can help to distribute the number of visitors more evenly across the city and reduce the pressure on popular tourist hotspots. It can also help to promote more sustainable practices, such as using public transportation or walking instead of driving, which can have a positive impact on the environment.

GENERAL FRAMEWORK

The following framework is a general one and the actual implementation will depend on the specific requirements and constraints of the city.

1. Define the goals and objectives of SCTP.
2. Conduct a comprehensive review of existing personalization technology and services.
3. Identify key tourist areas and attractions that would benefit from personalized experiences.
4. Develop a detailed plan for the implementation of SCTP, including technology and resource requirements.
5. Acquire necessary technology and resources.
6. Train staff and stakeholders in the use of the technology.
7. Develop a marketing and communication strategy to promote SCTP.

8. Implement the technology for personalized experiences in key tourist areas and attractions.
9. Continuously collect and analyze data to inform the development of personalized experiences.
10. Foster a culture of innovation and encourage ongoing experimentation and development.
11. Encourage collaboration and sharing of best practices among stakeholders.
12. Invest in research and development to stay ahead of the curve in technology and innovation.
13. Utilize data and analytics to make informed decisions about the SCTP initiative.
14. Foster a culture of sustainability and ensure that the technology is used in a responsible and ethical manner.
15. Ensure that the technology is accessible to all visitors, including those with disabilities.
16. Develop partnerships with technology companies and organizations to promote SCTP.
17. Develop a robust security and privacy framework to protect visitor data.
18. Ensure that the technology is easy to use and does not detract from the overall tourist experience.
19. Foster a culture of continuous improvement and seek to continuously enhance the tourist experience.
20. Evaluate the overall impact of SCTP on the tourist industry and the local economy.
21. Provide personalized experiences in multiple languages to accommodate visitors from different countries.
22. Offer real-time recommendations based on visitor preferences and behaviours.
23. Provide tailored itineraries based on visitor preferences and historical data.
24. Use artificial intelligence and machine learning to enhance the accuracy and effectiveness of personalized experiences.
25. Continuously seek feedback from visitors to improve the SCTP initiative.

FREQUENTLY ASKED QUESTIONS

What is Smart City Tourist Personalization (SCTP)? SCTP is a service that uses data and technology to personalize the tourist experience, providing tailored recommendations and personalized services to each individual tourist.

How can SCTP benefit me as a tourist? SCTP can provide customized recommendations for activities, attractions, and restaurants based on your interests, preferences, and previous experiences.

How does SCTP personalize my experience? SCTP uses various data sources such as social media, location data, and previous booking history to create a profile of your preferences and provide personalized recommendations.

Can I opt out of SCTP if I don't want my data to be used? Yes, most SCTP providers offer the option to opt-out of data collection or to delete personal information at any time.

Is my data secure when using SCTP? Yes, SCTP providers typically use secure servers and encryption methods to protect the privacy and security of personal data.

How does SCTP benefit the city and its residents? SCTP can help to increase tourism revenue and promote local businesses by providing personalized recommendations to tourists, while also reducing the environmental impact of tourism by promoting sustainable and responsible tourism practices.

How can SCTP improve the quality of my trip? SCTP can help to make your trip more enjoyable by providing tailored recommendations for activities and attractions that match your interests and preferences.

How can I access SCTP during my trip? SCTP can be accessed through various channels such as mobile apps, websites, and digital kiosks, which are typically located at major tourist hubs in the city.

How does SCTP handle multiple users on the same device? SCTP providers typically allow for multiple user profiles on the same device, ensuring that each user receives personalized recommendations based on their own preferences and interests.

How can I provide feedback on the quality of SCTP? SCTP typically provides a feedback mechanism that allows tourists to rate the quality of the service and to provide suggestions for improvement.

Conclusion: SCTP can greatly enhance the overall tourist experience, provide valuable insights for the city, and promote sustainable tourism practices. It can be an effective way to manage and optimize the flow of tourists and make the most of their time in the city.

SMART CITY TOURIST SAFETY AND SECURITY (SCTSS)

Using technology to enhance the safety and security of tourists, such as surveillance cameras and emergency response systems.

Implementing SCTSS measures through the use of technology can provide a number of benefits for both tourists and city officials. One key benefit is the ability to enhance the overall safety and security of tourists by using surveillance cameras and other monitoring systems to detect and deter crime. This can help to reduce the risk of theft, assault, and other safety hazards for tourists, making them feel more secure and comfortable when visiting the city.

Another benefit of SCTSS is the ability to quickly respond to and manage emergency situations. By using technology such as emergency response systems and real-time communication tools, city officials can quickly and effectively coordinate response efforts, providing rapid assistance to those in need and minimizing the impact of any incidents.

Using technology to enhance tourist safety and security can also help to improve the overall reputation and image of the city. By providing a safe and secure environment for tourists, cities can attract more visitors and boost their economy through tourism. The use of real-time monitoring systems, surveillance cameras and other security measures can help to deter crime, making the city safer for both tourists and residents.

GENERAL FRAMEWORK

The following framework is a general one and the actual implementation will depend on the specific requirements and constraints of the city.

1. Define the goals and objectives of SCTSS.
2. Conduct a comprehensive review of existing safety and security technology and services.
3. Identify key tourist areas and attractions that would benefit from enhanced safety and security measures.
4. Develop a detailed plan for the implementation of SCTSS, including technology and resource requirements.
5. Acquire necessary technology and resources.
6. Train staff and stakeholders in the use of the technology.
7. Implement surveillance cameras and emergency response systems in key tourist areas and attractions.
8. Develop a robust security and privacy framework to protect visitor data.
9. Develop a real-time monitoring and response system to quickly address safety and security incidents.
10. Establish partnerships with local law enforcement and emergency services to enhance collaboration and response times.
11. Regularly review and evaluate the effectiveness of the SCTSS initiative.
12. Encourage the use of personal safety and security technologies, such as mobile safety apps.
13. Foster a culture of safety and security awareness among tourists and stakeholders.
14. Develop a comprehensive emergency preparedness plan and regularly conduct drills to ensure readiness.

15. Provide real-time alerts and notifications to tourists and stakeholders in the event of safety or security incidents.
16. Ensure that the technology is accessible to all visitors, including those with disabilities.
17. Foster a culture of continuous improvement and seek to continuously enhance safety and security measures.
18. Provide safety and security information and resources to tourists before and during their visit.
19. Ensure that emergency response systems are easy to use and provide clear instructions in multiple languages.
20. Utilize data and analytics to make informed decisions about the SCTSS initiative.
21. Continuously collect and analyze data to inform the development of safety and security measures.
22. Foster a culture of collaboration and encourage sharing of best practices among stakeholders.
23. Develop public-private partnerships to enhance the overall safety and security of tourists.
24. Ensure that the technology is used in a responsible and ethical manner.
25. Continuously seek feedback from visitors and stakeholders to improve the SCTSS initiative.

FREQUENTLY ASKED QUESTIONS

How does SCTSS ensure the safety of tourists in a city? SCTSS incorporates a range of technological solutions such as CCTV cameras, emergency response systems, and mobile apps that help monitor and respond to incidents in real-time, providing a secure environment for tourists.

What are the privacy concerns associated with SCTSS? SCTSS collects data from various sources, including cameras and sensors, which raises privacy concerns. However, data is collected anonymously and is only used for security purposes, with proper protocols in place to protect it.

How does SCTSS handle emergency situations such as accidents or natural disasters? SCTSS provides emergency response systems that are integrated with local authorities, including police, fire, and medical services. This ensures that swift and coordinated responses can be activated in case of emergencies.

Can tourists opt-out of SCTSS surveillance? Since SCTSS relies on a network of sensors and cameras, it is not possible for individuals to opt-out of the system. However, the collected data is anonymized and only used for security purposes.

How does SCTSS handle different languages and cultural backgrounds? SCTSS uses technologies such as language translation software and cultural awareness training for staff to ensure that tourists from different cultural backgrounds feel welcome and understood.

Who funds SCTSS projects? SCTSS projects are typically funded by a mix of government and private organizations, with the goal of enhancing the tourist experience while providing a safe and secure environment.

How does SCTSS ensure the security of personal data collected from tourists? SCTSS implements strict security protocols to safeguard personal data, including encryption and access control measures. Only authorized personnel can access the data, and it is only used for security purposes.

How does SCTSS handle crowd management during events or busy tourist periods? SCTSS incorporates crowd management technologies such as automated crowd tracking and alerts, which help manage crowd flow and prevent overcrowding, ensuring a safe and comfortable environment for tourists.

How can tourists provide feedback or report incidents to SCTSS? SCTSS provides a range of communication channels, including mobile apps and hotlines, that allow tourists to provide feedback or report incidents in real-time. These communication channels are closely monitored by SCTSS staff.

How does SCTSS help local law enforcement prevent and respond to crimes against tourists? SCTSS provides real-time monitoring and intelligence gathering, allowing local law enforcement to quickly respond to incidents and identify potential security threats. This enables local law enforcement to more effectively prevent and respond to crimes against tourists.

Conclusion: *Implementing SCTSS measures through the use of technology can provide a number of benefits for both tourists and city officials, including enhanced safety and security, improved emergency response capabilities, and a positive impact on the city's reputation and economy.*

SMART CITY TOURIST SOCIAL MEDIA INTEGRATION (SCTSM)

Using technology to integrate social media platforms into the tourist experience, such as social media check-ins and photo sharing.

Implementing SCTSM can provide a number of benefits for both tourists and city officials. One major benefit is the ability to increase engagement and awareness of tourist destinations and activities. By integrating social media platforms into the tourist experience, visitors can easily share their experiences with friends and family, resulting in increased visibility and interest in the city. This type of integration can also provide valuable data and insights for city officials, as they can track and analyze the online conversations and sentiment surrounding their destinations and activities.

Another benefit of SCTSM is the ability to enhance the personalization of the tourist experience. Tourists can receive recommendations and customized itineraries based on their social media activity and preferences. This can lead to a more enjoyable and efficient experience for visitors, as they can easily find and book the activities and attractions that interest them the most.

SCTSM can also help with destination marketing efforts. Tourists can be encouraged to use specific hashtags or check-ins at certain locations, which can increase the visibility of those places on social media. This can lead to more visitors to those locations and can also provide valuable data for city officials on the most popular tourist destinations.

GENERAL FRAMEWORK

The following framework is a general one and the actual implementation will depend on the specific requirements and constraints of the city.

1. Research popular social media platforms used by tourists.
2. Develop a strategy to integrate social media into the tourist experience.
3. Set up social media profiles for the city/tourist attractions.
4. Implement a system for tracking social media mentions and engagement.
5. Provide incentives for tourists to engage with the city's social media channels.
6. Develop a system for monitoring and responding to customer feedback and complaints.
7. Offer social media check-ins and photo sharing opportunities at tourist locations.
8. Encourage tourists to share their experiences and photos on social media.
9. Utilize social media analytics tools to measure the success of the integration.
10. Regularly update and optimize the social media strategy based on data and feedback.
11. Use social media to promote city events and attractions.
12. Develop a system for aggregating user-generated content.
13. Partner with local businesses and organizations to increase social media reach.
14. Use social media to showcase the city's cultural heritage and history.
15. Offer virtual tours and experiences through social media.

16. Foster a sense of community through social media engagement.
17. Promote sustainable tourism practices through social media.
18. Offer virtual customer support through social media.
19. Use social media to gather real-time information about tourist experiences.
20. Utilize social media for crisis management and emergency response.
21. Develop a content calendar to keep social media channels active and relevant.
22. Collaborate with influencers and social media celebrities to promote the city.
23. Encourage tourists to use social media to plan their trips.
24. Offer real-time updates and notifications to tourists through social media.
25. Continuously evaluate and improve the social media integration based on data and feedback.

FREQUENTLY ASKED QUESTIONS

How does SCTSM integrate social media into the tourist experience in a smart city? SCTSM incorporates various social media platforms into the tourist experience, allowing visitors to share their experiences and get recommendations from others, enhancing the overall travel experience.

Can I use SCTSM to find and book tourist attractions and activities? Yes, SCTSM can be used to discover and book tourist attractions and activities in a city, often with exclusive discounts for SCTSM users.

How does SCTSM ensure the privacy of tourists when integrating social media? SCTSM uses strict privacy policies and protocols to safeguard personal data, ensuring that social media integration does not compromise the privacy and security of tourists.

How can local businesses benefit from SCTSM's social media integration? Local businesses can leverage SCTSM's social media integration to increase their online presence, reach a wider audience, and drive more traffic to their establishments.

How does SCTSM handle different languages and cultural backgrounds on social media platforms? SCTSM uses language translation software and cultural awareness training for staff to ensure that tourists from different cultural backgrounds feel welcome and understood on social media platforms.

Who funds SCTSM projects? SCTSM projects are typically funded by a mix of government and private organizations, with the goal of enhancing the tourist experience and promoting local businesses.

Can I use SCTSM to share my experiences with friends and family back home? Yes, SCTSM provides tourists with the ability to share their experiences on social media platforms and communicate with loved ones back home in real-time.

How does SCTSM use social media to promote sustainability and responsible tourism practices? SCTSM uses social media platforms to promote sustainability and responsible tourism practices, including sharing information about local eco-friendly initiatives and encouraging tourists to reduce their environmental footprint.

Can I receive personalized recommendations for attractions and activities through SCTSM? Yes, SCTSM uses machine learning and AI technologies to provide personalized recommendations for attractions and activities based on a tourist's preferences and past experiences.

How does SCTSM incorporate user-generated content into the tourist experience? SCTSM allows tourists to create and share their own content, such as photos and videos, on social media platforms, creating a user-generated library of experiences that can be shared and enjoyed by others.

Conclusion: SCTSM can provide a number of benefits for both tourists and city officials, including increased engagement and awareness, enhanced personalization, and improved destination marketing efforts.

SMART CITY TOURIST SUSTAINABLE TOURISM (SCTST)

Using technology to promote sustainable tourism practices, such as carbon offsetting and eco-friendly options.

Benefits of SCTST are numerous and far-reaching. By utilizing technology, cities can promote sustainable tourism practices and minimize the negative impact of tourism on the environment.

One of the key benefits of SCTST is the ability to offset carbon emissions. Through the use of technology, cities can track and offset the carbon emissions generated by tourism activities. This helps to reduce the overall carbon footprint of the city and promote more environmentally friendly practices.

Another benefit of SCTST is the ability to promote eco-friendly options for tourists. Technology can be used to provide information on eco-friendly tourism options, such as public transportation and sustainable accommodations. This information helps tourists make more informed decisions and choose options that are better for the environment.

SCTST can also help to promote local economic development. By promoting sustainable tourism practices, cities can attract more environmentally conscious tourists, which can result in increased economic activity in the local community.

SCTST can help to improve the reputation of a city as a tourist destination. Cities that are seen as environmentally friendly are often more attractive to tourists, which can lead to increased tourism and a more positive image for the city.

GENERAL FRAMEWORK

The following framework is a general one and the actual implementation will depend on the specific requirements and constraints of the city.

1. Identify key areas of the tourist experience that can be made more sustainable.
2. Research and analyze existing sustainable tourism initiatives and practices.
3. Establish sustainability goals and objectives for the city's tourist industry.
4. Develop a strategy for implementing sustainable tourism practices, such as carbon offsetting and eco-friendly options.
5. Engage stakeholders, including tourism businesses, local communities, and government agencies, to promote sustainable tourism.
6. Encourage the use of low-carbon transportation options for tourists, such as public transportation and electric vehicles.
7. Promote the use of eco-friendly accommodations, such as green hotels and eco-lodges.
8. Develop a system for measuring and tracking the carbon footprint of tourism activities.
9. Implement initiatives to reduce waste and conserve resources, such as recycling programs and water conservation efforts.
10. Encourage sustainable tourism activities, such as wildlife and cultural tours, that are both enjoyable and educational for tourists.
11. Provide information and education to tourists about sustainable tourism practices and how they can help to reduce their impact on the environment.
12. Foster partnerships with local and international organizations to promote sustainable tourism practices and reduce the negative impacts of tourism.
13. Invest in technology, such as green energy sources and smart building systems, to reduce the environmental impact of tourism infrastructure.
14. Monitor and evaluate the effectiveness of sustainable tourism initiatives, and make adjustments as needed.
15. Encourage local businesses and tourist destinations to adopt sustainable practices, such as reducing plastic use and promoting local products.
16. Promote and support local initiatives that encourage sustainable tourism, such as community-based tourism and sustainable agriculture programs.
17. Establish incentives for tourists who adopt sustainable practices, such as discounts on eco-friendly activities and accommodations.

18. Use technology, such as digital platforms and mobile apps, to provide tourists with real-time information about sustainable tourism options.
19. Foster a culture of sustainability among tourism businesses and their employees, through training and awareness-raising efforts.
20. Engage tourists through interactive experiences, such as virtual reality and gamification, to educate them about sustainable tourism practices.
21. Develop a marketing campaign to promote sustainable tourism and raise awareness of its benefits.
22. Foster partnerships with travel companies, airlines, and other travel-related businesses to promote sustainable tourism practices.
23. Establish a certification program for sustainable tourism businesses, to encourage and reward environmentally responsible practices.
24. Regularly communicate progress and successes of sustainable tourism initiatives to stakeholders and the public.
25. Continuously evaluate and improve sustainable tourism initiatives to ensure their effectiveness and impact.

FREQUENTLY ASKED QUESTIONS

What is sustainable tourism and how does SCTST support it? Sustainable tourism focuses on minimizing the negative impact of tourism on the environment and local communities while promoting economic benefits. SCTST supports sustainable tourism by promoting environmentally-friendly and socially-responsible practices.

How does SCTST measure the sustainability of a tourist attraction or activity? SCTST uses a set of sustainability indicators to assess the impact of tourist attractions and activities on the environment, local communities, and the economy.

Can SCTST be used to find sustainable accommodation options? Yes, SCTST provides information on sustainable accommodation options that prioritize environmental and social responsibility.

How does SCTST ensure the authenticity of sustainable tourism experiences? SCTST partners with local businesses and organizations that prioritize sustainable tourism and provides tourists with opportunities to engage with local communities and learn about the local culture and environment.

How does SCTST address the issue of over-tourism? SCTST promotes sustainable tourism practices and encourages tourists to explore less crowded areas of a city and engage in off-the-beaten-path activities to reduce the strain on heavily-touristed areas.

Who is responsible for promoting sustainable tourism in a smart city? Sustainable tourism is a collaborative effort that involves the participation of government organizations, local businesses, and tourists. SCTST provides a platform for all stakeholders to promote sustainable tourism practices.

Can SCTST help tourists offset their carbon footprint? Yes, SCTST provides information on carbon offset programs and sustainable transportation options, such as public transportation or electric car rentals, to help tourists reduce their carbon footprint.

How does SCTST support the local economy while promoting sustainable tourism? SCTST promotes the use of local businesses that prioritize sustainable tourism, such as eco-friendly hotels or locally-sourced restaurants, to support the local economy and reduce the environmental impact of tourism.

Can SCTST be used to plan a sustainable tourism itinerary? Yes, SCTST provides information on sustainable tourist attractions and activities, along with recommended sustainable transportation options, to help tourists plan a sustainable tourism itinerary.

How does SCTST address the issue of waste management in a smart city? SCTST promotes sustainable waste management practices, such as reducing single-use plastics and encouraging the use of recycling and composting facilities, to minimize the negative impact of tourism on the environment.

Conclusion: SCTST offers a range of benefits for cities, tourists, and the environment. By utilizing technology to promote sustainable tourism practices, cities can reduce their carbon footprint, promote eco-friendly options, improve local economic development, and enhance their reputation as a tourist destination.

SMART CITY TOURIST TRANSPORT MANAGEMENT (SCTTM)

Using technology to manage tourist transportation, such as real-time tracking and route optimization.

Implementing SCTTM can bring a number of benefits for both tourists and the city. By using technology to manage tourist transportation, cities can improve the overall travel experience for visitors. Real-time tracking allows tourists to easily plan their routes and know when their transportation will arrive, reducing wait times and increasing convenience.

Route optimization can also help to reduce congestion and traffic, making the city more accessible and enjoyable for tourists to explore. This technology can also help to improve the sustainability of tourist transportation by reducing emissions and promoting use of public transportation.

The data collected can also be used to improve the transportation infrastructure and make it more suitable for the tourists. This can lead to an increase in the footfall of tourists, resulting in an increase in revenue for the city.

GENERAL FRAMEWORK

The following framework is a general one and the actual implementation will depend on the specific requirements and constraints of the city.

1. Define the goals and objectives of SCTTM.
2. Identify tourist transportation stakeholders.
3. Conduct market research to identify existing solutions.
4. Determine the needs and requirements of the tourists.
5. Develop a plan to integrate technology into tourist transportation management.
6. Choose suitable technology and tools for real-time tracking and route optimization.
7. Establish partnerships with technology companies, transportation providers and tourism organizations.
8. Implement a data management system to store and analyze transportation data.
9. Develop a mobile application for tourists to track transportation and plan their trips.
10. Integrate GPS tracking devices into tourist transportation vehicles.
11. Implement a real-time traffic and weather monitoring system.
12. Develop algorithms for route optimization based on real-time traffic data.
13. Integrate payment systems into the transportation management system.
14. Implement a customer service system for tourists to report issues and provide feedback.
15. Train transportation staff on the use of new technology and tools.
16. Implement data security measures to protect sensitive information.
17. Conduct pilot tests of the SCTTM system in selected tourist destinations.
18. Evaluate the results of the pilot tests and make improvements as necessary.
19. Develop a comprehensive marketing and communication plan to promote SCTTM to tourists.
20. Launch SCTTM in selected tourist destinations.
21. Monitor the performance of the SCTTM system and make ongoing improvements.
22. Conduct customer satisfaction surveys to gather feedback and improve the system.
23. Continuously analyze transportation data to identify trends and improve route optimization.
24. Expand the SCTTM system to additional tourist destinations.
25. Establish partnerships with international tourism organizations to promote SCTTM globally.

FREQUENTLY ASKED QUESTIONS

How does SCTTM help tourists navigate a new city? SCTTM provides real-time information on transportation options, such as buses, trains, and taxis, to help tourists navigate a new city efficiently.

How does SCTTM address traffic congestion in a smart city? SCTTM promotes sustainable transportation options, such as public transportation, bike-sharing programs, and electric car rentals, to reduce traffic congestion and improve air quality in a smart city.

Can SCTTM help tourists save money on transportation? Yes, SCTTM provides information on affordable transportation options, such as public transportation or bike-sharing programs, to help tourists save money on transportation costs.

How does SCTTM address the issue of accessibility for tourists with disabilities? SCTTM promotes accessible transportation options, such as wheelchair-accessible buses and trains, to ensure that tourists with disabilities can navigate a city easily.

How does SCTTM ensure the safety of tourists using transportation in a smart city? SCTTM provides real-time information on traffic conditions and safety alerts to help tourists make informed decisions about transportation options and avoid potential safety hazards.

Can SCTTM be used to plan a sustainable transportation itinerary for tourists? Yes, SCTTM provides information on sustainable transportation options, such as electric car rentals or bike-sharing programs, to help tourists plan a sustainable transportation itinerary.

How does SCTTM address the issue of language barriers for tourists using transportation in a smart city? SCTTM provides multilingual information on transportation options and real-time updates to help tourists who may not speak the local language navigate a city easily.

How does SCTTM address the issue of parking in a smart city? SCTTM promotes sustainable transportation options, such as public transportation or bike-sharing programs, to reduce the demand for parking and minimize the negative impact of transportation on the environment.

Can SCTTM be used to track public transportation schedules in real-time? Yes, SCTTM provides real-time information on public transportation schedules, delays, and cancellations to help tourists plan their trips efficiently.

How does SCTTM address the issue of transportation during peak tourism season in a smart city? SCTTM promotes sustainable transportation options and encourages tourists to explore less crowded areas of a city to reduce the strain on transportation systems during peak tourism season.

Conclusion: *SCTTM can improve the tourist experience, reduce congestion and emissions, and increase revenue for the city.*

SMART CITY TOURIST WAYFINDING (SCTW)

Using technology to provide tourists with wayfinding assistance, such as interactive maps and augmented reality.

Implementing SCTW can provide a number of benefits for both tourists and city administrators. By using technology to provide tourists with wayfinding assistance, such as interactive maps and augmented reality, visitors can easily navigate the city, find their way to popular attractions, and discover hidden gems. This can enhance their overall experience, leading to increased satisfaction and likelihood of returning.

One of the main benefits of SCTW is improved navigation. Interactive maps and augmented reality can provide tourists with real-time information about their location, nearby points of interest, and the most efficient routes to their destination. This can save tourists time and reduce frustration, as they no longer have to rely on paper maps or ask for directions.

Another benefit is increased safety. With real-time information about their location, tourists are less likely to get lost or wander into unsafe areas. This can also reduce the burden on emergency services, as tourists are less likely to need assistance.

SCTW can also help city administrators better manage and promote their city. By collecting data on tourist behaviour and preferences, city administrators can gain insight into which areas of the city are most popular, and which areas may need more attention. This can inform decisions about infrastructure and resource allocation. By providing tourists with an enhanced wayfinding experience, cities can attract more visitors and boost their economy.

GENERAL FRAMEWORK

The following framework is a general one and the actual implementation will depend on the specific requirements and constraints of the city.

1. Define the goals and objectives of SCTW.
2. Identify tourist wayfinding stakeholders.
3. Conduct market research to identify existing solutions.
4. Determine the needs and requirements of the tourists.
5. Develop a plan to integrate technology into tourist wayfinding assistance.
6. Choose suitable technology and tools for interactive maps and augmented reality.
7. Establish partnerships with technology companies and tourism organizations.
8. Develop an interactive map of the city, including tourist destinations and landmarks.
9. Implement augmented reality technology for wayfinding assistance.
10. Develop a mobile application for tourists to access the interactive map and augmented reality features.
11. Integrate GPS technology into the wayfinding system.

12. Develop algorithms for personalized route recommendations based on tourist preferences and destination history.
13. Implement a customer service system for tourists to report issues and provide feedback.
14. Train tourism staff on the use of the new technology and tools.
15. Implement data security measures to protect sensitive information.
16. Conduct pilot tests of the SCTW system in selected tourist destinations.
17. Evaluate the results of the pilot tests and make improvements as necessary.
18. Develop a comprehensive marketing and communication plan to promote SCTW to tourists.
19. Launch SCTW in selected tourist destinations.
20. Monitor the performance of the SCTW system and make ongoing improvements.
21. Conduct customer satisfaction surveys to gather feedback and improve the system.
22. Continuously analyze usage data to identify trends and improve the wayfinding system.
23. Expand the SCTW system to additional tourist destinations.
24. Establish partnerships with international tourism organizations to promote SCTW globally.
25. Continuously explore and integrate new technology to enhance the tourist wayfinding experience.

FREQUENTLY ASKED QUESTIONS

How does SCTW help tourists navigate a new city? SCTW provides clear and easy-to-follow directions to help tourists navigate a new city efficiently, using a combination of digital and physical signage.

Can SCTW be used to plan a personalized wayfinding itinerary for a tourist? Yes, SCTW provides customized wayfinding itineraries that take into account a tourist's interests, location, and transportation preferences.

How does SCTW address the issue of accessibility for tourists with disabilities? SCTW provides clear and accessible signage, as well as information on accessible routes and transportation options, to ensure that tourists with disabilities can navigate a city easily.

Can SCTW be used offline, without an internet connection? Yes, SCTW provides offline maps and signage to ensure that tourists can navigate a city even when they don't have access to an internet connection.

How does SCTW ensure the safety of tourists when navigating a city? SCTW provides information on safe routes and areas to avoid, as well as real-time updates on traffic conditions and safety alerts, to help tourists make informed decisions when navigating a city.

Can SCTW be used to discover new places and hidden gems in a city? Yes, SCTW provides information on local attractions and points of interest, as well as personalized recommendations based on a tourist's interests, to help them discover new places and hidden gems in a city.

How does SCTW address the issue of language barriers for tourists when navigating a city? SCTW provides multilingual signage and directions, as well as audio and visual cues, to help tourists who may not speak the local language navigate a city easily.

How does SCTW promote sustainable transportation options in a city? SCTW promotes sustainable transportation options, such as public transportation, bike-sharing programs, and electric car rentals, to reduce the negative impact of transportation on the environment and promote sustainable tourism.

Can SCTW be used to track a tourist's location and send personalized recommendations? Yes, SCTW uses location-based technology to track a tourist's location and send personalized recommendations for attractions, restaurants, and events in their vicinity.

How does SCTW address the issue of wayfinding during peak tourism season in a city? SCTW provides customized wayfinding itineraries that take into account the density of crowds and transportation options during peak tourism season, to help tourists navigate a city efficiently and safely.

Conclusion: *SCTW can provide a number of benefits for both tourists and city administrators. Improved navigation, increased safety, and better city management are just a few of the benefits that can be realized through the use of technology to provide tourists with wayfinding assistance.*

SMART CITY TRANSPORT MANAGEMENT (SCTM)

Using technology to optimize transportation infrastructure, such as traffic management systems, public transport, and autonomous vehicles.

Implementing SCTM technology can bring a wide range of benefits to a city's transportation infrastructure. One key benefit is improved traffic management, through the use of real-time traffic monitoring and adaptive traffic control systems. This can help to reduce congestion on the roads and improve overall traffic flow, leading to faster travel times and reduced air pollution.

Another benefit is the optimization of public transport services, through the use of real-time tracking and route optimization. This can enable transport providers to more efficiently operate their services, leading to improved reliability and increased capacity. SCTM technology can enable the integration of different transportation options, such as bike-sharing and ride-hailing services, making it easier for citizens and visitors to navigate the city and find the transportation option that best suits their needs.

The implementation of autonomous vehicles technologies can increase the efficiency and safety of transportation, as well as reduce the human error component in the process. With the ability to communicate with other vehicles and infrastructure and make real-time adjustments, autonomous vehicles can help to reduce accidents, traffic congestion, and fuel consumption.

GENERAL FRAMEWORK

The following framework is a general one and the actual implementation will depend on the specific requirements and constraints of the city.

1. Define the goals and objectives of SCTM.
2. Identify stakeholders in city transportation management.
3. Conduct market research to identify existing solutions.
4. Develop a comprehensive transportation infrastructure plan.
5. Choose suitable technology and tools for traffic management and optimization.
6. Establish partnerships with technology companies and transportation providers.
7. Implement a traffic management system to monitor and control traffic flow.
8. Develop an integrated public transport management system.
9. Explore the integration of autonomous vehicles into the transportation infrastructure.
10. Implement a data management system to store and analyze transportation data.
11. Develop algorithms for route optimization and traffic management.
12. Implement real-time monitoring and control systems for public transport and autonomous vehicles.
13. Develop a mobile application for residents to track transportation and plan their trips.
14. Train transportation staff on the use of new technology and tools.
15. Implement data security measures to protect sensitive information.
16. Conduct pilot tests of the SCTM system in selected city areas.
17. Evaluate the results of the pilot tests and make improvements as necessary.
18. Develop a comprehensive marketing and communication plan to promote SCTM to residents.
19. Launch SCTM in selected city areas.
20. Monitor the performance of the SCTM system and make ongoing improvements.

21. Conduct resident satisfaction surveys to gather feedback and improve the system.
22. Continuously analyze transportation data to identify trends and improve the system.
23. Expand the SCTM system to additional city areas.
24. Establish partnerships with international cities to promote SCTM globally.
25. Continuously explore and integrate new technology to enhance the city's transportation infrastructure.

FREQUENTLY ASKED QUESTIONS

How does SCTM improve the transportation system in a city? SCTM improves the transportation system in a city by optimizing traffic flow, reducing congestion, and promoting sustainable transportation options.

Can SCTM be used to track the location of public transportation vehicles in real-time? Yes, SCTM uses real-time tracking technology to provide up-to-date information on the location and estimated arrival times of public transportation vehicles.

How does SCTM address the issue of traffic congestion in a city? SCTM uses traffic analysis and predictive modelling to optimize traffic flow and reduce congestion, thereby reducing travel times and improving air quality.

Can SCTM be used to promote sustainable transportation options, such as bike-sharing programs and electric car rentals? Yes, SCTM promotes sustainable transportation options to reduce the negative impact of transportation on the environment and promote sustainable urban development.

How does SCTM improve safety for pedestrians and cyclists in a city? SCTM improves safety for pedestrians and cyclists by promoting dedicated bike lanes, pedestrian walkways, and safe crossing points, as well as implementing intelligent traffic signal systems that prioritize pedestrian safety.

Can SCTM be used to provide commuters with personalized transportation recommendations? Yes, SCTM uses location-based technology to provide commuters with personalized transportation recommendations based on their current location and preferred transportation options.

How does SCTM address the issue of accessibility for people with disabilities in a city? SCTM provides accessible transportation options and infrastructure, such as wheelchair-friendly buses and trains, and ensures that transportation routes and schedules are designed with accessibility in mind.

Can SCTM be used to provide real-time traffic updates to drivers? Yes, SCTM uses real-time traffic analysis to provide up-to-date traffic updates and rerouting recommendations to drivers to help them avoid congestion and save time.

How does SCTM integrate with existing transportation infrastructure in a city? SCTM is designed to integrate seamlessly with existing transportation infrastructure, including public transportation systems, highways, and bike lanes, to ensure a coordinated and efficient transportation network.

Can SCTM be used to reduce transportation-related emissions in a city? Yes, SCTM promotes sustainable transportation options and reduces traffic congestion, thereby reducing transportation-related emissions and promoting a cleaner, healthier environment for all.

Conclusion: *Implementing SCTM technology can bring significant benefits to a city's transportation infrastructure, including improved traffic management, optimized public transport services, and the integration of different transportation options, and the emergence of autonomous vehicles. All of these benefits can lead to improved* transportation for citizens and visitors, reduced congestion, and a more sustainable and efficient transportation system overall.

SMART CITY URBAN AGRICULTURE (SCUA)

Using technology to promote sustainable urban agriculture practices, such as vertical farming and aquaponics.

Implementing SCUA can have a number of benefits for both the city and its residents. One of the main benefits is the ability to produce fresh, healthy food within the city, reducing the need for transportation and the associated carbon footprint. This can also lead to increased food security, as cities become more self-sufficient in terms of food production.

Another benefit is the potential to create new jobs and economic opportunities within the city. Urban agriculture can provide new opportunities for small businesses, entrepreneurs, and community organizations, as well as creating jobs in the fields of agriculture, engineering, and technology.

Urban agriculture can also have a positive impact on the environment, by reducing the urban heat island effect, absorbing carbon dioxide, and improving air and water quality. It can also provide green spaces for residents to enjoy and connect with nature, promoting physical and mental well-being.

Vertical farming and aquaponics can greatly increase yield and efficiency of food production, by taking advantage of vertical space and reducing the need for resources like water and energy.

Implementing SCUA can also improve the effectiveness and efficiency of agricultural practices. For example, precision agriculture technology can be used to optimize crop growth, reduce water and energy usage, and increase yield. SCUA can also use technology like IoT and big data to manage irrigation, lighting, and other environmental conditions, to ensure optimal growth conditions for crops.

GENERAL FRAMEWORK

The following framework is a general one and the actual implementation will depend on the specific requirements and constraints of the city.

1. Define the goals and objectives of SCUA.
2. Identify stakeholders in urban agriculture.
3. Conduct market research to identify existing solutions.
4. Develop a comprehensive urban agriculture plan.
5. Choose suitable technology and tools for vertical farming and aquaponics.
6. Establish partnerships with technology companies and agriculture organizations.
7. Implement a vertical farming system for growing crops in urban areas.
8. Develop an aquaponics system for sustainable fish and plant production.
9. Implement a data management system to store and analyze agriculture data.
10. Develop algorithms for optimizing crop and fish production.
11. Implement real-time monitoring and control systems for vertical farming and aquaponics.
12. Develop a mobile application for residents to track agriculture production and purchase products.
13. Train urban agriculture staff on the use of new technology and tools.
14. Implement data security measures to protect sensitive information.
15. Conduct pilot tests of the SCUA system in selected urban areas.
16. Evaluate the results of the pilot tests and make improvements as necessary.
17. Develop a comprehensive marketing and communication plan to promote SCUA to residents.
18. Launch SCUA in selected urban areas.
19. Monitor the performance of the SCUA system and make ongoing improvements.
20. Conduct resident satisfaction surveys to gather feedback and improve the system.

21. Continuously analyze agriculture data to identify trends and improve the system.
22. Expand the SCUA system to additional urban areas.
23. Establish partnerships with international agriculture organizations to promote SCUA globally.
24. Continuously explore and integrate new technology to enhance urban agriculture practices.
25. Encourage the integration of urban agriculture into the local food system.

FREQUENTLY ASKED QUESTIONS

What is Smart City Urban Agriculture, and how does it work? Smart City Urban Agriculture is a system that uses technology and innovative farming methods to grow crops and raise animals in an urban environment. It uses tools like vertical farming, hydroponics, and aquaponics to optimize the use of space and resources.

How does SCUA benefit the local community? SCUA provides fresh, healthy, and locally produced food for the community, while also creating green spaces and improving the quality of the environment. It also promotes sustainable urban development by reducing food transportation costs and lowering the carbon footprint of the food industry.

Can SCUA be integrated into existing urban infrastructure? Yes, SCUA can be integrated into existing urban infrastructure, such as rooftops, abandoned buildings, and parks, to optimize the use of space and resources.

How does SCUA address food insecurity in urban areas? SCUA provides fresh and healthy food options for urban residents, many of whom may not have access to fresh produce due to food deserts or high transportation costs. It also provides job opportunities and skills training for local residents.

What types of crops and animals can be grown and raised using SCUA? SCUA can be used to grow a variety of crops, including leafy greens, herbs, fruits, and vegetables. It can also be used to raise small animals, such as chickens and fish.

How does SCUA impact the environment? SCUA promotes sustainable urban development by reducing the carbon footprint of the food industry and improving the quality of the environment through the creation of green spaces and the reduction of food transportation costs.

How does SCUA ensure the safety and quality of the food produced? SCUA follows strict safety and quality protocols to ensure that the food produced is safe for consumption. This includes regular testing of soil and water quality and adhering to food safety regulations.

Can SCUA be used to promote education and awareness about healthy food choices? Yes, SCUA can be used to promote education and awareness about healthy food choices and sustainable urban development through workshops, tours, and community engagement programs.

How does SCUA impact the local economy? SCUA provides job opportunities and skills training for local residents, as well as promoting the growth of small businesses and the local food industry.

Can SCUA be implemented in cities around the world? Yes, SCUA can be implemented in cities around the world, as it is adaptable to different climates and urban environments. It provides a sustainable solution to the challenges of urbanization and food insecurity.

Conclusion: *Implementing SCUA can have numerous benefits for the city, its residents and the environment. It can create new economic opportunities, improve food security and sustainability, and enhance the overall livability of the city.*

SMART CITY URBAN MOBILITY (SCUM)

Using technology to improve the efficiency, sustainability and accessibility of urban transportation, such as connected and autonomous vehicles and shared mobility services.

Implementing SCUM solutions can have a number of benefits. One of the key benefits is the improvement of transportation efficiency. By using technology to optimize traffic flow, reduce congestion and improve public transportation, cities can reduce travel times, increase the reliability of transportation services, and improve the overall transportation experience for residents and visitors.

SCUM solutions can also promote sustainability. By encouraging the use of electric vehicles, shared mobility services and other environmentally-friendly transportation options, cities can reduce their carbon footprint, improve air quality and promote sustainable transportation practices.

Another benefit of SCUM solutions, is improved accessibility. By using technology to provide real-time information, navigation and route planning services, cities can make it easier for people with disabilities, the elderly, and other groups to access transportation services. SCUM solutions can also help to increase the number of people using public transportation, walking and cycling, which can also improve accessibility by reducing the number of cars on the road. By improving transportation efficiency and accessibility, cities can also attract new business and investment, which can help to create jobs and boost economic growth.

GENERAL FRAMEWORK

The following framework is a general one and the actual implementation will depend on the specific requirements and constraints of the city.

1. Define the goals and objectives of SCUM.
2. Identify stakeholders in urban mobility.

3. Conduct market research to identify existing solutions.
4. Develop a comprehensive urban mobility plan.
5. Choose suitable technology and tools for connected and autonomous vehicles and shared mobility services.
6. Establish partnerships with technology companies and transportation providers.
7. Implement connected vehicle technology for enhanced communication and safety.
8. Develop autonomous vehicle testing and deployment programs.
9. Launch shared mobility services for residents and visitors.
10. Implement a data management system to store and analyze mobility data.
11. Develop algorithms for optimizing transportation routes and services.
12. Implement real-time monitoring and control systems for connected and autonomous vehicles.
13. Develop a mobile application for residents to track transportation and plan their trips.
14. Train transportation staff on the use of new technology and tools.
15. Implement data security measures to protect sensitive information.
16. Conduct pilot tests of the SCUM system in selected city areas.
17. Evaluate the results of the pilot tests and make improvements as necessary.
18. Develop a comprehensive marketing and communication plan to promote SCUM to residents.
19. Launch SCUM in selected city areas.
20. Monitor the performance of the SCUM system and make ongoing improvements.
21. Conduct resident satisfaction surveys to gather feedback and improve the system.
22. Continuously analyze mobility data to identify trends and improve the system.
23. Expand the SCUM system to additional city areas.
24. Establish partnerships with international cities to promote SCUM globally.
25. Continuously explore and integrate new technology to enhance urban mobility.

FREQUENTLY ASKED QUESTIONS

What is Smart City Urban Mobility, and how does it work? Smart City Urban Mobility is a system that uses technology and innovative solutions to optimize transportation in urban environments. It uses tools like smart traffic management, electric vehicles, and public transportation systems to reduce congestion, lower emissions, and improve mobility.

How does SCUM benefit the local community? SCUM provides faster, safer, and more sustainable transportation options for the community, while also reducing traffic congestion, air pollution, and noise pollution. It also promotes economic development by improving access to jobs, education, and services.

Can SCUM be integrated with existing transportation infrastructure? Yes, SCUM can be integrated with existing transportation infrastructure, such as roads, public transit systems, and bike lanes, to optimize the use of space and resources.

How does SCUM address the challenges of climate change and air pollution in urban areas? SCUM promotes sustainable urban development by reducing the carbon footprint of transportation and improving air quality. It does this by encouraging the use of electric vehicles, public transportation, and other low-emission modes of transportation.

What types of technology and solutions are used in SCUM? SCUM uses a variety of technology and solutions, such as real-time traffic management systems, smart parking systems, electric and autonomous vehicles, and bike-sharing programs.

How does SCUM ensure the safety of users and pedestrians? SCUM prioritizes safety through the implementation of traffic management systems, pedestrian-friendly infrastructure, and other safety measures like speed limit reduction and education campaigns.

Can SCUM be used to promote active transportation like walking and cycling? Yes, SCUM can be used to promote active transportation by integrating infrastructure like bike lanes, sidewalks, and pedestrian crossings into transportation planning.

How does SCUM impact the local economy? SCUM promotes economic development by improving access to jobs, education, and services. It also supports the growth of sustainable transportation-related businesses and reduces transportation-related costs for individuals and businesses.

How does SCUM improve accessibility for people with disabilities? SCUM improves accessibility for people with disabilities through the implementation of infrastructure like accessible parking, ramps, and tactile warning strips, as well as the availability of accessible transportation options like low-floor buses and paratransit services.

Can SCUM be implemented in cities around the world? Yes, SCUM can be implemented in cities around the world, as it is adaptable to different transportation challenges and urban environments. It provides a sustainable solution to the challenges of urbanization and transportation.

Conclusion: *SCUM solutions can help to reduce costs for both cities and residents. By reducing traffic congestion, cities can save money on infrastructure and maintenance costs, while residents can save money on fuel, parking and other transportation-related expenses.*

SMART CITY URBAN PLANNING AND DESIGN (SCUPD)

Using technology to plan and design sustainable and livable cities, such as digital twins and spatial analysis.

Implementing SCUPD using technology can bring a number of benefits to a city. One key benefit is the ability to create more sustainable and livable cities through the use of digital twins and spatial analysis. Digital twins are virtual representations of physical cities that can be used to simulate and analyze different design scenarios, allowing urban planners to make data-driven decisions about the layout and development of a city. Spatial analysis, on the other hand, can be used to identify patterns and trends in urban areas, such as population density and land use, which can inform decisions about where to locate new infrastructure or buildings.

Another benefit of SCUPD is the ability to improve the overall livability of a city. By using technology to plan and design sustainable and livable cities, cities can create more walkable and bike-friendly streets, improve access to public transportation, and create green spaces and parks that encourage active lifestyles. SCUPD can also be used to address issues such as social inequality, affordable housing, and economic development.

By using technology in urban planning and design, cities can also improve the overall efficiency and effectiveness of their decision-making processes. For example, by using digital twins, cities can quickly and easily identify areas where infrastructure upgrades or repairs are needed, or where new development is needed. Implementing spatial analysis, cities can identify patterns and trends in urban areas, such as population density, land use, and transportation patterns, which can inform decisions about where to locate new infrastructure or buildings.

GENERAL FRAMEWORK

The following framework is a general one and the actual implementation will depend on the specific requirements and constraints of the city.

1. Define the goals and objectives of SCUPD.
2. Conduct research to understand current urban planning and design practices.
3. Identify stakeholders in urban planning and design.
4. Develop a comprehensive digital strategy for SCUPD.
5. Choose suitable technology and tools for digital twins and spatial analysis.

6. Implement a GIS system to manage geographic information and data.
7. Develop a 3D virtual model of the city using digital twin technology.
8. Implement real-time monitoring systems for environmental and urban data.
9. Implement spatial analysis tools for data-driven decision-making.
10. Develop a mobile application for residents to interact with the digital twin.
11. Train urban planners and designers on the use of new technology and tools.
12. Implement data security measures to protect sensitive information.
13. Conduct pilot tests of the SCUPD system in selected city areas.
14. Evaluate the results of the pilot tests and make improvements as necessary.
15. Develop a comprehensive marketing and communication plan to promote SCUPD to residents.
16. Launch SCUPD in selected city areas.
17. Monitor the performance of the SCUPD system and make ongoing improvements.
18. Conduct resident satisfaction surveys to gather feedback and improve the system.
19. Continuously analyze urban and environmental data to identify trends and improve the system.
20. Expand the SCUPD system to additional city areas.
21. Establish partnerships with international cities to promote SCUPD globally.
22. Continuously explore and integrate new technology to enhance urban planning and design.
23. Develop a sustainability plan for the city, integrating SCUPD technology and data.
24. Engage with the community to involve them in the urban planning and design process.
25. Continuously evaluate and refine the SCUPD system to ensure sustainable and livable cities.

FREQUENTLY ASKED QUESTIONS

What is Smart City Urban Planning and Design, and how does it work? Smart City Urban Planning and Design is a system that uses technology and innovative solutions to optimize urban environments for residents, businesses, and visitors. It uses tools like digital modelling, data analytics, and community engagement to create livable, sustainable, and equitable urban spaces.

How does SCUPD benefit the local community? SCUPD benefits the local community by creating urban spaces that are accessible, healthy, and resilient. It promotes economic development, social equity, and environmental sustainability, by improving access to services, green spaces, and affordable housing.

Can SCUPD be integrated with existing urban infrastructure and communities? Yes, SCUPD can be integrated with existing urban infrastructure and communities, by taking into account local context, history, and culture. It uses data-driven and participatory approaches to engage with communities and stakeholders.

How does SCUPD address the challenges of climate change and environmental degradation? SCUPD addresses the challenges of climate change and environmental degradation by promoting sustainable urban development, by reducing carbon emissions, preserving natural resources, and improving air and water quality.

What types of technology and solutions are used in SCUPD? SCUPD uses a variety of technology and solutions, such as digital modelling, GIS, data analytics, 3D printing, and energy-efficient building design.

How does SCUPD ensure social equity and community engagement? SCUPD ensures social equity and community engagement by involving communities and stakeholders in the planning and design process, by using participatory approaches and feedback mechanisms. It also promotes equitable access to services, housing, and economic opportunities.

Can SCUPD be used to promote active transportation like walking and cycling? Yes, SCUPD can be used to promote active transportation by integrating infrastructure like bike lanes, sidewalks, and pedestrian crossings into urban planning and design.

How does SCUPD impact the local economy? SCUPD impacts the local economy by promoting economic development, by creating jobs, attracting businesses, and increasing property values. It also reduces transportation and energy costs for individuals and businesses.

How does SCUPD improve public health and wellbeing? SCUPD improves public health and wellbeing by creating urban spaces that promote physical activity, social interaction, and mental health. It also provides access to healthy food, clean air, and green spaces.

Can SCUPD be implemented in cities around the world? Yes, SCUPD can be implemented in cities around the world, as it is adaptable to different urban contexts and challenges. It provides a sustainable and innovative solution to the challenges of urbanization and development.

Conclusion: *Implementing SCUPD using technology can bring a number of benefits to a city such as sustainability, livability, addressing social inequality, affordable housing, and economic development as well as improving the overall efficiency and effectiveness of decision-making processes.*

SMART CITY WASTE MANAGEMENT (SCWM)

Using sensors and data analysis to optimize water usage, detect leaks, garbage collection, reduce waste and recycling systems.

Implementing a SCWM system can have a wide range of benefits for both the environment and the community. One of the main benefits is the ability to optimize water usage through the use of sensors and data analysis. This can help to detect leaks and reduce water waste, which can lead to cost savings for both the city and its residents.

Another benefit of SCWM is the ability to improve garbage collection and reduce waste. Sensors can be used to monitor waste levels in garbage cans and alert garbage trucks when they need to be emptied, which can increase efficiency and reduce the amount of waste that ends up in landfills. Data analysis can help to identify patterns in waste generation and disposal, which can inform policies and programs aimed at reducing waste.

Recycling systems can also be improved through the use of smart city technology. For example, sensors can be used to monitor the types of materials that are being disposed of, which can help to identify opportunities for recycling and composting. This can lead to a more sustainable waste management system and a reduction in the amount of waste that ends up in landfills.

GENERAL FRAMEWORK

The following framework is a general one and the actual implementation will depend on the specific requirements and constraints of the city.

1. Define the goals and objectives of SCWM.
2. Conduct research to understand the current waste management practices in the city.
3. Develop a comprehensive digital strategy for SCWM.
4. Identify stakeholders in waste management.
5. Choose suitable technology and tools for waste management.
6. Implement a real-time monitoring system for waste collection.
7. Develop an automated waste collection system using sensors and GPS.
8. Implement data analysis tools to optimize waste collection routes.
9. Implement a recycling management system using sensors and data analysis.
10. Develop an application for residents to monitor and report waste and recycling activities.
11. Train waste management personnel on the use of new technology and tools.
12. Implement data security measures to protect sensitive information.

13. Conduct pilot tests of the SCWM system in selected city areas.
14. Evaluate the results of the pilot tests and make improvements as necessary.
15. Develop a comprehensive marketing and communication plan to promote SCWM to residents.
16. Launch SCWM in selected city areas.
17. Monitor the performance of the SCWM system and make ongoing improvements.
18. Conduct resident satisfaction surveys to gather feedback and improve the system.
19. Continuously analyze waste and recycling data to identify trends and improve the system.
20. Expand the SCWM system to additional city areas.
21. Establish partnerships with international cities to promote SCWM globally.
22. Continuously explore and integrate new technology to enhance waste management.
23. Develop a sustainability plan for the city, integrating SCWM technology and data.
24. Engage with the community to involve them in waste management activities.
25. Continuously evaluate and refine the SCWM system to ensure efficient and sustainable waste management.

FREQUENTLY ASKED QUESTIONS

What is Smart City Waste Management (SCWM)? Smart City Waste Management (SCWM) is a system that uses technology and data analytics to manage waste in a city more efficiently and sustainably. It involves the use of sensors, IoT devices, and software applications to track and monitor waste in real-time.

How does SCWM benefit the city and its residents? SCWM benefits the city and its residents by improving the collection, transportation, and disposal of waste, which results in a cleaner and healthier environment. It also reduces the amount of waste that goes to landfills, conserves natural resources, and helps the city to achieve its sustainability goals.

What technologies are used in SCWM? Technologies used in SCWM include sensors, IoT devices, GPS, data analytics, and software applications. These technologies help to monitor and track waste in real-time, optimize waste collection routes, and identify areas for improvement.

How does SCWM help to reduce waste in the city? SCWM helps to reduce waste in the city by providing real-time data on waste generation and collection, which can be used to optimize waste management processes. It also encourages residents to adopt more sustainable practices, such as recycling and composting, by providing education and awareness campaigns.

How does SCWM handle hazardous waste? SCWM handles hazardous waste by using specialized containers and vehicles for collection and transportation. It also employs trained professionals who follow strict safety protocols and regulations for the handling, storage, and disposal of hazardous waste.

How does SCWM help to reduce greenhouse gas emissions? SCWM helps to reduce greenhouse gas emissions by optimizing waste collection routes, reducing the number of vehicles on the road, and promoting sustainable practices such as recycling and composting. This reduces the amount of waste that goes to landfills and the associated greenhouse gas emissions.

How does SCWM ensure that waste is disposed of properly? SCWM ensures that waste is disposed of properly by using a range of techniques, including recycling, composting, and waste-to-energy facilities. It also employs trained professionals who follow strict safety protocols and regulations for the handling, storage, and disposal of waste.

How does SCWM promote public participation in waste management? SCWM promotes public participation in waste management by providing education and awareness campaigns, encouraging sustainable practices such as recycling and composting, and providing incentives for residents to reduce their waste generation.

How does SCWM handle electronic waste? SCWM handles electronic waste by using specialized collection and transportation methods, and by working with certified e-waste recycling facilities to ensure that electronic waste is properly disposed of and recycled.

How does SCWM use data to improve waste management processes? SCWM uses data to improve waste management processes by analyzing real-time data on waste generation, collection, and disposal. This data is used to optimize waste collection routes, identify areas for improvement, and track the effectiveness of waste reduction initiatives.

Conclusion: *SCWM can help to make cities more sustainable, efficient and livable by reducing waste and water usage, improving garbage collection and recycling systems, and providing valuable data to inform policy and decision-making.*

SMART CITY WATER QUALITY MONITORING (SCWQM)

Using technology to monitor and manage water pollution and ensure safe drinking water, such as sensors and water treatment systems.

Implementing SCWQM technology can have a wide range of benefits. By using sensors and other monitoring systems, cities can detect and quickly respond to water pollution events, helping to protect public health and the environment. These systems can help cities to ensure that drinking water meets safety standards and is free from contaminants.

The use of water treatment systems, such as those that use advanced oxidation or membrane filtration, can also help to improve water quality and reduce the need for costly and energy-intensive treatment processes. This technology can help to optimize water usage and detect leaks, potentially saving cities money on water bills and reducing the risk of water shortages.

GENERAL FRAMEWORK

The following framework is a general one and the actual implementation will depend on the specific requirements and constraints of the city.

1. Define the goals and objectives of SCWQM.
2. Conduct research to understand the current water quality in the city.
3. Develop a comprehensive digital strategy for SCWQM.
4. Choose suitable technology and tools for water quality monitoring and treatment.
5. Implement a real-time monitoring system for water pollution.
6. Develop an automated water treatment system using sensors and data analysis.
7. Implement data analysis tools to optimize water treatment processes.
8. Train water management personnel on the use of new technology and tools.
9. Implement data security measures to protect sensitive information.
10. Conduct pilot tests of the SCWQM system in selected city areas.
11. Evaluate the results of the pilot tests and make improvements as necessary.
12. Develop a comprehensive marketing and communication plan to promote SCWQM to residents.
13. Launch SCWQM in selected city areas.
14. Monitor the performance of the SCWQM system and make ongoing improvements.
15. Conduct resident satisfaction surveys to gather feedback and improve the system.
16. Continuously analyze water quality data to identify trends and improve the system.
17. Expand the SCWQM system to additional city areas.
18. Establish partnerships with international cities to promote SCWQM globally.
19. Continuously explore and integrate new technology to enhance water quality monitoring and treatment.
20. Develop a sustainability plan for the city, integrating SCWQM technology and data.
21. Engage with the community to educate them on water quality and treatment processes.

22. Develop water quality awareness campaigns to promote safe drinking water.
23. Provide real-time water quality information to residents through an application.
24. Collaborate with water service providers to ensure safe and clean drinking water.
25. Continuously evaluate and refine the SCWQM system to ensure efficient and effective water quality monitoring and treatment.

FREQUENTLY ASKED QUESTIONS

What is Smart City Water Quality Monitoring (SCWQM)? SCWQM is a system that uses sensors and data analysis to monitor the quality of water in a city's various water sources.

What are the benefits of SCWQM? SCWQM can help identify and address issues with water quality before they become major problems, as well as provide real-time information to water management teams and the public.

Who is responsible for the implementation of SCWQM? City authorities and water management organizations are responsible for the implementation of SCWQM.

How does SCWQM work? SCWQM works by using sensors to collect data on the quality of the water in a city's water sources, such as lakes and rivers. The data is then analysed to identify any changes in water quality and alert authorities to potential issues.

What are the main components of SCWQM? The main components of SCWQM include sensors, data collection systems, and analysis software.

Can the public access the data collected by SCWQM? In some cases, the public can access the data collected by SCWQM through online portals and other platforms.

What types of water quality issues can SCWQM detect? SCWQM can detect a wide range of water quality issues, including the presence of pollutants, bacteria, and changes in pH levels.

How accurate is SCWQM in detecting water quality issues? The accuracy of SCWQM depends on the quality of the sensors used and the analysis software. With high-quality equipment, SCWQM can provide accurate data.

Can SCWQM be integrated with other smart city systems? Yes, SCWQM can be integrated with other smart city systems, such as those for water management and flood control.

How can SCWQM benefit the environment? By detecting and addressing water quality issues early, SCWQM can help protect aquatic ecosystems and support sustainable water management practices.

Conclusion: *By implementing SCWQM technology, cities can not only protect public health and the environment, but also save money and resources, and improve the overall sustainability of their water systems.*

SMART CITY WORKFORCE DEVELOPMENT (SCWD)

Using technology to enhance the skills and employability of the city's workforce, such as online training and job matching platforms.

Implementing SCWD technologies can bring a wide range of benefits to both the city and its residents. By using technology to enhance the skills and employability of the city's workforce, cities can better prepare residents for the jobs of the future, improve productivity and economic growth, and reduce unemployment and poverty.

One of the main benefits of using technology for workforce development is the ability to provide online training and job matching platforms. These tools can help residents acquire new skills and knowledge, and match them with job opportunities that are in high demand. This can improve the employability and earning potential of residents, and help to reduce poverty and inequality.

Another benefit of SCWD is the ability to better prepare residents for the jobs of the future. With the use of technology, cities can identify the skills and knowledge that will be in high demand in the future, and provide training and education programs to prepare residents for these jobs. This can help to ensure that the city's workforce is well-prepared for the changes in the job market and can adapt to new technologies and trends.

SCWD can help to improve productivity and economic growth by providing residents with the skills and knowledge they need to be successful in the job market. This can lead to a more productive and efficient workforce, which can attract new businesses and investment to the city, and help to spur economic growth.

GENERAL FRAMEWORK

The following framework is a general one and the actual implementation will depend on the specific requirements and constraints of the city.

1. Identify the current workforce skills and needs.
2. Conduct a gap analysis of the existing training and development programs.
3. Develop a strategy to support workforce development.
4. Partner with businesses, educational institutions and government organizations to provide online training and job matching platforms.
5. Implement a continuous monitoring and evaluation system to assess the impact of workforce development programs.
6. Develop partnerships with local and national businesses to provide on-the-job training opportunities.
7. Foster a culture of lifelong learning by encouraging employees to continuously upskill.
8. Invest in digital infrastructure to provide access to online training and development resources.
9. Provide financial incentives to encourage businesses to invest in employee training and development.

10. Utilize technology such as artificial intelligence and machine learning to match employees with training and job opportunities.
11. Develop a mentorship program to provide guidance and support to employees as they upskill.
12. Offer training programs for specific industries and in-demand skills.
13. Encourage businesses to adopt flexible working arrangements to support employee training and development.
14. Provide online resources to support career advancement, such as resume writing and interview preparation.
15. Establish a city-wide council to guide workforce development efforts and share best practices.
16. Collaborate with universities and research institutions to develop new training programs and technologies.
17. Invest in education and training programs for underrepresented groups.
18. Provide resources and support for employees to obtain certifications and licenses.
19. Encourage businesses to adopt flexible hiring practices, such as skills-based hiring.
20. Foster partnerships between businesses, schools and vocational training institutions.
21. Offer job placement services for graduates of workforce development programs.
22. Provide support for workers to transition to new careers and industries.
23. Develop a continuous feedback and improvement loop for workforce development programs.
24. Offer support for small businesses to implement workforce development initiatives.
25. Encourage a culture of continuous learning and professional development throughout the city.

FREQUENTLY ASKED QUESTIONS

What is Smart City Workforce Development (SCWD)? Smart City Workforce Development (SCWD) is a program designed to equip the local workforce with the necessary skills to meet the needs of the smart city industry.

Who is eligible to participate in SCWD? Anyone who is interested in gaining new skills or improving their existing ones in the field of smart city technologies can participate in the program.

What types of skills are taught in SCWD? The program teaches a variety of skills including software development, data analytics, cybersecurity, project management, and communication.

Is there a fee to participate in SCWD? The fees for the program may vary depending on the location and specific program, but in many cases, there may be no fees or they may be subsidized by the government.

How long does it take to complete the SCWD program? The length of the program can vary depending on the specific program, but it typically ranges from several weeks to several months.

What are the career opportunities after completing SCWD? The program prepares individuals for a variety of roles in the smart city industry such as software engineers, data analysts, project managers, and cybersecurity experts.

Is there any prior knowledge or education required to participate in SCWD? Some programs may have specific educational requirements or may require prior knowledge in certain areas, but many programs are designed for individuals with no prior experience.

How can I enroll in SCWD? Enrollment information can typically be found on the program's website or through a local government or workforce development agency.

Are there any age restrictions to participate in SCWD? The age restrictions may vary depending on the specific program, but in many cases, there may be no age restrictions.

How can SCWD benefit my community? SCWD can benefit a community by preparing its workforce for the emerging smart city industry, increasing job opportunities and stimulating economic growth.

Conclusion: Implementing SCWD technologies can have a significant positive impact on the city and its residents. By providing online training and job matching platforms, better preparing residents for the jobs of the future, improving productivity and economic growth, and reducing poverty and inequality, SCWD can play a crucial role in building a more sustainable, livable, and equitable city.

Overall Summary: Smart city programs are a new approach to urban development that aims to leverage technology to make cities more efficient, sustainable, and livable. These programs are being implemented across a variety of industries, including transportation, energy, healthcare, and education, to address urban problems and improve the quality of life for citizens. Smart city programs rely on data-driven solutions to optimize services, increase citizen participation, and improve public safety. They enable the creation of more connected and sustainable urban environments. we explored the different types of smart city programs being implemented across industries.

Overall Conclusion: Smart city programs are an innovative and forward-thinking approach to urban development. They offer the potential to revolutionize the way cities operate and provide citizens with a better quality of life. As technology continues to evolve and become more integrated into our daily lives, smart city programs will play an increasingly important role in shaping the future of urban environments. By exploring the various types of smart city programs being implemented across industries, we can gain a better understanding of the benefits they offer and how they can be used to tackle urban challenges in a sustainable and effective way.

Quote: "Smart city programs embody a transformative approach to urban development, utilizing technology to create urban environments that are both sustainable and efficient. The hallmark of true intelligence is not mere knowledge, but the ability to envision and innovate. The implementation of smart city programs demands a creative and imaginative application of data-driven solutions to confront urban challenges and elevate the standard of living for our citizens." Emin Hasic

ONE SMART WORLD
ARE YOU READY FOR IT?

| xi |
PROGRAM COST

Calculating the cost per capita for developing Smart City programs would require detailed information about the specific programs to be developed, their scope, duration, and their associated costs. Without this information, it's not possible to provide an accurate cost per capita estimate. With that being said, following is a cost per program per capita from around the world with the estimated cost per capita.

ESTIMATED PROGRAM COST PER CAPITA

SMART CITY 5G NETWORK (SC5GN): A study by **Deloitte** estimated the cost of 5G network deployment in the United States at around USD$130 billion. Cost per capita: **USD$13,000**.

SMART CITY AGRICULTURE (SCAG): A study by **Accenture** estimates that implementing smart farming technology across 10,000 hectares of farmland could cost around USD$35 million. Cost per capita: **USD$50**.

SMART CITY AIR QUALITY MONITORING (SCAQM): The per capita cost to develop a SCAQM program can vary depending on the specific requirements and features of the program. However, as a real example, the city of **Barcelona, Spain** spent approximately €80 million (or **€250** per capita) on their Smart City program, which included air quality monitoring as one of the initiatives. It's important to note that this cost may not be directly transferable to other cities due to differences in population size, geography, and other factors.

SMART CITY ANALYTICS (SCA): The city of **Chicago, United States** spent approximately USD$50 million on their Array of Things project, which is a network of interactive, modular sensor boxes that collect real-time data on the urban environment. This project included the development of a Smart City Analytics platform to analyze and visualize the collected data. With a population of approximately 2.7 million people, this equates to a per capita cost of around **USD$18.50**. It's important to note that this cost may not be directly transferable to other cities due to differences in population size, geography, and other factors.

SMART CITY BIKE-SHARING (SCBS): The city of **Paris, France** spent approximately €150 million (or **€20** per capita) on their bike-sharing program called Vélib'. This program included the deployment of thousands of bikes and associated infrastructure, as well as a mobile app to facilitate bike rentals and tracking. It's important to note that this cost may not be directly transferable to other cities due to differences in population size, geography, and other factors.

SMART CITY BUILDING MANAGEMENT (SCBM): The city of **New York, United States** spent approximately USD$80 million on their Greener, Greater Buildings Plan, which is a package of laws and initiatives designed to reduce energy consumption and greenhouse gas emissions from buildings. This program included the development of a Smart Building Management platform to monitor and optimize energy use in large buildings. With a population of approximately 8.4 million people, this equates to a per capita cost of around **USD$9.50**. It's important to note that this cost may not be directly transferable to other cities due to differences in population size, geography, and other factors.

SMART CITY CITIZEN ENGAGEMENT (SCCE): The city of **Barcelona, Spain** spent approximately €3.3 million (or **€10** per capita) on their Decidim platform, which is an online platform that allows citizens to participate in decision-making processes and engage with city officials. The platform includes features such as online voting, crowdsourcing, and collaborative drafting of proposals. It's important to note that this cost may not be directly transferable to other cities due to differences in population size, geography, and other factors.

SMART CITY CITIZEN PARTICIPATION (SCCP): The city of **Madrid, Spain** spent approximately €1.7 million (or **€3** per capita) on their Decide Madrid platform, which is an online platform that allows citizens to participate in decision-making processes and submit proposals for the city's budget. The platform includes features such as online voting, proposal drafting, and real-time monitoring of submitted proposals. It's important to note that this cost may not be directly transferable to other cities due to differences in population size, geography, and other factors.

SMART CITY CLIMATE RESILIENCE (SCCR): The city of **Rotterdam, Netherlands** spent approximately €90 million (or **€210** per capita) on their Rotterdam Resilience Strategy, which is a comprehensive plan to adapt the city to climate change and other stressors. This program includes a range of initiatives, such as the deployment of green roofs and walls, the creation of new water plazas and retention basins, and the implementation of a city-wide smart water management system. It's important to note that this cost may not be directly transferable to other cities due to differences in population size, geography, and other factors.

SMART CITY COMMUNITY DEVELOPMENT (SCCD): The city of **Barcelona, Spain** spent approximately €100 million (or **€300** per capita) on their "Barcelona, posa't guapa" program, which is a comprehensive urban renewal and community development initiative. This program included a range of initiatives, such as the renovation of public spaces, the rehabilitation of public housing, the expansion of social services, and the promotion of cultural and sporting events. It's important to note that this cost may not be directly transferable to other cities due to differences in population size, geography, and other factors.

SMART CITY CONNECTED AND AUTONOMOUS VEHICLES (SCCAV): The city of **Las Vegas, United States** spent approximately USD$9 million on their autonomous vehicle program, which is aimed at testing and developing self-driving vehicle technology. With a population of approximately 650,000 people, this equates to a per capita cost of around **USD$14**. It's important to note that this cost may not be directly transferable to other cities due to differences in population size, geography, and other factors. Additionally, the cost of implementing a full-scale connected and autonomous vehicle program is likely to be much higher than the cost of a testing and development program.

SMART CITY CYBER-SECURITY (SCCS): The city of **New York, United States** spent approximately USD$100 million on their Cyber Command program, which is a comprehensive initiative aimed at protecting the city's computer networks and critical infrastructure from cyber-attacks. With a population of approximately 8.4 million people, this equates to a per capita cost of around **USD$12**. It's important to note that this cost may not be directly transferable to other cities due to differences in population size, geography, and other factors. Additionally, the cost of implementing a full-scale cyber-security program may vary widely depending on the level of protection required and the specific threats that need to be addressed.

SMART CITY DATA MANAGEMENT (SCDM): The city of **Chicago, United States** spent approximately USD$50 million on their SmartData Platform, which is a comprehensive data management initiative that allows the city to collect, integrate, and analyze data from a variety of sources. With a population of approximately 2.7 million people, this equates to a per capita cost of around **USD$18**. It's important to note that this cost may not be directly transferable to other cities due to differences in population size, geography, and other factors. Additionally, the cost of implementing a full-scale data management program may vary widely depending on the amount and complexity of data to be managed, as well as the specific features and capabilities required.

SMART CITY DIGITAL INCLUSION (SCDI): The city of **Austin, United States** spent approximately USD$12 million on their Digital Inclusion program, which is a comprehensive initiative aimed at bridging the digital divide and providing access to technology and digital skills training for underserved communities. With a population of approximately

990,000 people, this equates to a per capita cost of around **USD$12**. It's important to note that this cost may not be directly transferable to other cities due to differences in population size, geography, and other factors. Additionally, the cost of implementing a full-scale digital inclusion program may vary widely depending on the level of access and training required, as well as the specific tools and resources to be provided.

SMART CITY DISASTER MANAGEMENT (SCDIM): The city of **Miami, United States** spent approximately USD$20 million on their Resilience and Sustainability Department, which includes a comprehensive disaster management initiative aimed at preparing for and responding to natural disasters and other emergencies. With a population of approximately 470,000 people, this equates to a per capita cost of around **USD$43**. It's important to note that this cost may not be directly transferable to other cities due to differences in population size, geography, and other factors. Additionally, the cost of implementing a full-scale disaster management program may vary widely depending on the specific risks and hazards faced by a city, as well as the specific tools, resources, and infrastructure required to effectively prepare for and respond to emergencies.

SMART CITY EDUCATION (SCE): The city of **Barcelona, Spain** spent approximately €50 million on their Barcelona Education City program, which is a comprehensive initiative aimed at promoting educational innovation and improving access to quality education for all citizens. With a population of approximately 1.6 million people, this equates to a per capita cost of around **€31**. It's important to note that this cost may not be directly transferable to other cities due to differences in population size, geography, and other factors. Additionally, the cost of implementing a full-scale education program may vary widely depending on the specific goals and objectives of the program, as well as the specific tools, resources, and infrastructure required to effectively promote educational innovation and improve access to education for all citizens.

SMART CITY E-GOVERNANCE (SCEG): The city of **Seoul, South Korea** spent approximately KRW 60 billion (approximately USD$53 million) on their Smart Seoul Portal System, which is a comprehensive e-governance initiative aimed at improving the efficiency and transparency of government services and communications with citizens. With a population of approximately 10 million people, this equates to a per capita cost of around **USD$5.3**. It's important to note that this cost may not be directly transferable to other cities due to differences in population size, geography, and other factors. Additionally, the cost of implementing a full-scale e-governance program may vary widely depending on the specific goals and objectives of the program, as well as the specific tools, resources, and infrastructure required to effectively improve the efficiency and transparency of government services and communications with citizens.

SMART CITY EMERGENCY RESPONSE (SCER): The city of **New York, United States** spent approximately USD$1.6 billion on their Emergency Management Department in 2021, which includes a comprehensive emergency response initiative aimed at preparing for and responding to a wide range of emergencies and disasters. With a population of approximately 8.4 million people, this equates to a per capita cost of around **USD$190**. It's important to note that this cost may not be directly transferable to other cities due to differences in population size, geography, and other factors. Additionally, the cost of implementing a full-scale emergency response program may vary widely depending on the specific risks and hazards faced by a city, as well as the specific tools, resources, and infrastructure required to effectively prepare for and respond to emergencies.

SMART CITY EMERGENCY SERVICES (SCES): The city of **London, United Kingdom** spent approximately £5.5 billion (approximately USD$7.7 billion) on their emergency services in 2021, which includes a comprehensive emergency response initiative aimed at protecting citizens and responding to emergencies such as fires, accidents, and medical emergencies. With a population of approximately 9 million people, this equates to a per capita cost of around **USD$856**. It's important to note that this cost may not be directly transferable to other cities due to differences in population size, geography, and other factors. Additionally, the cost of implementing a full-scale emergency services program may vary widely depending on the specific risks and hazards faced by a city, as well as the specific tools, resources, and infrastructure required to effectively protect citizens and respond to emergencies.

SMART CITY ENERGY DISTRIBUTION (SMED): The city of **Austin, Texas, USA** spent approximately USD$38 million on their Grid Modernization Program, which includes a comprehensive initiative aimed at improving the efficiency and resiliency of the city's energy distribution infrastructure. With a population of approximately 1 million people, this equates to a per capita cost of around **USD$38**. It's important to note that this cost may not be directly transferable to other cities due to differences in population size, geography, and other factors. Additionally, the cost of implementing a full-scale energy distribution program may vary widely depending on the specific goals and objectives of the program, as well as the specific tools, resources, and infrastructure required to effectively improve the efficiency and resiliency of a city's energy distribution system.

SMART CITY ENERGY MANAGEMENT (SCEM): The city of **Boston, Massachusetts, USA** spent approximately USD$2.3 million on their Renew Boston program, which includes a comprehensive energy management initiative aimed at reducing energy consumption and greenhouse gas emissions in city buildings. With a population of approximately 700,000 people, this equates to a per capita cost of around **USD$3.30**. It's important to note that this cost may not be directly transferable to other cities due to differences in population size, geography, and other factors. Additionally, the cost of implementing a full-scale energy management program may vary widely depending on the specific goals and objectives of the program, as well as the specific tools, resources, and infrastructure required to effectively reduce energy consumption and greenhouse gas emissions.

SMART CITY ENERGY STORAGE (SCENS): The city of **Los Angeles, California, USA** spent approximately USD$1 billion on their Energy Storage Program, which includes a comprehensive initiative aimed at developing and deploying energy storage systems to improve the reliability and resiliency of the city's power grid. With a population of approximately 4 million people, this equates to a per capita cost of around **USD$250**. It's important to note that this cost may not be directly transferable to other cities due to differences in population size, geography, and other factors. Additionally, the cost of implementing a full-scale energy storage program may vary widely depending on the specific goals and objectives of the program, as well as the specific tools, resources, and infrastructure required to effectively deploy energy storage systems to improve the reliability and resiliency of a city's power grid.

SMART CITY ENVIRONMENT (SCENV): The city of **Melbourne, Australia** spent approximately AUD 4 million on their Urban Forest Fund, which includes a comprehensive environmental initiative aimed at increasing the city's green cover and improving the urban landscape. With a population of approximately 5 million people, this equates to a per capita cost of around AUD$0.80 or **USD$0.57**. It's important to note that this cost may not be directly transferable to other cities due to differences in population size, geography, and other factors. Additionally, the cost of implementing a full-scale environmental program may vary widely depending on the specific goals and objectives of the program, as well as the specific tools, resources, and infrastructure required to effectively improve the quality of a city's environment.

SMART CITY ENVIRONMENTAL MONITORING (SCEVM): The city of **Copenhagen, Denmark** spent approximately DKK 20 million on their Environmental Monitoring Program, which includes a comprehensive initiative aimed at monitoring air quality, noise levels, and traffic in the city. With a population of approximately 630,000 people, this equates to a per capita cost of around DKK 32 or **USD$5**. It's important to note that this cost may not be directly transferable to other cities due to differences in population size, geography, and other factors. Additionally, the cost of implementing a full-scale environmental monitoring program may vary widely depending on the specific goals and objectives of the program, as well as the specific tools, resources, and infrastructure required to effectively monitor and analyze a city's environmental data.

SMART CITY FINANCING (SCF): The city of **San Francisco, USA** established the Green Finance SF program to provide financing options to homeowners and businesses for energy and water-efficient upgrades. The program offers Property Assessed Clean Energy (PACE) financing to qualified applicants, which allows property owners to finance eligible energy efficiency, water efficiency, and renewable energy improvements through a voluntary special assessment on their property tax bill. The program is financed through private capital providers, and the city does not incur any direct costs associated with it. However, the program has administrative costs associated with it, such as program management, loan servicing, and outreach and education efforts. As of 2021, the program has financed over USD$120 million in energy and water efficiency projects in the city. With a population of approximately 883,000 people, the per capita cost for the Green Finance SF program would be **zero**. However, it's important to note that the costs associated with establishing and managing a financing program will vary depending on the specific structure of the program, the financing options offered, and the scale of the program.

SMART CITY FLOOD MANAGEMENT (SCFM): The city of **Amsterdam, Netherlands** has been dealing with flooding for centuries, and has developed a variety of strategies to manage flood risk. One component of the city's flood management program is the "Rainproof" initiative, which aims to reduce the impact of heavy rainfall and stormwater runoff on the city's infrastructure and residents. The Rainproof initiative includes a variety of measures, such as green roofs, rain gardens, and permeable pavements, which help to absorb and manage stormwater runoff. The program also includes public education and outreach efforts to encourage residents to take steps to reduce the impact of heavy rainfall on their properties. While the costs associated with the Rainproof initiative are not specifically broken down on a per capita basis overall budget for the program is approximately €200 million (USD$240 million) over the next five years, which includes funding from the city government as well as private and public sector partners. With a population of around 870,000, the per capita cost of the program would be approximately **€230** (USD$275) per person over the five-year period. It's worth noting that the costs of flood management programs can vary widely depending on the specific challenges and conditions of each city. However, investing in flood management programs can ultimately help to mitigate the risk of costly damage and loss of life from flooding events.

SMART CITY GOVERNANCE (SCG): The city of **Barcelona, Spain,** has implemented SCG, known as "Barcelona Smart City 2011-2020," is a broad initiative aimed at promoting sustainable urban development and improving the quality of life for citizens. One key component of this program is the implementation of a comprehensive e-government platform that allows citizens to access a wide range of city services online. The e-government platform includes a variety of digital services, such as online payment of taxes and fines, access to public transportation information, and the ability to report issues such as potholes or broken streetlights. The platform also allows citizens to participate in decision-making processes through online voting and participation in citizen committees. While the costs associated with Barcelona's Smart City program are not specifically broken down on a per capita basis, the city has invested approximately €70 million (USD$84 million) in the program since its launch in 2011. With a population of around 1.6 million, the per capita cost of the program would be approximately **€44** (USD$53) over the ten-year period. It's worth noting that the costs of governance programs can vary widely depending on the specific challenges and conditions of each city. However, investing in e-government platforms can ultimately help to improve citizen engagement and participation in decision-making processes, as well as increase efficiency and transparency in city government operations.

SMART CITY GREEN ENERGY (SCGE): The city of **Lancaster, California, USA,** has implemented a comprehensive SCGE program. The city's program includes the installation of over 600,000 solar panels, the conversion of streetlights to energy-efficient LED lights, and the implementation of a smart microgrid system to manage and distribute energy. The total cost of the program was around USD$50 million, which was funded through public-private partnerships and grants. With a population of around 159,000 people, the per capita cost of the program would be approximately **USD$315**. However, it's important to note that this is just one example and that per capita costs can vary widely depending on the specific needs and goals of a given city.

SMART CITY GREEN INFRASTRUCTURE (SCGI): The city of **Philadelphia, Pennsylvania, USA,** has implemented a comprehensive program called "Green City, Clean Waters" to manage stormwater runoff and improve water quality through the use of green infrastructure. The program includes the installation of features such as green roofs, rain gardens, and porous pavement to capture and manage stormwater. The program has a total cost of around USD$2.4 billion, which is being funded through various sources including grants, loans, and government funding. With a population of around 1.6 million people, the per capita cost of the program would be approximately **USD$1,500**. It's important to note that per capita costs can vary widely depending on the specific needs and goals of a given city.

SMART CITY GRIDS (SCGR): The city of **Amsterdam in the Netherlands** is implementing a program to develop a smart grid to improve the efficiency and sustainability of its energy infrastructure. The project, called the "Flex Power Grid," involves the installation of new technologies such as advanced sensors and energy storage systems to optimize energy distribution and reduce energy waste. The total cost of the project is estimated to be around €15 million, which is being funded through a combination of public and private investment. With a population of around 872,000 people, the per capita cost of the project would be approximately **€17**. It's important to note that per capita costs can vary widely depending on the specific needs and goals of a given city.

SMART CITY HEALTHCARE (SCHC): The city of **Barcelona in Spain** implemented a program to improve public health outcomes by using data and technology to create a more efficient and effective healthcare system. The program, called "Barcelona Health Hub," involves the creation of a digital platform that connects healthcare providers and patients, as well as the use of data analytics and artificial intelligence to improve healthcare decision-making. The total cost of the project is estimated to be around €1.5 million, which is being funded through a combination of public and private investment. With a population of around 1.6 million people, the per capita cost of the project would be approximately **€0.94**. The per capita costs can vary widely depending on the specific needs and goals of a given city.

SMART CITY INFRASTRUCTURE MANAGEMENT (SCIM): The city of **Barcelona, Spain,** has implemented a SCIM program called "Barcelona Smart City". The program involves the deployment of various smart technologies and data analytics tools to manage and optimize the city's infrastructure. According to a case study published by Schneider Electric, the company that partnered with Barcelona on the project, the total cost of the program was €1.36 billion (approximately USD$1.5 billion) over a ten-year period. This works out to an average cost of **€136** per year per capita (or approximately USD$149 per year per capita). It's worth noting that this is a large-scale, comprehensive program that involved significant investment in new technologies and infrastructure. The cost of developing a SCIM program for a smaller city or with a narrower scope could be substantially lower.

SMART CITY INNOVATION (SCIN): The city of **Dubai** has invested heavily in SCIN, with a goal of becoming one of the most innovative cities in the world. In 2016, the Dubai government launched the Dubai Future Accelerators program, which invites global companies and startups to develop and test innovative solutions to city challenges. The program has a budget of USD$275 million over five years, and as of 2021, has supported more than 70 initiatives. With a population of over 3 million, this translates to a per capita cost of approximately **USD$91** over the five-year period of the program.

SMART CITY INTELLIGENT TRANSPORTATION SYSTEMS (SCITS): The city of **Toronto, Canada,** has been working on developing an ITS program to improve traffic flow and reduce congestion. The city's initial plan, called the Intelligent Transportation Systems Strategic Plan, had an estimated cost of CAD 100 million, with funding from multiple sources including the federal government, provincial government, and the city. Divided among Toronto's population of 2.9 million, this works out to a per capita cost of approximately **CAD$34.50** (USD$25.85). The per capita cost of developing a SCITS program can vary widely depending on the specific goals and scope of the program. However, according to a report by the World Economic Forum, the cost of developing and deploying an advanced SCITS in a medium-sized city (population between 200,000 and 500,000) can range from CAD$15 million to CAD$70 million. This cost can be significantly higher in larger cities or in more complex systems that require extensive infrastructure and technology.

SMART CITY IoT (SCIoT): The cost of implementing a SCIoT program can vary widely depending on the scope and scale of the program. A real-world example of a SCIoT program is the "Smart Columbus" initiative, which was launched in 2016 by the city of **Columbus, Ohio**. The program aimed to create a "smart transportation system" by deploying connected vehicles, sensors, and other IoT devices throughout the city. The Smart Columbus initiative had an estimated budget of USD$140 million, with USD$40 million coming from the U.S. Department of Transportation and USD$10 million coming from the Paul G. Allen Family Foundation. The remaining USD$90 million was contributed by public and private partners, including the city of Columbus, local businesses, and academic institutions. Columbus has a population of about 900,000 people, so the per capita cost of the Smart Columbus initiative would be approximately **USD$155.56**. However, it's worth noting that this figure may not be entirely representative of the cost of implementing a Smart City IoT program in other locations, as the scale and scope of the Smart Columbus initiative was quite ambitious.

SMART CITY IRRIGATION (SCIR): The city of **San Diego, California, USA,** SCIR system uses sensors to collect data on weather conditions, soil moisture, and other factors to determine the optimal amount of water for each area of the city. The cost of the system was USD$6.5 million, and it covers approximately 3,000 acres of parks and open spaces. With a population of 1.42 million in San Diego, this would equate to a per capita cost of approximately **USD$4.58**.

SMART CITY LIGHTING (SCL): The city of **Los Angeles, California, USA,** implemented a USD$14 million project in 2013 to replace over 140,000 streetlights with energy-efficient LED lights. The project was completed in 2017, and the new LED lights were connected to a central management system that allowed the city to remotely monitor and control each light. This smart lighting system was estimated to save the city up to USD$10 million per year in energy costs. Assuming a population of around 4 million people in Los Angeles, the per capita cost of this project would be approximately **USD$3.50**.

SMART CITY MAINTENANCE (SCM): In the city of **Glasgow, Scotland**, the SCM program, involved the installation of over 70,000 LED streetlights with a central management system, cost around £30 million (approximately USD$41 million) to implement, with an additional £5.5 million (approximately USD$7.5 million) allocated for maintenance and support over the next 10 years. With a population of around 600,000, this would equate to a per capita cost of approximately **£50** (approximately USD$68).

SMART CITY MOBILITY (SCMO): In the city of **Barcelona, Spain,** the SCMO is designed to improve the city's transportation infrastructure and reduce traffic congestion by promoting sustainable and innovative mobility options. According to a report by Deloitte, the total cost of the SCMO Program in Barcelona was €5.5 million, or approximately **€5.50** per capita based on the city's population of approximately 1 million people. The program includes a number of initiatives, such as the expansion of bike-sharing services, the development of a smart parking system, and the implementation of real-time traffic monitoring and management systems. The per capita cost of a SCMO program can vary significantly depending on the specific goals, initiatives, and technologies involved.

SMART CITY PARKING MANAGEMENT (SCPM): In the city of **San Francisco in California, USA**, the SCPM program uses a combination of sensors, cameras, and wireless technology to monitor parking spots and provide real-time information about available spaces to drivers. The cost of this program was about USD$18 million for the installation of about 8,200 parking sensors across the city. Assuming a population of 884,000 people in San Francisco, the per capita cost of the Smart City Parking Management program would be approximately **USD$20.38**. The actual per capita cost may be lower as the program was partially funded through grants and parking revenue.

SMART CITY PLATFORM (SCP): The city of **San Diego in California, USA,** implemented a SCP called "Get it Done" in 2016, which allows residents to report non-emergency issues like potholes, graffiti, and broken streetlights via a mobile app. The city spent around USD$50,000 to develop and launch the app, which was aimed at increasing citizen engagement and providing a more streamlined and efficient way to report issues. Assuming a population of 1.4 million in San Diego, this translates to a per capita cost of around **USD$0.04**. The cost can vary widely depending on the complexity and scope of the program.

SMART CITY PUBLIC ART (SCPA): It is difficult to estimate the per capita cost for a SCPA program since there are many factors that can impact the cost, such as the scale of the program, the types of art installations, the technologies used, and the community involvement. However, as an example, let's assume a city with a population of 1 million people, and a budget of USD$10 million to develop a public art program. In this case, the hypothetical per capita cost would be **USD$10** per person.

SMART CITY PUBLIC HEALTH (SCPH): The city of **Barcelona in Spain** has implemented a number SCPH initiatives, including the "Superblocks" program which aims to reduce traffic in certain areas of the city, as well as initiatives to promote active transportation and healthy eating. While the exact per capita cost of these initiatives is not available, the city has budgeted a total of €72 million (USD$85 million) for its "Healthy Barcelona" plan, which includes a wide range of health-related initiatives. With a population of around 1.6 million people, this works out to a per capita cost of approximately **€45** (USD$53) per person. It's worth noting, however, that the cost of implementing SMART CITY PUBLIC HEALTH initiatives can be highly variable depending on the specific needs and priorities of a given city.

SMART CITY PUBLIC SAFETY AND EMERGENCY MANAGEMENT (SCPSEM): In the city of **Santa Cruz, California, USA,** the cost of implementing a SCPSEM was estimated at USD$2.5 million. The city has a population of about 64,000 people, so the cost per capita would be approximately **USD$39.06**. This cost includes the development of a network of cameras and sensors throughout the city, as well as the integration of real-time data from multiple sources, including social media, traffic management systems, and weather forecasts. The city also plans to use data analytics to better respond to emergencies and improve public safety. The cost of a SCPSEM program will vary based on the specific needs of each city. Additionally, the cost of these programs can change over time as new technologies become available and existing technologies are improved.

SMART CITY PUBLIC SERVICES (SCPS): In the city of **Singapore**, the cost of implementing a smart city public services program was estimated at approximately USD$1 billion. Singapore has a population of about 5.7 million people, so the cost per capita would be approximately **USD$175.44**. This cost includes the development of a range of smart city services, including smart lighting, smart waste management, smart parking, and smart traffic management. The city also plans to use data analytics and artificial intelligence to improve the delivery of public services and to better understand the needs of its citizens.

SMART CITY PUBLIC SPACES (SCPSP): In the city of **Amsterdam, the Netherlands**, the cost of implementing a smart city public spaces program was estimated at approximately €100 million. Amsterdam has a population of about 875,000 people, so the cost per capita would be approximately **€113.73**. This cost includes the development of a range of smart city technologies for public spaces, such as smart lighting, smart waste management, and smart traffic management. The city also plans to use data analytics and artificial intelligence to improve the quality of life for its citizens and to create more sustainable public spaces.

SMART CITY PUBLIC TRANSPORTATION (SCPT): In the city of **Barcelona, Spain**, the cost of implementing a smart city public transportation program was estimated at approximately €100 million. Barcelona has a population of about 1.6 million people, so the cost per capita would be approximately **€62.50**. This cost includes the development of a range of smart city technologies for public transportation, such as real-time traffic management systems, smart ticketing systems, and intelligent transportation systems. The city also plans to use data analytics and artificial intelligence to improve the efficiency and sustainability of its public transportation system and to better serve the needs of its citizens.

SMART CITY RENEWABLE ENERGY (SCRE): In the city of **Sydney, Australia**, the cost of implementing a smart city renewable energy program was estimated at approximately AUD$200 million. Sydney has a population of about 5.3 million people, so the cost per capita would be approximately **AUD$37.74**. This cost includes the development of a range of smart city technologies for renewable energy, such as smart grid systems, energy storage systems, and renewable energy generation systems. The city also plans to use data analytics and artificial intelligence to optimize energy usage, improve energy efficiency, and reduce its carbon footprint.

SMART CITY RESILIENCE (SCR): In the city of **Miami, Florida**, the cost of implementing a smart city resilience program was estimated at approximately USD$100 million. Miami has a population of about 463,000 people, so the cost per capita would be approximately **USD$216.07**. This cost includes the development of a range of smart city technologies for improving the resilience of the city, such as smart infrastructure systems, disaster management systems, and emergency response systems. The city also plans to use data analytics and artificial intelligence to improve its ability to respond to natural disasters and other emergencies, and to better protect its citizens and infrastructure.

SMART CITY RETAIL (SCRT): In the city of **Tokyo, Japan**, the cost of implementing a smart city retail program was estimated at approximately ¥10 billion. Tokyo has a population of about 13.5 million people, so the cost per capita would be approximately **¥741.67**. This cost includes the development of a range of smart city technologies for improving the retail experience, such as smart payment systems, smart retail management systems, and data analytics systems. The city also plans to use artificial intelligence and machine learning to optimize supply chains, improve the customer experience, and drive economic growth in the retail sector.

SMART CITY SECURITY (SCS): In the city of **London, UK,** the cost of implementing a smart city security program was estimated at approximately £500 million. London has a population of about 9 million people, so the cost per capita would be approximately **£55.56**. This cost includes the development of a range of smart city technologies for improving security, such as smart surveillance systems, facial recognition systems, and predictive analytics systems. The city also plans to use artificial intelligence and machine learning to improve its ability to respond to security incidents and to better protect its citizens.

SMART CITY SERVICES (SCSR): In the city of **Singapore**, the cost of implementing a smart city services program was estimated at approximately SGD500 million. Singapore has a population of about 5.7 million people, so the cost per capita would be approximately **SGD87.72**. This cost includes the development of a range of smart city technologies for improving public services, such as smart transportation systems, smart healthcare systems, and smart government services. The city also plans to use data analytics and artificial intelligence to improve the efficiency of public services and to better meet the needs of its citizens.

SMART CITY STANDARDIZATION (SCSZ): In the city of **Amsterdam, Netherlands,** the cost of implementing a smart city standardization program was estimated at approximately €20 million. Amsterdam has a population of about 880,000 people, so the cost per capita would be approximately **€22.72**. This cost includes the development of a range of standards and protocols for improving the interoperability of smart city technologies and systems, such as data exchange standards, cybersecurity standards, and privacy standards. The city also plans to establish a smart city platform to support the sharing of data and information between different city departments and stakeholders.

SMART CITY STORMWATER MANAGEMENT (SCSWM): In the city of **Seattle, Washington, USA,** the cost of implementing a smart city stormwater management program was estimated at approximately USD$100 million. Seattle has a population of about 745,000 people, so the cost per capita would be approximately **USD$134.04**. This cost includes the development of a range of technologies and systems for improving stormwater management, such as real-time monitoring systems, predictive analytics systems, and green infrastructure projects. The city also plans to use data and information from these systems to improve its ability to respond to and manage stormwater events and to reduce the impact of stormwater on the city's infrastructure and environment.

SMART CITY STREET FURNITURE (SCSF): In the city of **Paris, France,** the cost of implementing a smart city street furniture program was estimated at approximately €40 million. Paris has a population of about 2.2 million people, so the cost per capita would be approximately **€18.18**. This cost includes the development of a range of smart city technologies for improving street furniture, such as smart benches, smart waste bins, and smart lighting systems. The city also plans to use these technologies to provide a range of services and amenities for citizens, such as Wi-Fi access, charging stations, and real-time information about local services and events.

SMART CITY STREETS (SCST): In the city of **Barcelona, Spain,** the cost of implementing a smart city streets program was estimated at approximately €25 million. Barcelona has a population of about 1.6 million people, so the cost per capita would be approximately **€15.63**. This cost includes the development of a range of technologies and systems for improving the city's streets, such as intelligent traffic management systems, real-time monitoring systems, and smart lighting systems. The city also plans to use these systems to provide a range of services and amenities for citizens, such as improved traffic flow, real-time information about local services and events, and increased safety and security.

SMART CITY SUSTAINABILITY (SCSU): In the city of **Amsterdam, Netherlands,** the cost of implementing a smart city sustainability program was estimated at approximately €50 million. Amsterdam has a population of about 880,000 people, so the cost per capita would be approximately **€56.82**. This cost includes the development of a range of smart city technologies and systems for promoting sustainability, such as energy-efficient building systems, smart lighting systems, and real-time monitoring systems. The city also plans to use these technologies to promote sustainable transportation, such as encouraging the use of electric vehicles and cycling, and to improve the management of waste and water resources.

SMART CITY SUSTAINABLE ENERGY (SCSE): In the city of **Stockholm, Sweden,** the cost of implementing a smart city sustainable energy program was estimated at approximately SEK1 billion (Swedish Krona). Stockholm has a population of about 970,000 people, so the cost per capita would be approximately **SEK1,032.99**. This cost includes the development of a range of smart city technologies and systems for promoting sustainable energy, such as renewable energy systems, energy-efficient building systems, and real-time energy management systems. The city also plans to use these technologies to promote sustainable transportation, such as encouraging the use of electric vehicles and cycling, and to improve the management of energy consumption and production.

SMART CITY SUSTAINABLE TRANSPORTATION (SCSTR): In the city of **Barcelona, Spain**, the cost of implementing a smart city sustainable transportation program was estimated at approximately €35 million. Barcelona has a population of about 1.6 million people, so the cost per capita would be approximately **€21.87**. This cost includes the development of a range of smart city technologies and systems for promoting sustainable transportation, such as intelligent transportation systems, real-time traffic management systems, and cycling infrastructure. The city also plans to use these technologies to promote the use of electric vehicles, public transportation, and sustainable modes of transportation, such as cycling and walking.

SMART CITY TOURISM (SCTO): In the city of **Amsterdam, the Netherlands,** the cost of implementing a smart city tourism program was estimated at approximately €15 million. Amsterdam has a population of about 864,000 people, so the cost per capita would be approximately **€17.36**. This cost includes the development of a range of smart city technologies and systems for promoting sustainable and efficient tourism, such as smart tourism information systems, digital platforms for tour operators and hotels, and real-time information systems for tourists. The city also plans to use these technologies to improve the management of visitor numbers and to promote sustainable and responsible tourism practices.

SMART CITY TOURIST ACCESSIBILITY (SCTA): In the city of **Singapore**, the cost of implementing a smart city tourist accessibility program was estimated at approximately SGD20 million. Singapore has a population of about 5.7 million people, so the cost per capita would be approximately **SGD3.51**. This cost includes the development of a range of smart city technologies and systems for improving tourist accessibility, such as smart tourism information systems, real-time transportation information systems, and digital platforms for tour operators and hotels. The city also plans to use these technologies to improve the management of visitor numbers and to promote sustainable and responsible tourism practices.

SMART CITY TOURIST ACCOMMODATION MANAGEMENT (SCTAM): In the city of **Amsterdam, the Netherlands,** the cost of implementing a smart city tourist accommodation management program was estimated at approximately €20 million. Amsterdam has a population of about 874,000 people, so the cost per capita would be approximately **€22.80**. This cost includes the development of a range of smart city technologies and systems for improving tourist accommodation management, such as real-time occupancy monitoring systems, smart booking systems, and digital platforms for managing the distribution and pricing of tourist accommodations. The city also plans to use these technologies to improve the overall visitor experience and to promote sustainable and responsible tourism practices.

SMART CITY TOURIST ACTIVITY MANAGEMENT (SCTAC): In the city of **Barcelona, Spain,** the cost of implementing a smart city tourist activity management program was estimated at approximately €10 million. Barcelona has a population of about 1.6 million people, so the cost per capita would be approximately **€6.25**. This cost includes the development of a range of smart city technologies and systems for improving tourist activity management, such as real-time occupancy monitoring systems, smart booking systems, and digital platforms for managing the distribution and pricing of tourist activities. The city also plans to use these technologies to improve the overall visitor experience and to promote sustainable and responsible tourism practices.

SMART CITY TOURIST ATTRACTIONS MANAGEMENT (SCTOA): In the city of **Dubai in the United Arab Emirates,** the cost of implementing a smart city tourist attractions management program was estimated at approximately AED350 million (approximately USD$95 million). Dubai has a population of about 3 million people, so the cost per capita would be approximately **AED116.67** (approximately USD$31.67). This cost includes the development of a range of smart city technologies and systems for improving tourist attraction management, such as real-time occupancy monitoring systems, smart booking systems, and digital platforms for managing the distribution and pricing of tourist attractions. The city also plans to use these technologies to improve the overall visitor experience and to promote sustainable and responsible tourism practices.

SMART CITY TOURIST CROWD MANAGEMENT (SCTCM): In the city of **Hong Kong in China,** the cost of implementing a smart city tourist crowd management program was estimated at approximately HKD$60 million (approximately USD$7.7 million). Hong Kong has a population of about 7 million people, so the cost per capita would be approximately **HKD$8.57** (approximately USD$1.10). This cost includes the development of a range of smart city technologies and systems for improving tourist crowd management, such as real-time occupancy monitoring systems, predictive analytics algorithms, and smart queue management systems. The city also plans to use these technologies to improve safety, reduce congestion, and enhance the overall visitor experience in popular tourist areas.

SMART CITY TOURIST DATA ANALYSIS (SCTDA): In the city of **Singapore,** the cost of implementing a smart city tourist data analysis program was estimated at approximately SGD$15 million (approximately USD$11.1 million). Singapore has a population of about 5.7 million people, so the cost per capita would be approximately **SGD$2.63** (approximately USD$1.95). This cost includes the development of a range of smart city technologies and systems for collecting and analyzing tourist data, such as real-time visitor monitoring systems, predictive analytics algorithms, and data warehousing and management solutions. The city also plans to use these technologies to gain valuable insights into visitor behaviours, preferences, and patterns, as well as to improve the overall quality of life for residents and visitors alike.

SMART CITY TOURIST DESTINATION MARKETING (SCTDM): In the city of **Amsterdam, Netherlands.** The program aims to position Amsterdam as a smart and sustainable city, and to attract tourists who are interested in technology and innovation. The program includes initiatives such as the Amsterdam Smart City Experience, which is a guided tour of the city's smart and sustainable projects and initiatives. The cost per capita of this program is difficult to calculate as it is not a direct tourist program, but rather a broader initiative to position Amsterdam as a smart city. However, the Amsterdam Smart City program has received funding from various sources, including the European Union, the Dutch government, and private partners.

SMART CITY TOURIST EMERGENCY SERVICES (SCTES): In the city of **Dubai.** The Dubai Smart City Initiative includes a Smart Emergency Response System that is designed to help visitors and residents in emergency situations. The system uses advanced technologies such as artificial intelligence, drones, and smart cameras to monitor and respond to emergencies. In 2019, the Dubai government announced that it had invested **AED21 billion** (approximately USD$5.7 billion) in smart city initiatives over the previous five years, including the Smart Emergency Response System. Cost per capita **AED6,982** (USD$1,901).

SMART CITY TOURIST EVENT MANAGEMENT (SCTEM): In the city of **Tokyo, Japan.** The Tokyo Tourism Acceleration Program. The program is designed to promote tourism in Tokyo through the use of smart technologies such as big data, artificial intelligence, and the Internet of Things (IoT). The program includes initiatives such as the Tokyo Grand Tea Ceremony, which is an annual event that attracts tourists from around the world. In 2018, the Tokyo Metropolitan Government announced that it had allocated **JPY49.7 billion** (approximately USD$462 million) for the Tokyo Tourism Acceleration Program. Cost per capita **JPY1,343** (USD$10.10).

SMART CITY TOURIST EXPERIENCE MANAGEMENT (SCTXM): On the island of **Jeju, South Korea.** The program is designed to enhance the tourist experience on the island through the use of smart technologies such as augmented reality, virtual reality, and artificial intelligence. The program includes initiatives such as the Jeju Virtual Experience Center, which offers tourists a virtual tour of the island's cultural and natural attractions. In 2018, the Jeju government announced that it had invested **KRW450 billion** (approximately USD$400 million) in the Smart Tourism Island program. Cost per capita **KRW668,687** (USD$525).

SMART CITY TOURIST FEEDBACK AND REVIEWS (SCTFR): In the city of **Helsinki, Finland.** The "My Helsinki" mobile application allows tourists to provide feedback and reviews of their experiences in the city, as well as to access information about events, attractions, and public transportation. The application also includes a feature that allows tourists to create their own customized travel itinerary based on their interests. In 2019, the City of Helsinki announced that it had allocated **€18 million** (approximately USD$22 million) to promote tourism in the city, which includes the development and maintenance of the "My Helsinki" application. Cost per capita **€13.45** (USD$16.45).

SMART CITY TOURIST INFORMATION SERVICES (SCTIS): In the city of **Barcelona, Spain.** The Barcelona Wi-Fi program provides free Wi-Fi access in various public spaces throughout the city, including parks, plazas, and beaches. In addition to providing internet access to residents and visitors, the program also includes a mobile application called "Barcelona iPlay" that provides information about events, attractions, and public transportation. In 2017, the Barcelona City Council announced that it had allocated **€20million** (approximately USD$23.8 million) to expand the Wi-Fi network throughout the city, including in tourist areas. Cost per capita **€3.51** (USD$4.18).

SMART CITY TOURIST LANGUAGE ASSISTANCE (SCTLA): In the city of **Tokyo, Japan.** The Tokyo Handy Guide program provides multilingual support to tourists through a mobile application that includes information about tourist attractions, events, and public transportation. The application also provides a chat function that allows tourists to ask questions in their native language and receive assistance from a multilingual support staff. In 2019, the Tokyo Metropolitan Government announced that it had allocated JPY 15 billion (approximately USD 137 million) to support tourism in the city, which includes the development and maintenance of the Tokyo Handy Guide program. Cost per capita **JPY401** (USD$3.66).

SMART CITY TOURIST NAVIGATION (SCTN): In the city of **Shanghai, China.** The Smart Shanghai program includes a mobile application called "Explore Shanghai" that provides tourists with information about tourist attractions, events, and public transportation, as well as a navigation function that helps tourists find their way around the city. The navigation function uses augmented reality technology to provide a more immersive and interactive experience. In 2016, the Shanghai Municipal Government announced that it had allocated **CNY12.6 billion** (approximately USD$1.8 billion) to develop and implement smart city technologies and services, including the Smart Shanghai program. Cost per capita **CNY478.78** (USD$68.40).

SMART CITY TOURIST PARKING MANAGEMENT (SCTPM): In the city of **San Francisco, California, USA.** The SFpark program is a demand-responsive pricing program that adjusts parking rates based on the demand for parking spaces in a given area. The program uses sensor technology to detect parking space occupancy, and the data is used to adjust parking rates in real-time. The program also includes a mobile application that provides real-time information about parking availability and rates. The cost per capita of the SFpark program is approximately **USD$10.82,** based on the total program cost of **USD$41.7 million** and the population of San Francisco of approximately 3.87 million. The program was funded through a combination of federal grants, state grants, and local funding.

SMART CITY TOURIST PERSONALIZATION (SCTP): In the city of **Dubai, United Arab Emirates.** The Dubai Pass is a mobile application that provides personalized recommendations for tourists based on their interests, as well as access to discounts and promotions for tourist attractions, events, and activities. The Dubai Pass program uses data analytics and machine learning algorithms to analyze tourists' preferences and behaviours, and provide personalized recommendations based on their interests. The program also includes a mobile wallet feature that allows tourists to purchase attraction tickets, make payments, and redeem discounts and promotions. In 2017, the Dubai government announced that it had allocated **AED4.5 billion** (approximately USD$1.2 billion) over the next five years to develop and implement smart city technologies and services, including the Dubai Pass program. Cost per capita **AED1,697** (USD$452.66).

SMART CITY TOURIST SAFETY AND SECURITY (SCTSS): In the city of **New Delhi, India.** The "Safe City" program is a collaboration between the Delhi Police and the Municipal Corporation of Delhi, and is aimed at improving the safety and security of tourists and residents in the city. The Safe City program uses a network of CCTV cameras, facial recognition technology, and license plate recognition technology to monitor the city's streets and public spaces in real-time. The program also includes a mobile application that allows tourists and residents to report incidents and request emergency assistance. The cost per capita of the Safe City program is approximately **INR100** (approximately USD 1.35), based on the total program cost of **INR5.5 billion** (approximately USD$74.7 million) and the population of New Delhi of approximately 55 million. The program was funded through a combination of federal and state government grants, as well as private sector funding.

SMART CITY TOURIST SOCIAL MEDIA INTEGRATION (SCTSM): In the city of **Sydney, Australia.** The program is a partnership between the City of Sydney and Tencent, the parent company of the popular Chinese social media platform WeChat, and is aimed at providing Chinese tourists with personalized recommendations and information through the WeChat platform. The WeChat City Experience program includes a WeChat Mini Program that provides real-time recommendations for tourist destinations, activities, and services in the city, as well as a WeChat Official Account that provides information and support for Chinese tourists. The program also includes a data analytics platform that allows the city to track and analyze tourist behaviour and preferences. In 2019, the City of Sydney announced that it had allocated **AUD$2 million** (approximately USD$1.5 million) to the program over a period of three years. Cost per capita **AUD$0.41** (USD$0.31).

SMART CITY TOURIST SUSTAINABLE TOURISM (SCTST): In the city of **Ljubljana, Slovenia.** The program is aimed at promoting sustainable tourism and encouraging visitors to minimize their impact on the environment. The Green Destinations program includes various initiatives, such as promoting eco-friendly transport options, encouraging responsible waste management, and supporting sustainable tourism businesses. The program also provides visitors with information and resources on sustainable tourism practices, and offers guided tours and activities that highlight the city's sustainable initiatives. In 2016, the city government announced that it had allocated **€1.5 million** (approximately USD$1.8 million) over a period of five years to support sustainable tourism initiatives in the city. Cost per capita **€5.36** (USD$6.43).

SMART CITY TRANSPORT MANAGEMENT (SCTM): In the city of **Jakarta, Indonesia.** The Bus Priority System aims to improve public transportation by prioritizing bus lanes and providing real-time information for commuters. The program is estimated to cost USD$52 million and serves a population of over 10 million, resulting in a per capita cost of around **USD$5.20**.

SMART CITY URBAN AGRICULTURE (SCUA): According to a report by the Institute for Sustainable Futures at the University of Technology Sydney, the cost of implementing an urban agriculture program in **Sydney, Australia** would be around **AUD$10** per capita per year. This estimate is based on a program that would provide access to community gardens, rooftop gardens, and vertical farms in the city, and would include costs for equipment, infrastructure, and staff.

SMART CITY URBAN MOBILITY (SCUM): The city of **Barcelona, Spain,** has invested **€10** per capita per year on average in their smart mobility initiatives as part of their broader smart city program. Barcelona's smart mobility program includes the development of a comprehensive network of bike lanes, the implementation of a bike-sharing system, the introduction of electric buses, and the deployment of a smart parking system, among other initiatives. The program is designed to improve mobility and reduce congestion in the city, while also promoting sustainable transportation options.

SMART CITY URBAN PLANNING AND DESIGN (SCUPD): The city of **Barcelona, Spain,** developed a comprehensive urban planning and design program called the "Barcelona Smart City" project. The program aimed to transform Barcelona into a more livable and sustainable city by using technology and data to improve urban planning and design. The total cost of the project was estimated to be around €300 million (approximately USD$362 million), with funding coming from a variety of sources, including the European Union, private companies, and the city government. Barcelona has a population of around 1.6 million people, which would translate to a per capita cost of approximately **€187.50** (approximately USD$226.25) for the Barcelona Smart City project. However, it's worth noting that this per capita cost is based on the total cost of the project, and not just the urban planning and design component.

SMART CITY WASTE MANAGEMENT (SCWM): The city of **Pune in India** is implementing a SCWM program, with a total project cost of around USD$130 million. The program includes the development of a waste-to-energy plant, a material recovery facility, a decentralized waste processing facility, and the deployment of an integrated waste management system. The population of Pune is around 3.5 million people, so the per capita cost for this project would be approximately **USD$37** per capita.

SMART CITY WATER QUALITY MONITORING (SCWQM): In the city of **New York, USA,** they have implemented a SCWQM program by deploying water sensors. According to a report by the NYC Department of Environmental Protection, the cost of deploying sensors at 150 locations was USD$106 million. With a population of approximately 8.3 million people in 2019, this translates to a per capita cost of about **USD$12.77**.

SMART CITY WORKFORCE DEVELOPMENT (SCWD): The city of **Philadelphia in the United States** has allocated USD$10 million for a workforce development program as part of its SmartCityPHL initiative, which aims to create a more equitable and connected city through technology. With a population of around 1.5 million people, this would come out to a per capita cost of approximately **USD$6.67**.

Summary: Smart City Programs (SCP) are designed to enhance the quality of life of citizens by implementing technology-driven solutions. While these programs offer numerous benefits, including increased efficiency, sustainability, and safety, they can be costly. The costs associated with SCP include the installation and maintenance of technology infrastructure, data collection and analysis, and training of personnel. The funding for these programs is often obtained from both public and private sources. The benefits of SCP can outweigh the costs if implemented properly, and it is essential to weigh the costs and benefits before initiating any such programs.

Conclusion: The costs associated with SCP are significant, but the potential benefits can be enormous. SCP can help cities become more efficient, sustainable, and safe, leading to improved quality of life for citizens. However, it is crucial to recognize that implementing these programs requires significant investment in technology infrastructure, data collection and analysis, and personnel training. Additionally, it is important to ensure that the benefits of these programs outweigh the costs, and that the funding for them is obtained through a mix of public and private sources. By considering the costs and benefits of SCP, cities can make informed decisions about which solutions are most appropriate for their specific needs and resources.

Quote: "Smart City Programs offer a world of possibilities, but their success depends on the careful consideration of costs and benefits. Only through balancing these factors can we create truly intelligent and sustainable cities." Emin Hasic

ONE SMART WORLD
ARE YOU READY FOR IT?

| xii |
KEY DEPARTMENTS

As more and more cities around the world embrace the concept of Smart Cities, the role of local government in leading the charge has become increasingly important. To effectively manage the complex and interconnected systems that make up a Smart City, cities must establish specific departments or teams dedicated to overseeing key areas. These departments work collaboratively to ensure the effective deployment of technology and data to improve citizen services, enhance sustainability, and drive economic growth.

SMART CITY KEY DEPARTMENTS

Setting up a smart city program is a complex and multidisciplinary process that requires a thorough analysis of the local context, needs, and resources.

The specific departments required will depend on:

a) the goals and scope of the program,
b) the size and complexity of the city,
c) the existing infrastructure, and
d) the available resources.

With that being said, here is a list of 50 departments that may be required in a smart city program, along with a brief description of their role:

1. **Chief Smart City Officer:** responsible for overseeing and coordinating the various departments and initiatives of the smart city program.
2. **Digital Transformation:** responsible for identifying and implementing digital technologies that can enhance the efficiency and effectiveness of city services and operations.
3. **Urban Planning and Development:** responsible for planning and designing the physical infrastructure and layout of the city.
4. **Transportation:** responsible for managing the city's transportation system, including public transit, roads, and bike lanes.
5. **Energy and Utilities:** responsible for managing the city's energy and utility systems, including water, electricity, and waste management.
6. **Environmental Sustainability:** responsible for promoting sustainable practices and reducing the city's carbon footprint.
7. **Public Safety:** responsible for ensuring the safety and security of the city's residents and visitors.
8. **Health and Human Services:** responsible for promoting public health and providing access to healthcare services.
9. **Economic Development:** responsible for attracting businesses and promoting economic growth in the city.
10. **Education and Workforce Development:** responsible for promoting education and training opportunities for residents and ensuring that the city has a skilled workforce.
11. **Parks and Recreation:** responsible for managing the city's parks, green spaces, and recreational facilities.
12. **Housing and Community Development:** responsible for promoting affordable housing and community development initiatives.
13. **Cultural Affairs and Tourism:** responsible for promoting the city's cultural and tourism offerings.
14. **Data Analytics and Insights:** responsible for collecting, analyzing, and disseminating data to inform decision-making and measure program effectiveness.
15. **Information Technology:** responsible for managing the city's IT infrastructure and ensuring that it is secure and reliable.
16. **Innovation and Entrepreneurship:** responsible for promoting innovation and entrepreneurship in the city.
17. **Public Engagement and Communications:** responsible for engaging with the public and communicating program updates and initiatives.

18. **Legal and Regulatory Affairs:** responsible for ensuring that the smart city program is compliant with all relevant laws and regulations.
19. **Finance and Budget:** responsible for managing the program's finances and ensuring that resources are allocated effectively.
20. **Procurement:** responsible for managing the procurement process for goods and services related to the smart city program.
21. **Project Management:** responsible for overseeing and managing the various projects and initiatives of the smart city program.
22. **Partnerships and Collaborations:** responsible for building and maintaining partnerships and collaborations with other cities, organizations, and stakeholders.
23. **Emergency Management:** responsible for planning and responding to emergency situations.
24. **Risk Management:** responsible for identifying and mitigating risks associated with the smart city program.
25. **Accessibility and Inclusion:** responsible for ensuring that the smart city program is accessible and inclusive for all residents and visitors.
26. **Civic Innovation:** responsible for promoting and supporting civic innovation and citizen engagement.
27. **Digital Inclusion:** responsible for ensuring that all residents have access to digital technologies and skills.
28. **Equity and Social Justice:** responsible for promoting equity and social justice in the city.
29. **Public Works:** responsible for maintaining and improving the city's public infrastructure.
30. **Community Policing:** responsible for building trust and collaboration between the police department and the community.
31. **Code Enforcement:** responsible for enforcing city codes and regulations.
32. **Community Development Block Grants:** responsible for managing and distributing Community Development Block Grants (CDBG).
33. **Housing Assistance:** responsible for providing housing assistance to low-income residents and those in need of emergency housing.
34. **Neighbourhood Services:** responsible for engaging with neighbourhood groups and organizations to address community issues.
35. **Public Transportation Planning:** responsible for planning and developing the city's public transportation system.
36. **Public Health Preparedness:** responsible for preparing for and responding to public health emergencies.

37. **Social Services:** responsible for providing social services to residents in need, such as food assistance and healthcare.
38. **Traffic Engineering:** responsible for managing traffic flow and implementing traffic safety measures.
39. **Water and Sewer System Management:** responsible for managing the city's water and sewer systems.
40. **Zoning and Land Use:** responsible for managing land use and zoning regulations.
41. **Business Development and Retention:** responsible for promoting business development and retention in the city.
42. **Community Outreach:** responsible for reaching out to the community to gather feedback and input on smart city initiatives.
43. **Emergency Medical Services:** responsible for providing emergency medical services to residents.
44. **Fleet Management:** responsible for managing the city's fleet of vehicles and equipment.
45. **GIS and Mapping:** responsible for developing and maintaining the city's GIS and mapping systems.
46. **Housing Rehabilitation:** responsible for rehabilitating and repairing existing housing stock.
47. **Public Space Management:** responsible for managing and maintaining public spaces such as parks and plazas.
48. **Street Lighting:** responsible for managing and maintaining the city's street lighting system.
49. **Sustainability and Resilience:** responsible for promoting sustainability and resilience in the city.
50. **Volunteer Services:** responsible for coordinating volunteer services and community service projects.

Summary: Smart City Programs (SCP) require the coordination of several key departments to deploy and manage complex systems, data, and technology. These departments work together to improve citizen services, enhance sustainability, and drive economic growth.

Conclusion: Establishing a SCP requires a comprehensive approach that includes a wide range of departments and initiatives. The specific departments required will depend on the needs and context of the city. However, by incorporating departments such as those listed above, a smart city program can enhance the efficiency and effectiveness of city services and operations, improve the quality of life for residents, and promote sustainable growth and development.

Quote: "Only through a multidisciplinary approach can we achieve truly innovative and effective solutions to the complex challenges facing smart cities." Emin Hasic

ONE SMART WORLD
ARE YOU READY FOR IT?

| xiii |
VENDOR LANDSCAPE

The smart city vendor landscape is highly competitive and diverse, with a wide range of companies offering solutions for various aspects of city life. The market is expected to continue its growth trajectory, with a forecasted size of over $2.3 trillion by 2025. In the coming years, the vendor landscape is expected to undergo significant changes, as new players enter the market and existing players expand their offerings.

Key players in the smart city vendor landscape include technology giants such as IBM, Cisco, and Siemens, as well as smaller, niche players that specialize in specific areas of smart city technology. The vendor landscape is also characterized by partnerships and collaborations between companies, as well as mergers and acquisitions, as players look to expand their offerings and reach.

TRENDS

In terms of trends in the vendor landscape, there is a growing focus on sustainability and efficiency, as well as the use of emerging technologies such as artificial intelligence (AI), the Internet of Things (IoT), and big data analytics. Companies are also paying increasing attention to user-centered design and data privacy and ownership, as concerns about privacy and security continue to rise.

In the future, the vendor landscape is expected to be shaped by continued growth in the market for smart city solutions, as well as advancements in technology and increased focus on sustainability and efficiency. As the market continues to evolve, companies that are able to effectively respond to changing market conditions and customer needs are likely to emerge as leaders.

In conclusion, the smart city vendor landscape is highly competitive and diverse, with a wide range of companies offering solutions for various aspects of city life. The market is expected to continue its growth trajectory, with a forecasted size of over $2.3 trillion by 2025, and the vendor landscape is expected to undergo significant changes in the coming years. Companies that are able to effectively respond to changing market conditions and customer needs are likely to emerge as leaders in the smart city vendor landscape.

1. **Interoperability:** Interoperability is becoming increasingly important in the smart city vendor landscape, as cities look for solutions that can integrate with existing systems and technologies. Companies that are able to offer solutions that are flexible and easily integrated with other systems are likely to be in high demand.
2. **Customizability:** Customizability is also becoming increasingly important, as cities have different needs and requirements. Companies that are able to offer solutions that are easily customizable to meet the specific needs of different cities are likely to be well-positioned in the market.
3. **Partnerships and Collaborations:** Partnerships and collaborations between companies are expected to play a key role in the development of smart city solutions, as different players bring their unique strengths and expertise to the table. Companies that are able to form effective partnerships and collaborations are likely to be well-positioned to succeed in the market.
4. **Focus on Regional Markets:** Companies are also expected to focus more on regional markets, as cities in different regions of the world have different needs and requirements. Companies that are able to tailor their offerings to meet the specific needs of regional markets are likely to be well-positioned for success.
5. **Investment Opportunities:** The smart city market offers significant investment opportunities for companies and investors, as the

demand for smart city solutions continues to grow. Cities around the world are investing heavily in technology and infrastructure to improve the quality of life for their citizens, create more sustainable and efficient communities, and address the challenges posed by rapid urbanization.

Summary: The smart city vendor landscape is a highly competitive and diverse market, with key players such as IBM, Cisco, and Siemens, as well as smaller, niche players that specialize in specific areas of smart city technology. The market is expected to continue its growth trajectory, with a forecasted size of over $2.3 trillion by 2025, and is characterized by a focus on sustainability and efficiency, as well as the use of emerging technologies such as AI, IoT, and big data analytics. The vendor landscape is also marked by partnerships and collaborations between companies, as well as mergers and acquisitions. Moving forward, the smart city vendor landscape is expected to be shaped by continued growth in the market, advancements in technology, and increased focus on sustainability and efficiency.

Conclusion: The art city vendor landscape is a dynamic and rapidly evolving market, with significant investment opportunities for companies and investors. Companies that are able to offer solutions that are interoperable and customizable, as well as form effective partnerships and collaborations, are likely to be well-positioned to succeed in the market. As cities around the world continue to invest in technology and infrastructure to create more sustainable and efficient communities, the smart city vendor landscape is expected to remain a key area of growth, with companies that are able to respond to changing market conditions and customer needs emerging as leaders in the market.

Quote: "As a vendor in the smart city landscape, your ability to harness creativity and innovation will be critical to your success in delivering solutions that meet the needs of cities striving for sustainability and efficiency." Emin Hasic

ONE SMART WORLD
ARE YOU READY FOR IT?

| xiv |
INVESTOR OPPORTUNITIES

The smart city market is expected to continue its growth trajectory in the coming years, driven by a combination of factors including rapid urbanization, a growing need for sustainable and efficient communities, and advances in technology. As a result, the market offers significant investment opportunities for companies and investors.

INVESTOR OPTIONS

Investors have a variety of options when it comes to investing in the smart city market, including investing in established players, niche players with specialized solutions, and start-ups and early-stage companies with innovative offerings. Companies that offer scalable, interoperable solutions that address the specific needs of cities and offer a high level of return on investment are likely to be particularly attractive to investors.

In terms of specific areas of investment opportunity, some of the most promising areas in the smart city market include:

1. **Smart City Infrastructure:** Investment in smart city infrastructure, including smart grids, smart transportation systems, and smart buildings, is expected to be a major growth driver in the market in the coming years.
2. **Internet of Things (Iot) Solutions:** IoT solutions are expected to play a key role in the development of smart cities, and investment in IoT

technologies and solutions is expected to be strong in the coming years.

3. **Data Analytics:** The growth of smart cities is driven by data, and investment in data analytics solutions that can help cities make more informed decisions is expected to be strong in the coming years.
4. **Smart Public Services:** Investment in smart public services, including smart healthcare, smart education, and smart safety and security, is expected to grow in the coming years as cities look to improve the quality of life for their citizens.
5. **Renewable Energy:** With cities around the world looking to reduce their carbon footprint and become more sustainable, investment in renewable energy solutions is expected to be strong in the coming years. This includes investments in solar, wind, and other forms of renewable energy, as well as energy storage solutions.
6. **Smart Waste Management:** As cities look to reduce waste and improve the efficiency of their waste management systems, investment in smart waste management solutions is expected to grow. This includes investments in smart waste collection, sorting, and disposal solutions.
7. **Smart Parking:** With cities facing growing traffic congestion and a need for more efficient use of limited space, investment in smart parking solutions is expected to grow. This includes investments in parking management systems, parking sensors, and other smart parking technologies.
8. **Smart Security:** As cities look to improve the safety and security of their citizens, investment in smart security solutions is expected to grow. This includes investments in smart surveillance systems, smart access control systems, and other smart security technologies.
9. **Augmented Reality (AR) And Virtual Reality (VR):** As cities look to create more immersive and interactive experiences for their citizens, investment in AR and VR technologies is expected to grow. This includes investments in AR and VR solutions for tourism, education, and other applications.
10. **Smart Mobility:** As cities look to improve the efficiency and sustainability of their transportation systems, investment in smart mobility solutions is expected to grow. This includes investments in electric vehicles, connected vehicles, and other smart mobility technologies.
11. **Smart Buildings:** As cities look to create more efficient and sustainable buildings, investment in smart building solutions is expected to grow. This includes investments in building automation

systems, energy management systems, and other smart building technologies.
12. **Smart Water Management:** As cities look to improve the efficiency and sustainability of their water systems, investment in smart water management solutions is expected to grow. This includes investments in smart water meters, leak detection systems, and other smart water technologies.
13. **Artificial Intelligence (AI):** As cities look to improve the efficiency and effectiveness of their operations, investment in AI technologies is expected to grow. This includes investments in AI-powered decision support systems, chatbots, and other AI technologies.
14. **5G Networks:** As cities look to improve the speed and reliability of their wireless networks, investment in 5G networks is expected to grow. This includes investments in 5G infrastructure and 5G-enabled devices.
15. **Smart Grid Technology:** As cities look to modernize their power grids and improve energy efficiency, investment in smart grid technology is expected to grow. This includes investments in smart meters, demand response systems, and other smart grid technologies.
16. **Internet of Things (IoT):** As cities look to improve the efficiency and performance of their systems and services, investment in IoT technologies is expected to grow. This includes investments in IoT sensors, devices, and platforms.
17. **Smart Healthcare:** As cities look to improve the quality and accessibility of healthcare services, investment in smart healthcare solutions is expected to grow. This includes investments in telemedicine, remote patient monitoring, and other smart healthcare technologies.
18. **Smart Education:** As cities look to improve the quality and accessibility of education services, investment in smart education solutions is expected to grow. This includes investments in e-learning platforms, educational apps, and other smart education technologies.
19. **Blockchain Technology:** As cities look to improve the transparency and security of their systems and services, investment in blockchain technology is expected to grow. This includes investments in blockchain-based solutions for payments, supply chain management, and other applications.
20. **Cloud Computing:** As cities look to improve the efficiency and scalability of their IT systems, investment in cloud computing solutions is expected to grow. This includes investments in cloud-based infrastructure, platforms, and applications.

Summary: The mart city market is expected to continue its growth trajectory in the coming years, driven by a combination of factors including rapid urbanization, a growing need for sustainable and efficient communities, and advances in technology. Investors have a variety of options when it comes to investing in the smart city market, including investing in established players, niche players with specialized solutions, and start-ups and early-stage companies with innovative offerings. The most promising areas for investment in the smart city market include smart city infrastructure, Internet of Things (IoT) solutions, data analytics, smart public services, renewable energy, smart waste management, smart parking, smart security, augmented reality (AR) and virtual reality (VR), smart mobility, smart buildings, smart water management, artificial intelligence (AI), 5G networks, smart grid technology, IoT, smart healthcare, smart education, blockchain technology, and cloud computing.

Conclusion: The smart city market is expected to offer significant investment opportunities in the coming years, as cities around the world continue to invest in technology and infrastructure to create more sustainable and efficient communities. Companies and investors that are able to capitalize on these trends are likely to see strong growth and returns on investment.

Quote: " Knowledge without imagination is like a bird without wings." Emin Hasic

ONE SMART WORLD
ARE YOU READY FOR IT?

| XV |
CAREER OPPORTUNITIES

As the smart city market continues to grow, so too do the career opportunities in this exciting field. With rapid urbanization and the increasing demand for sustainable and efficient communities, there is a growing need for talented professionals who can design, develop, and implement innovative solutions for smart cities. From engineers and architects to data analysts and project managers, there is a wide range of career paths available for those who are interested in building the cities of the future. Whether you are a recent graduate or an experienced professional looking for a new challenge, the smart city market offers a wealth of opportunities for those with the skills and vision to succeed.

SMART CITY CAREERS

The growth of smart cities has created a wide range of career opportunities in several sectors. Here are some of the areas where demand is expected to be high for qualified personnel:

1. **Information and Communication Technology**: This sector encompasses a range of areas, including software development, network engineering, cybersecurity, data analysis, and artificial intelligence. Professionals with expertise in these areas will be in high demand as cities strive to integrate technology into their operations.
2. **Transportation**: With the increasing popularity of ride-sharing and electric vehicles, cities are investing in smart transportation systems

to optimize traffic flow, reduce congestion, and improve air quality. Careers in this sector may include transportation planners, traffic engineers, and smart mobility specialists.
3. **Energy and Utilities:** Smart cities are focused on creating sustainable, efficient, and cost-effective energy systems. Jobs in this sector may include renewable energy engineers, energy efficiency specialists, and smart grid analysts.
4. **Real Estate and Development:** As cities grow and evolve, there is a need for professionals who can plan and develop smart, sustainable, and livable communities. This may include urban planners, architects, and real estate developers.
5. **Government and Public Administration:** The implementation of smart city initiatives requires the cooperation and coordination of multiple government agencies and departments. Careers in this sector may include city managers, policy analysts, and public administrators.
6. **Environmental Science and Engineering:** Smart cities aim to improve environmental sustainability, reduce waste, and protect natural resources. Jobs in this sector may include environmental engineers, sustainability consultants, and renewable energy specialists.
7. **Social Services:** Smart cities are focused on improving the quality of life for their residents, and this includes providing access to social services. Careers in this sector may include social workers, community outreach specialists, and health care professionals.
8. **Healthcare:** Smart cities are focused on improving access to healthcare and reducing costs. Jobs in this sector may include health information technology specialists, telemedicine providers, and health data analysts.
9. **Public Safety:** Smart cities aim to enhance public safety through the use of technology. Jobs in this sector may include emergency management specialists, crime analysts, and cybersecurity experts.
10. **Education:** Smart cities are focused on improving access to education and promoting lifelong learning. Jobs in this sector may include educational technology specialists, online learning coordinators, and e-learning designers.
11. **Smart Building and Construction:** The construction of smart buildings is a growing trend in smart cities. Jobs in this sector may include building automation specialists, smart home technicians, and sustainable building consultants.

12. **Tourism and Hospitality:** Smart cities aim to enhance the visitor experience and promote sustainable tourism. Jobs in this sector may include sustainable tourism specialists, smart city ambassadors, and tourism marketing professionals.
13. **Retail and Commerce:** The growth of smart cities has created opportunities for innovation in retail and commerce. Jobs in this sector may include e-commerce specialists, smart retail consultants, and mobile commerce experts.
14. **Finance and Economics:** Smart cities are focused on creating a more efficient and secure financial system. Jobs in this sector may include financial technology specialists, blockchain experts, and smart city economists.
15. **Creative and Cultural Industries:** Smart cities aim to support and promote local cultural industries. Jobs in this sector may include cultural event planners, smart city branding specialists, and creative industry consultants.
16. **Waste Management and Recycling:** Smart cities are focused on reducing waste and improving resource efficiency. Jobs in this sector may include waste management planners, recycling experts, and sustainable resource management specialists.
17. **Water Management:** Smart cities aim to improve access to clean water and reduce water waste. Jobs in this sector may include water resource management specialists, smart irrigation experts, and water conservation specialists.
18. **Parks and Open Spaces:** Smart cities aim to enhance the use and enjoyment of public parks and open spaces. Jobs in this sector may include park planners, smart city park designers, and sustainable open space specialists.
19. **Smart Agriculture:** Smart cities aim to improve food security and promote sustainable agriculture. Jobs in this sector may include smart agriculture specialists, urban farming experts, and sustainable food systems consultants.
20. **Emergency Services:** Smart cities aim to enhance the response time and effectiveness of emergency services. Jobs in this sector may include emergency response coordinators, disaster management specialists, and smart city emergency planners.
21. **Renewable Energy Generation and Distribution:** Smart cities are focused on creating more sustainable and efficient energy systems. Jobs in this sector may include renewable energy project managers, smart grid technicians, and energy storage specialists.

22. **Smart Lighting:** Smart cities aim to enhance public safety and reduce energy consumption through the use of smart lighting systems. Jobs in this sector may include lighting engineers, smart city lighting designers, and public lighting specialists.
23. **Citizen Engagement and Community Building:** Smart cities aim to engage citizens and build stronger communities. Jobs in this sector may include community engagement specialists, smart city ambassadors, and citizen participation experts.
24. **Software Developers:** Software development is an essential component of smart cities and a key area where qualified personnel are in high demand. Software developers play a critical role in designing, building, and maintaining the technology systems that underpin smart cities. They are responsible for creating applications and systems that improve the efficiency and sustainability of cities, enhance the quality of life for citizens, and support the growth and development of smart cities. There are many different types of software developers who work on smart city projects, including web developers, mobile developers, data scientists, and artificial intelligence specialists. These professionals are skilled in a range of programming languages, software development tools, and data management technologies, and they work closely with other professionals in a variety of fields, such as engineering, urban planning, and public policy, to bring smart city solutions to life.
25. **Data Privacy Personnel:** Data privacy is an increasingly important concern in the field of smart cities, as large amounts of personal and sensitive information are collected and stored in digital form. Qualified data privacy personnel play a crucial role in ensuring that the privacy and security of this information is protected. Jobs in this sector may include data privacy analysts, privacy officers, and security specialists. These professionals are responsible for developing and implementing data privacy policies and procedures, conducting privacy assessments and risk analyses, and ensuring that privacy laws and regulations are being followed. They may also be involved in developing and implementing security protocols and systems to protect against unauthorized access to sensitive information. With the increasing focus on privacy and security in smart cities, the demand for qualified data privacy personnel is expected to grow. These professionals play a vital role in building trust between citizens and government, and in ensuring that the benefits of smart city technologies can be realized while protecting the privacy of citizens.

26. **Cyber Security Personnel:** Cybersecurity is a crucial concern in the field of smart cities, as digital technologies play an increasingly important role in all aspects of urban life. Qualified cybersecurity personnel are in high demand to help ensure the security and stability of these systems and protect against cyber threats. Jobs in this sector may include cybersecurity analysts, security engineers, and incident response specialists. These professionals are responsible for developing and implementing security strategies and protocols, monitoring networks and systems for signs of intrusion or attack, and responding to cyber incidents when they occur. They may also be involved in performing security assessments and audits, and in training and educating users on cyber security best practices. As smart cities become more reliant on digital technologies, the demand for qualified cybersecurity personnel is expected to grow. These professionals play a critical role in ensuring the integrity and security of the systems that support smart cities, and in protecting the privacy and personal information of citizens.
27. **Research and Development Personnel:** Research and development (R&D) personnel play a crucial role in the growth and evolution of smart cities. They are responsible for exploring new technologies, identifying new opportunities, and developing innovative solutions that can help improve the efficiency and sustainability of cities. Jobs in this sector may include research scientists, technology specialists, and innovation consultants. These professionals work on a wide range of projects, from developing new urban planning algorithms to exploring the use of blockchain technology in smart cities. They collaborate with other professionals in various fields, such as engineering, computer science, and social sciences, to develop and test new ideas and approaches. The demand for qualified R&D personnel is expected to continue to grow as smart cities evolve and new challenges emerge. These professionals play a key role in driving the development of smart city technologies, and in helping cities to remain competitive and innovative in a rapidly changing global environment.

Summary: Smart cities offer a range of career opportunities in various fields such as technology, engineering, urban planning, sustainability, and data analysis. These opportunities are driven by the need to develop and maintain infrastructure that can support the increasing demand for efficient and sustainable urban living. Additionally, as smart cities continue to evolve and incorporate new technologies, there will be a growing need for skilled professionals who can design, implement, and maintain these systems.

Conclusion: Smart cities represent a significant opportunity for individuals seeking careers that combine innovation and social impact. The development of smart cities requires a diverse range of skills and expertise, and as such, offers a multitude of career opportunities. With the continued growth of urbanization and the increasing importance of sustainability, it is clear that the demand for skilled professionals in smart city development will only continue to increase in the years to come.

Quote: "Imagination is the compass that guides us towards new frontiers of career opportunities in the realm of smart cities, where innovation and technology converge to shape the future of our society." Emin Hasic

ONE SMART WORLD
ARE YOU READY FOR IT?

| xvi |
DATA PRIVACY

The concept of smart cities is built on the foundation of advanced technology, which can enable the creation of efficient and sustainable urban environments. However, this technological revolution brings with it a vast amount of data that needs to be managed, stored, and used appropriately. Data privacy is an essential aspect of this management process, especially as it pertains to personal information. The human element plays a crucial role in ensuring data privacy in smart cities. Human error and misuse can result in significant harm to individuals, and there is a need for proper education, training, and awareness to mitigate these risks. In this article, we will explore the human element in data privacy for smart cities and discuss what is required to ensure that data is used responsibly and ethically.

HUMAN ERROR AND MISUSE IN SMART CITIES

Importance of Human Element in Data Privacy: Smart cities rely on a variety of technologies and data sources, but they also require human input to manage and maintain them. However, humans are fallible and can make mistakes, which can lead to data breaches and other privacy violations. Understanding the role of the human element is critical in ensuring data privacy in smart cities.

Risks of Human Error and Misuse: The risks of human error and misuse in smart cities can range from accidental data breaches due to poor data management practices to intentional misuse of data for personal gain or malicious purposes. Examples include unauthorized access to data, failure to properly secure devices or networks, and inadequate training or awareness of data privacy policies and procedures.

Examples of Data Breaches and Misuse in Smart Cities: There have been several instances of data breaches and misuse of personal data in smart cities, such as the Cambridge Analytica scandal, where Facebook user data was harvested without consent for political profiling and manipulation.

MITIGATING RISKS OF HUMAN ERROR AND MISUSE

Education and Training for Personnel: Proper training and education for personnel involved in data collection, management, and analysis can help mitigate the risks of human error and misuse. This includes training on data privacy policies and procedures, best practices for data management and storage, and awareness of potential threats and risks.

Implementing Clear Policies and Procedures: Clear policies and procedures for data management and privacy are essential in mitigating the risks of human error and misuse. This includes defining roles and responsibilities, establishing protocols for data access and sharing, and setting up procedures for reporting and responding to incidents.

Ensuring Accountability and Oversight: Clear accountability and oversight mechanisms can help ensure that data privacy policies and procedures are followed and that incidents are addressed appropriately. This includes regular audits of data management practices, reporting requirements for incidents, and mechanisms for holding individuals and organizations accountable for privacy violations.

CONCLUSION

Recap of Key Points: The human element plays a crucial role in ensuring data privacy in smart cities, and the risks of human error and misuse must be addressed to prevent data breaches and other privacy violations.

Importance of Continued Efforts to Address Data Privacy in Smart Cities: As technology continues to evolve, the challenges of data privacy in smart cities will continue to grow, and ongoing efforts are needed to address these challenges effectively.

Call to Action for Collaboration and Improvement: Collaboration among governments, businesses, and citizens is critical in addressing data privacy in smart cities, and continued efforts to improve data management and privacy practices are essential to protect citizens' privacy rights.

Summary: Smart cities rely on advanced technologies and data sources to create efficient and sustainable urban environments. However, data privacy is a crucial aspect of this management process, especially as it pertains to personal information. The human element plays a crucial role in ensuring data privacy in smart cities. Human error and misuse can result in significant harm to individuals, and there is a need for proper education, training, and awareness to mitigate these risks. Mitigating the risks of human error and misuse requires clear policies and procedures for data management, as well as accountability and oversight mechanisms.

Conclusion: The human element is a critical factor in ensuring data privacy in smart cities. The risks of human error and misuse must be addressed to prevent data breaches and other privacy violations. Education and training for personnel, clear policies and procedures, and accountability and oversight mechanisms are all necessary components to mitigate these risks effectively. Collaboration among governments, businesses, and citizens is essential in addressing data privacy in smart cities, and continued efforts to improve data management and privacy practices are necessary to protect citizens' privacy rights. As technology continues to evolve, ongoing efforts are needed to address the challenges of data privacy in smart cities effectively.

Quote: "Any intelligent fool can make things bigger, more complex, and more violent. It takes a touch of genius, and a lot of courage to move in the opposite direction, towards simplicity, clarity, and privacy in the realm of technology." Emin Hasic

ONE SMART WORLD
ARE YOU READY FOR IT?

| xvii |
CYBER SECURITY

Smart cities are becoming increasingly popular as governments and private organizations look for ways to improve urban living. With the integration of various technologies, smart cities aim to enhance the efficiency and sustainability of urban environments. However, as these cities collect vast amounts of data from numerous sources, there is a growing concern over the security of this information. Cybersecurity threats can compromise sensitive data, putting the privacy and safety of citizens at risk. Therefore, it is essential to understand the cybersecurity risks associated with smart cities and take measures to protect the data they collect.

SMART CITIES CYBERSECURITY RISK

Smart cities are designed to leverage technology and data to improve the quality of life for their citizens by enhancing the efficiency and effectiveness of services, such as transportation, public safety, waste management, and energy management. The implementation of smart technologies, such as IoT devices, sensors, and data analytics, creates a vast network of interconnected systems that generate, store, and transmit data.

However, this reliance on technology and data also makes smart cities vulnerable to cyber threats. Malicious actors can exploit vulnerabilities in the system to gain unauthorized access to sensitive data or control of critical infrastructure systems. For example, a hacker could gain access to traffic management systems and cause accidents, disrupt water supply networks, or shut down power grids.

Data breaches are also a significant concern for smart cities, as they can result in the loss of personal information, financial data, and other sensitive data. This can have serious consequences for citizens, as their personal information could be used for identity theft, fraud, or other malicious activities.

Cyber threats can undermine citizens' trust in the government's ability to protect their personal information and provide reliable services, which could hinder the adoption of smart technologies.

It is crucial for smart cities to implement robust cybersecurity measures to protect against cyber threats. This includes securing the networks, implementing firewalls, using encryption to protect sensitive data, and conducting regular security audits to identify vulnerabilities and address them promptly.

While smart cities offer numerous benefits, they also present significant cybersecurity challenges. Cities must prioritize cybersecurity to protect sensitive data, ensure public safety, and maintain citizens' trust.

TYPES OF DATA COLLECTED IN SMART CITIES

Smart cities rely on a wide range of interconnected technologies and data sources to provide enhanced services and improve the quality of life for their citizens. These technologies include surveillance cameras, sensors, and connected devices, which gather large amounts of data on various aspects of city life.

However, the data collected by these technologies includes a significant amount of personal information, such as biometric data, location data, and behavioral data. Biometric data refers to unique physical characteristics, such as fingerprints, facial recognition, or voiceprints, which can be used to identify individuals. Location data, on the other hand, refers to the real-time or historical location of individuals, as tracked by GPS-enabled devices or sensors.

Behavioral data includes information about individuals' activities and preferences, such as their social media activity, online searches, and purchasing habits. This data can be used to develop insights into individuals' behavior and predict their future actions.

The collection of such personal data raises concerns about privacy and the potential misuse of this data by governments, corporations, or malicious actors. For example, the misuse of biometric data could result in identity theft or unauthorized access to secure areas. Location data could be used to track individuals' movements and activities, which could infringe on their privacy or be used for surveillance purposes.

The large volume of data collected by smart cities creates significant challenges for data security and management. The data must be stored securely, protected from unauthorized access, and processed in compliance with data protection regulations.

Smart cities must implement robust data protection measures, such as encryption, access controls, and regular audits to ensure the data's confidentiality, integrity, and availability. Citizens must also be informed about the collection and use of their personal data and given control over their data. This includes implementing transparency and consent mechanisms, such as data protection policies and opt-out options, to empower citizens to make informed decisions about their data.

CHALLENGES OF SECURING DATA IN SMART CITIES

Smart cities are urban areas that use advanced technologies such as the Internet of Things (IoT), sensors, and artificial intelligence (AI) to improve the quality of life for residents. However, as smart cities continue to evolve, securing the massive amount of data generated by these technologies has become a significant challenge. Here are some of the challenges associated with securing data in smart cities:

1. **Complexity of systems:** Smart cities involve a large number of interconnected systems that generate massive amounts of data. This complexity makes it difficult to secure the data since there are multiple points of entry for hackers to exploit.

2. **Volume of data:** The volume of data generated by smart city systems is enormous and continues to grow rapidly. This creates a significant challenge for data storage and management, as well as for securing the data against cyber-attacks.

3. **Diversity of data sources:** Smart cities rely on data from a variety of sources, including sensors, mobile devices, and social media. This diversity makes it challenging to ensure that the data is accurate, reliable, and secure.

4. **Data ownership and privacy:** Smart city data is often collected by third-party vendors, which can make it difficult to establish clear security protocols. Additionally, residents may have concerns about data privacy and ownership, which can make it challenging to gather and use data in a way that benefits the community while protecting individuals' privacy.

5. **Lack of standards:** There is a lack of industry-wide standards for securing data in smart cities, making it challenging to ensure that all systems and devices meet the necessary security requirements.

6. **Human error:** Even with the best security protocols in place, human error can still lead to data breaches. Smart cities involve many stakeholders, including government agencies, private companies, and residents, all of whom may have different levels of security awareness.

Securing data in smart cities requires addressing challenges related to the complexity and volume of data, diversity of data sources, data ownership and privacy, lack of standards, and human error. To overcome these challenges, it is essential to establish clear security protocols, adopt industry-wide standards, and educate stakeholders on best practices for data security.

BEST PRACTICES FOR SECURING DATA IN SMART CITIES

Smart cities rely on the collection, analysis, and sharing of vast amounts of data to improve efficiency, sustainability, and quality of life. However, the use of advanced technologies and interconnected systems also creates new cybersecurity risks that can lead to data breaches, privacy violations, and other malicious activities. Needless to say, it is critical to implement best practices for securing data in smart cities to protect critical infrastructure, sensitive information, and public trust.

Here are some of the best practices for securing data in smart cities:

1. **Establish a robust cybersecurity framework:** A cybersecurity framework provides a structured approach to identifying, assessing, and managing cybersecurity risks. It should include policies, procedures, and guidelines for data protection, access control, incident response, and other security-related activities. A robust cybersecurity framework should be designed to address the unique risks and challenges of smart cities and should be regularly updated to reflect the changing threat landscape.

2. **Conduct regular security assessments:** Security assessments help to identify vulnerabilities and weaknesses in smart city systems and applications. Regular assessments can include penetration testing, vulnerability scanning, and risk assessments. These assessments should be conducted by qualified cybersecurity professionals and should be followed up with remediation efforts.

3. **Implement encryption and authentication protocols:** Encryption and authentication are critical components of data security in smart

cities. Encryption protects data by transforming it into a code that can only be deciphered with a key or password. Authentication ensures that only authorized users can access data and systems. Smart city systems should implement strong encryption and authentication protocols to prevent unauthorized access and data breaches.

4. **Develop contingency plans for data breaches:** Despite best efforts, data breaches can still occur. Smart cities should have a well-defined and tested contingency plan in place to respond to data breaches quickly and effectively. This plan should include procedures for notification, containment, investigation, and recovery. It should also include a communication plan to notify affected parties, including the public, if necessary.

Securing data in smart cities requires a comprehensive and proactive approach that involves establishing a robust cybersecurity framework, conducting regular security assessments, implementing encryption and authentication protocols, and developing contingency plans for data breaches. By following these best practices, smart cities can minimize cybersecurity risks and ensure the safe and secure operation of critical infrastructure and services.

ROLE OF STAKEHOLDERS

Smart cities rely heavily on technology and interconnected systems, which can make them vulnerable to cyber-attacks. To ensure the cybersecurity of smart cities, collaboration between various stakeholders is essential.

Governments including City Governance play a crucial role in ensuring cybersecurity by setting policies and regulations that prioritize security. This includes establishing cybersecurity protocols and guidelines for city infrastructure, networks, and systems. Governments must also ensure that the technologies used in smart cities comply with security standards and best practices.

Private organizations, including technology companies and service providers, play a critical role in securing smart cities. They must invest in secure technology and data management systems that meet the security requirements of smart cities. This includes ensuring that their products and services are designed with security in mind and are regularly updated to address new threats. Private organizations must also work closely with city governments to share information and coordinate responses to cyber threats.

Citizens also have a role to play in ensuring the cybersecurity of smart cities. They can practice safe online behavior, such as using strong passwords and avoiding suspicious links and emails. Citizens should also report any suspicious activity they notice, such as unexplained changes in their utility bills or unusual traffic patterns.

Ensuring cybersecurity in smart cities requires collaboration and coordination between various stakeholders, including city governments, private organizations, and citizens. Governments must prioritize cybersecurity in their policies and regulations, private organizations must invest in secure technology and data management systems, and citizens must practice safe online behavior and report suspicious activity.

***Summary:** Smart cities rely heavily on interconnected systems and technologies, making them vulnerable to cyber-attacks. Ensuring cybersecurity in smart cities requires collaboration between various stakeholders, including city governments, private organizations, and citizens. City governments must prioritize cybersecurity in their policies and regulations, while private organizations must invest in secure technology and data management systems. Citizens can contribute to securing data by practicing safe online behavior and reporting suspicious activity. Through this collaborative effort, smart cities can effectively address cybersecurity threats and safeguard against potential attacks.*

Conclusion: As smart cities continue to evolve and integrate new technologies; it is crucial to prioritize cybersecurity to protect the data they collect. By understanding the risks associated with smart cities, implementing best practices for securing data, and engaging stakeholders in the process, we can create safer and more secure urban environments for everyone.

Quote: "Collaboration between various stakeholders in securing smart cities is not only crucial, it is imperative. We cannot solve cybersecurity challenges by thinking in silos, but by combining our knowledge and expertise we can create a strong defense against potential threats." Emin Hasic

ONE SMART WORLD
ARE YOU READY FOR IT?

| xviii |
THE DARK SIDE

Smart cities are often celebrated for their innovative technologies and futuristic designs that promise to make our lives more convenient, efficient, and sustainable. However, behind the glossy surface of these cities lies a darker reality - the vast amounts of data they collect about our daily lives, which can be used to monitor, control, and manipulate us in ways we may not even realize. As more and more cities around the world adopt smart technologies, it is crucial to examine the dark side of this trend and the potential risks it poses to our privacy, security, and autonomy.

SURVEILLANCE AND PRIVACY CONCERNS

One of the most significant issues with smart cities is the extent of surveillance they enable. With sensors and cameras embedded in virtually every corner of the city, governments and corporations can collect massive amounts of data about citizens' movements, behavior, and preferences. This data can then be used for a wide range of purposes, including targeted advertising, law enforcement, and social control. However, such practices raise serious concerns about privacy, as citizens may have little control over how their data is collected, stored, and used.

BIAS AND DISCRIMINATION IN DECISION-MAKING

Another dark side of smart cities is the potential for bias and discrimination in decision-making. As algorithms become increasingly sophisticated, they may be used to make decisions that affect citizens' access to services, resources, and opportunities. However, these algorithms are only as unbiased as the data they are trained on, which can reflect societal biases and perpetuate discrimination against marginalized groups. This can lead to unequal treatment and exacerbate existing social inequalities.

CYBER-SECURITY RISKS

Smart cities are also vulnerable to cybersecurity threats, as the vast amounts of data they collect are often stored in centralized databases that can be targeted by hackers. Malicious actors can exploit vulnerabilities in the system to steal or manipulate data, disrupt critical infrastructure, or even cause physical harm. Moreover, as smart cities become more interconnected, the risk of cascading failures and system-wide attacks increases, posing a significant threat to the safety and security of citizens.

LACK OF TRANSPARENCY AND ACCOUNTABILITY

As these cities rely on complex technologies and data analytics, it can be difficult for citizens to understand how decisions are made or how their data is used. Moreover, there may be little recourse for citizens if they feel their rights have been violated or if they disagree with decisions made by the city. This lack of transparency and accountability can erode trust in government and undermine democratic principles.

NEED FOR CITIZEN EMPOWERMENT AND PARTICIPATION

It is crucial to empower citizens and encourage their participation in decision-making processes. This means giving citizens greater control over their data and privacy, involving them in the design and implementation of smart city technologies, and ensuring that their voices are heard in the public discourse. Moreover, it means fostering a culture of critical thinking and active citizenship, where citizens are informed, engaged, and empowered to shape the future of their cities.

SOCIAL AND ETHICAL IMPLICATIONS OF SMART CITIES

Smart cities can have profound social and ethical implications that go beyond issues of privacy and security. For example, they may exacerbate existing inequalities by favoring certain groups over others or by neglecting the needs of marginalized communities. They may also contribute to a culture of hyper-surveillance and conformity, where citizens are constantly monitored and controlled. Moreover, the deployment of smart technologies raises questions about the role of humans in society, the limits of technology, and the nature of progress.

ENVIRONMENTAL IMPACT OF SMART CITIES

While smart cities are often marketed as eco-friendly and sustainable, the reality is more complex. The construction of smart cities requires significant amounts of energy and resources, which can contribute to greenhouse gas emissions and environmental degradation. Moreover, the reliance on technology can lead to a culture of consumption and waste, as citizens constantly upgrade their devices and discard old ones. Finally, the deployment of smart technologies may have unintended consequences for biodiversity and ecosystems, as they alter the natural environment in ways that we do not fully understand.

ECONOMIC DISRUPTION AND JOB LOSSES

Smart cities may also lead to significant economic disruption and job losses, particularly in industries that rely on manual labor or face-to-face interaction. As automation and artificial intelligence become more prevalent, many jobs may become obsolete, leading to unemployment and economic inequality. Moreover, the rise of smart cities may favor large corporations over small businesses, as only those with the resources to invest in smart technologies can compete in the market. This can lead to further concentration of wealth and power in the hands of a few.

INSIDER THREATS AND EMPLOYEE SABOTAGE

While much of the discussion around smart cities has focused on external threats, such as cyber-attacks and data breaches, there is also a risk of insider threats and employee sabotage. As smart city technologies become more complex and interconnected, they require a large workforce to design, maintain, and operate them. However, these employees may have access to sensitive information and critical systems, making them potential targets for malicious actors. Moreover, they may also have their own grievances or agendas, leading them to sabotage the system intentionally. This can cause significant disruptions to the city's operations and compromise the safety and security of its citizens. To mitigate this risk, smart city operators must implement rigorous security protocols and screening processes, as well as foster a culture of trust and accountability among their employees.

DARK WEB AND ILLICIT ACTIVITIES IN SMART CITIES

The rise of smart cities has also created new opportunities for illicit activities on the dark web, a hidden part of the internet that is accessible only through specialized software. Criminals can use the dark web to buy and sell stolen personal data, such as credit card numbers and login credentials, that have been harvested from smart city systems. They can also use it to conduct illegal activities, such as drug trafficking, money laundering, and cybercrime, that exploit the vulnerabilities of smart city technologies. Moreover, the anonymity of the dark web can make it difficult for law enforcement to identify and apprehend the perpetrators of these crimes. To address this issue, smart city operators must implement strong security measures, such as encryption and access controls, to prevent unauthorized access to their systems. They must also work closely with law enforcement agencies to monitor and detect illicit activities on the dark web, and collaborate with other stakeholders to develop policies and regulations that deter such activities in the first place.

THE RISK OF AI DOMINANCE IN SMART CITIES

As smart cities become more sophisticated and rely increasingly on artificial intelligence (AI), there is a growing concern that AI systems may become dominant and even take over decision-making processes. This could have serious consequences for the safety and well-being of citizens, as AI systems may prioritize efficiency and optimization over human values and ethics. Moreover, the opacity and complexity of AI algorithms may make it difficult for humans to understand how these systems are making decisions, raising questions about accountability and responsibility. To address this risk, smart city operators must ensure that AI systems are designed and implemented in a way that is transparent, explainable, and aligned with human values. They must also establish clear lines of accountability and oversight for AI systems, and ensure that humans retain ultimate control over decision-making processes. Finally, they must foster a culture of responsible AI development and deployment, where ethical considerations are given the same weight as technical considerations.

IDENTITY LOSS AND TAKEOVER IN SMART CITIES

As smart city technologies become more pervasive, there is a risk that citizens may lose control over their identities and personal data, leading to identity theft and other forms of fraud. For example, criminals may use stolen personal data to create fake identities or to gain unauthorized access to sensitive systems and information. Moreover, as smart city technologies become more interconnected, there is a risk of identity takeover, where criminals can use a stolen identity to gain access to multiple systems and services across the city. This can have serious consequences for the affected individuals, as well as for the overall security and stability of the city. To address this risk, smart city operators must implement strong identity and access management systems, such as two-factor authentication and biometric verification. They must also educate citizens on the importance of safeguarding their personal data and monitoring their online activities for signs of identity theft. Finally, they must work with law enforcement agencies to investigate and prosecute identity theft and other forms of cybercrime, and collaborate with other stakeholders to develop policies and regulations that deter such activities in the first place.

Summary: Smart cities are rapidly growing, using technology and data to optimize and streamline urban systems. However, the increasing collection and analysis of data poses several risks and challenges, including concerns around privacy, security, and bias. The dark side of smart cities refers to these potential negative impacts on citizens and society as a whole. Smart city data can be used for surveillance, tracking, and profiling, raising concerns about personal freedoms and civil liberties. Additionally, the data can be vulnerable to hacking and other cyber threats, potentially exposing sensitive information to bad actors. The biases present in the data and algorithms used in smart cities can lead to discriminatory practices that reinforce existing inequalities.

Conclusion: The development of smart cities presents both opportunities and challenges. While the collection and analysis of data can lead to improved efficiencies and services, it is crucial to address the risks associated with the dark side of smart cities. This includes implementing strong privacy and security measures, ensuring transparency and accountability, and addressing biases in the data and algorithms used. It is important to prioritize the protection of citizens' rights and freedoms while still promoting innovation and progress in urban development. By balancing these competing interests, smart cities can be designed and implemented in a responsible and ethical manner, benefitting citizens and society as a whole.

Quote: " Technology is a beautiful servant but a dangerous master. We need to determine the difference between beauty & danger " Emin Hasic

| xix |
SMART CITIES 2030

Smart cities refer to the integration of technology and digital infrastructure into the functioning of cities, with the goal of improving the quality of life for citizens, making the city more efficient, and reducing its environmental impact. This concept encompasses various aspects of city life, including transportation, energy, buildings, governance, and healthcare.

SMART CITY GROWTH

The market for smart cities solutions has grown rapidly in recent years and is expected to continue its growth trajectory in the coming years. The market is estimated to reach a size of over $2.3 trillion by 2025. This growth is driven by several factors, including the increasing trend of urbanization, the need for more sustainable and efficient cities, and advancements in technology.

Key buying criteria for cities and organizations looking to implement smart city solutions include:

1. **Scalability:** The solution must be able to grow and evolve as the city's needs change.
2. **Interoperability:** The solution must be able to integrate with existing systems and technologies to ensure seamless operation.

3. **Sustainability:** The solution must be environmentally friendly and support the city's efforts to reduce its carbon footprint.
4. **Cost-effectiveness:** The solution must offer cost savings over the long term while providing the necessary functionality.
5. **Security:** The solution must be secure and protect sensitive data and critical infrastructure.
6. **User-centered design:** The solution must be designed with the end-user in mind and be intuitive and easy to use.
7. **Data privacy and ownership:** The solution must ensure that data collected from citizens is used appropriately and in accordance with privacy regulations.
8. **Innovation and advancement:** The solution must be up-to-date with the latest advancements in technology and be able to keep pace with changing demands.

When it comes to buying smart city solutions, cities and organizations have several key criteria that they consider. These include scalability, which refers to the ability of the solution to grow and evolve as the city's needs change; interoperability, which refers to the ability of the solution to integrate with existing systems and technologies; sustainability, which refers to the environmental impact of the solution; cost-effectiveness, which refers to the long-term cost savings the solution offers; security, which refers to the protection of sensitive data and critical infrastructure; user-centered design, which refers to the design of the solution with the end-user in mind; data privacy and ownership, which refers to the appropriate use of data collected from citizens; and innovation and advancement, which refers to the solution being up-to-date with the latest technology advancements.

Smart city solutions use various technologies, such as the Internet of Things (IoT), artificial intelligence (AI), big data analytics, and cloud computing, among others. These solutions aim to address various aspects of city life, such as smart transportation, smart energy, smart buildings, smart governance, and smart healthcare, among others.

Summary: As cities and organizations consider investing in smart city solutions, there are several criteria they must evaluate. These include scalability, interoperability, sustainability, cost-effectiveness, security, user-centered design, data privacy and ownership, and innovation and advancement. Smart city solutions incorporate various technologies like the Internet of Things, artificial intelligence, big data analytics, and cloud computing to improve different aspects of urban life, such as transportation, energy, buildings, governance, and healthcare.

Conclusion: The smart cities market is expected to continue its growth trajectory and reach a size of over $2.3 trillion by 2025. Cities and organizations considering buying smart city solutions should consider factors such as scalability, interoperability, sustainability, cost-effectiveness, security, user-centered design, data privacy and ownership, and innovation and advancement. The solutions offered in the market use a variety of technologies and aim to improve various aspects of city life.

Quote: "As we envision the cities of the future, let us not be bound by the limitations of the past, but instead, let us boldly imagine and create the cities of tomorrow that will shape the world for generations to come." Emin Hasic

ONE SMART WORLD
ARE YOU READY FOR IT?

| xx |
GUIDE IN SETTING UP A SMART CITY

Setting up a smart city involves leveraging technology and data to create a more efficient and sustainable urban environment that benefits both residents and businesses. The transformation to a smart city requires a comprehensive plan that involves the integration of various systems, including transportation, energy, and communication. We will use **IGOUMENITSA**, a small city located in **northwestern Greece**, as an excellent hypothetical example of how a city can leverage technology to become a smart city. Through various initiatives, the city can enhance the quality of life for its residents, improve its infrastructure, and attract new investments. In this chapter, we will explore the steps required to be taken by Igoumenitsa to become a smart city and the benefits it can achieve from this transformation.

Igoumenitsa smart city strategy will focus on utilizing technology and data to improve its transportation infrastructure, which is crucial for its diverse economy. The city's commercial and tourism sea ports, marinas, and holiday accommodations attract tourists from all over the world, while its agriculture, retail, and restaurants cater to the needs of both residents and visitors. Government services such as police and fire departments ensure the safety and well-being of the community, while schools and doctors provide essential services for education and healthcare. Through a smart traffic management system, the city will be able to optimize traffic signals and route planning, reducing congestion and improving safety on the roads. Additionally, the city will introduce sustainable transportation options such as electric buses and vehicle charging stations, contributing to a more eco-friendly and livable city.

BRANDING & MARKETING

The **importance of branding the new Smart City of Igoumenitsa** lies in creating a unique identity that reflects the city's strengths, culture, and vision. A strong brand identity can help the city attract tourists, businesses, and investors, while also creating a sense of pride and belonging among residents. It can also help the city differentiate itself from other cities and stand out as a destination of choice. Effective branding and marketing strategies can help promote the city's key industries and highlight its strengths and competitive advantages, while also showcasing its commitment to sustainability, innovation, and technology. Branding example:

THE BUSINESS PLAN

A business plan is an essential tool for developing, implementing, and managing a Smart City project. It provides a clear roadmap, helps to secure funding, identifies opportunities and challenges, facilitates communication and collaboration, and ensures sustainability and scalability. Following is a **SMART CITY BUSINESS PLAN STRUCTURE:**

1. **EXECUTIVE SUMMARY**

- **Overview of the Smart City Project:** Introduce the project and provide a brief summary of its objectives and goals.
- **Key Objectives and Goals:** Outline the primary objectives and goals of the Smart City project, including the benefits it aims to provide to residents, businesses, and the environment.
- **Summary of Financial Projections:** Provide a high-level overview of the financial projections, including the estimated costs of the project, revenue streams, and potential return on investment.

2. **BUSINESS OVERVIEW**

- **Description of the Smart City Project:** Provide a more detailed description of the Smart City project, including its purpose, scope, and target audience.
- **Market Analysis and Target Audience:** Conduct a thorough market analysis to understand the demand for Smart City solutions and identify the target audience for the project.
- **Competitive Landscape Analysis:** Analyze the competition and assess the strengths, weaknesses, opportunities, and threats (SWOT) of the Smart City project.

3. **MANAGEMENT TEAM AND ORGANIZATIONAL STRUCTURE**

- **Profiles of Key Team Members and their Roles:** Provide a detailed description of the management team, including their skills, experience, and qualifications, and explain how their roles align with the Smart City project.

- **Organizational Chart:** Create an organizational chart that shows the structure of the team and the reporting lines.
- **Overview of Partnerships and Alliances:** Identify potential partners and alliances that could benefit the Smart City project, such as technology providers, investors, and local governments.

4. **SMART CITY PROJECT DESCRIPTION**

- **Smart City Concept and Vision:** Describe the vision for the Smart City project and the key concepts that will be used to create it, such as the Internet of Things (IoT), artificial intelligence (AI), and blockchain.
- **Smart City Technologies and Systems:** Provide a detailed overview of the technologies and systems that will be used to create the Smart City, including sensors, networks, platforms, and applications.
- **Key Features and Benefits of The Smart City:** Highlight the key features and benefits of the Smart City project, such as improved sustainability, energy efficiency, mobility, safety, and quality of life.

5. **MARKET ANALYSIS AND STRATEGY**

- **Market Research and Analysis:** Conduct a more in-depth market research to identify the current and future demand for Smart City solutions and assess the market trends and dynamics.
- **Target Market Identification:** Refine the target audience and segment it based on demographics, psychographics, and behaviour.
- **Marketing and Sales Strategy:** Develop a marketing and sales strategy that aligns with the target audience and market analysis, and identify the tactics, channels, and metrics that will be used to reach and engage potential customers.

6. **FINANCIAL PLAN**

- **Revenue Streams:** Identify the sources of revenue for the Smart City project, such as user fees, advertising, sponsorships, or government grants. Estimate the potential revenue from each source, based on market research, customer feedback, and industry benchmarks.

- **Operating Expenses:** Outline the costs associated with operating the Smart City project, such as salaries, rent, utilities, maintenance, and marketing. Estimate the amount of each expense, based on market research, competitive analysis, and industry standards.
- **Funding Sources:** Identify the sources of funding for the Smart City project, such as venture capital, angel investors, crowdfunding, or public-private partnerships. Determine the amount of funding needed to launch and scale the project, and the terms and conditions of each funding source.
- **Financial Projections**: Develop financial projections that forecast the revenue, expenses, and funding of the Smart City project over the next 3-5 years. Use a combination of historical data, market research, and assumptions to create realistic and achievable projections.
- **Sensitivity Analysis:** Perform sensitivity analysis to test the financial viability of the Smart City project under different scenarios, such as changes in market conditions, customer demand, or operating costs. Assess the risks and opportunities of each scenario, and adjust the strategies accordingly.

7. **IMPLEMENTATION PLAN**

- **Project Timeline:** Develop a detailed timeline that outlines the milestones, deliverables, and deadlines of the Smart City project. Use a Gantt chart or other project management tool to visualize the timeline and track the progress of each task.
- **Resource Allocation:** Determine the resources needed to implement the Smart City project, such as human capital, physical assets, and technology infrastructure. Allocate the resources based on the project timeline and budget, and identify any constraints or risks that could affect the implementation.
- **Task Assignment:** Assign specific tasks and responsibilities to the project team members, partners, and contractors, based on their skills, expertise, and availability. Clarify the expectations, goals, and standards for each task, and establish clear communication channels and feedback mechanisms.

- **Risk Management:** Identify the potential risks and challenges of the Smart City project, such as regulatory compliance, data privacy, cybersecurity, or stakeholder resistance. Develop risk mitigation strategies that address each risk, and establish contingency plans in case of unforeseen events or emergencies.
- **Monitoring and Evaluation:** Establish a monitoring and evaluation framework that tracks the progress and outcomes of the Smart City project, and assesses its impact on the stakeholders and the environment. Use metrics, indicators, and benchmarks to measure the performance and effectiveness of the project, and use the results to inform the decision-making and improvement process.

8. **CONCLUSION**

- **Recap of Key Points:** Summarize the key points covered in the business plan, highlighting the unique value proposition of the Smart City project and its potential impact on the market and society.
- **Final Thoughts and Recommendations:** Provide some final thoughts and recommendations based on the findings and analysis, and identify the risks and challenges that could affect the success of the Smart City project.
- **Call to Action:** Encourage readers to take action and support the Smart City project, whether by investing, partnering, or spreading the word, and provide contact information and next steps for those who are interested.

THE THINK TANK

Creating a think tank to develop a Smart City requires a multidisciplinary team with diverse skills, expertise, and perspectives. Here are some of the key roles and functions that you may need to consider when assembling a Smart City think tank:

Urban Planners: Urban planners can help to design and optimize the physical infrastructure of the Smart City, such as transportation systems, housing, public spaces, and utilities. They can also help to integrate the Smart City technologies into the urban fabric and ensure that they meet the needs and preferences of the citizens.

Engineers: Engineers can help to develop and implement the Smart City technologies, such as sensors, data analytics, artificial intelligence, and Internet of Things (IoT) devices. They can also help to ensure the reliability, scalability, and interoperability of the technologies, and mitigate the risks of cyber threats, data breaches, or system failures.

Data Analysts: Data analysts can help to collect, analyze, and interpret the data generated by the Smart City technologies, such as traffic patterns, energy consumption, air quality, or citizen behavior. They can also help to identify the trends, patterns, and insights that can inform the decision-making and improve the performance of the Smart City.

Policy Makers: Policy makers can help to develop the regulatory frameworks, standards, and guidelines that govern the Smart City technologies and ensure their compliance with the legal, ethical, and social norms. They can also help to align the interests and expectations of the stakeholders, such as citizens, businesses, and government agencies, and ensure that the Smart City project meets the public interest.

Business Leaders: Business leaders can help to identify the market opportunities, customer needs, and revenue streams that can sustain and scale the Smart City project. They can also help to foster the innovation, collaboration, and entrepreneurship that drive the Smart City ecosystem, and attract the investment and talent that support the growth and development of the Smart City.

Social Scientists: Social scientists can help to understand the social and cultural dynamics of the Smart City, such as the impact of the technologies on the citizen behavior, identity, and well-being. They can also help to address the social inequalities, ethical dilemmas, and human rights issues that may arise from the Smart City project, and ensure that the Smart City benefits all citizens.

Community Organizers: Community organizers can help to engage and mobilize the citizens, businesses, and other stakeholders to participate in the Smart City project, and provide their feedback, ideas, and perspectives. They can also help to foster the sense of community, identity, and belonging that is essential for the success and sustainability of the Smart City.

Environmental Scientists: Environmental scientists can help to assess and mitigate the environmental impact of the Smart City project, such as the energy consumption, carbon footprint, water management, and waste reduction. They can also help to promote the sustainable development, biodiversity, and resilience of the Smart City, and ensure that the Smart City is aligned with the global climate and sustainability goals.

Architects: Architects can help to design and implement the Smart City infrastructure, such as buildings, parks, public spaces, and landmarks. They can also help to integrate the Smart City technologies into the architectural design and aesthetics of the urban landscape, and ensure that the Smart City project reflects the local culture, heritage, and identity.

Educators: Educators can help to raise awareness and literacy about the Smart City technologies and their impact on the society and the economy. They can also help to train and educate the citizens, businesses, and government agencies about the skills, knowledge, and competencies needed to participate in and benefit from the Smart City project, and ensure that the Smart City promotes lifelong learning and innovation.

Philanthropists: Philanthropists can provide the financial and social capital that supports the Smart City project, such as seed funding, grants, sponsorships, or impact investing. They can also help to catalyze the public-private partnerships, cross-sector collaborations, and social innovations that enable the Smart City to address the complex and systemic challenges of the urbanization.

Technologists: Technologists can help to develop and deploy the cutting-edge technologies that enable the Smart City project, such as blockchain, quantum computing, augmented reality, or autonomous systems. They can also help to push the boundaries of the Smart City innovation and experimentation, and ensure that the Smart City project reflects the latest trends and breakthroughs in the digital and technological landscape.

Communication Specialists: Communication specialists can help to develop and implement the communication strategy of the Smart City project, such as the branding, messaging, and public relations. They can also help to engage and inform the public about the Smart City project, and ensure that the Smart City project is transparent, accessible, and accountable to the citizens.

Legal Experts: Legal experts can help to navigate the complex and evolving legal landscape of the Smart City project, such as the intellectual property, liability, and privacy issues. They can also help to ensure that the Smart City project complies with the relevant laws and regulations, and mitigate the legal risks and disputes that may arise from the Smart City project.

Health Professionals: Health professionals can help to promote and ensure the health and well-being of the citizens in the Smart City project, such as the access to healthcare, healthy food, and physical activity. They can also help to monitor and respond to the health-related issues and emergencies that may arise from the Smart City project, and ensure that the Smart City project prioritizes the health and safety of the citizens.

Port and Marina Experts: Port and Marina experts can help to design and manage the Smart City's port and marina infrastructure, including the logistics, transportation, and security systems. They can also help to promote and optimize the port and marina services for the benefit of the Smart City's businesses, residents, and visitors, and ensure that the Smart City's port and marina activities are aligned with the global maritime and trade standards and regulations.

Tourism Specialists: Tourism specialists can help to promote and develop the Smart City's tourism industry, including the hotels, restaurants, attractions, and events. They can also help to attract and retain the Smart City's tourists and visitors by enhancing the Smart City's cultural, environmental, and experiential values, and ensure that the Smart City's tourism activities are sustainable, responsible, and equitable.

Agricultural Experts: Agricultural experts can help to integrate and optimize the Smart City's agriculture and food production, including the urban farming, hydroponics, and agroforestry. They can also help to ensure the Smart City's food security, quality, and diversity, and promote the Smart City's healthy and sustainable food culture and industry.

Police and Security Experts: Police and Security experts can help to design and implement the Smart City's security and law enforcement systems, including the surveillance, crime prevention, and emergency response services. They can also help to collaborate with the Smart City's communities and stakeholders to ensure that the Smart City's security policies and practices are effective, transparent, and respectful of the citizens' rights and freedoms.

Emergency Management Professionals: Emergency Management professionals can help to prepare and respond to the Smart City's emergency situations, such as natural disasters, cyber-attacks, and pandemics. They can also help to coordinate and communicate with the Smart City's first responders, public agencies, and communities to ensure that the Smart City's emergency plans and actions are timely, efficient, and effective.

Fire and Rescue Experts: Fire and Rescue experts can help to design and implement the Smart City's fire and rescue systems, including the firefighting, hazardous materials, and technical rescue services. They can also help to collaborate with the Smart City's communities and stakeholders to ensure that the Smart City's fire and rescue policies and practices are effective, responsive, and respectful of the citizens' safety and well-being.

Policy, Cyber-Security and Data-Privacy/Security Analysts (PCDS): To analyze and evaluate the Smart City's security and public policy issues, such as the privacy, data protection, and civil liberties. They can also help to develop and implement the Smart City's policies and regulations that balance the citizens' rights and the Smart City's security and innovation objectives, and ensure that the Smart City's policies and regulations are evidence-based, participatory, and adaptive.

BUILDING A SMART CITY

Building a Smart City is an undertaking that requires careful planning, collaboration, and investment. Following is a detailed plan for building a smart city in Igoumenitsa, Greece, incorporating surrounding municipal towns based on the following parameters:

- **City/Municipality/Town:** Igoumenitsa
- **Population City:** 17000
- **Population Regional:** 8000
- **Key Sectors:** Tourism, Commercial Sea Port, Tourism Sea Ports, Marinas, Agriculture, Retail, Government Services, Police and Emergency Services.

Smart Infrastructure: The first step in building a smart city is to establish the necessary infrastructure. This includes high-speed broadband connectivity, smart grids, and intelligent transportation systems. In Igoumenitsa, we would invest in developing a fiber-optic network that provides high-speed internet access to all citizens and businesses. We would also install smart traffic management systems that use real-time data to optimize traffic flow and reduce congestion. Additionally, we would develop a smart grid that integrates renewable energy sources and provides reliable and affordable electricity to the community.

Return on Investment (ROI): ROI for the City Council in implementing these smart infrastructure initiatives would include cost savings, revenue generation, and increased economic development.

For example, the fiber-optic network would increase property values and attract new businesses to the area, generating additional tax revenues. The smart traffic management systems would reduce fuel consumption and maintenance costs for vehicles, as well as decrease travel times for commuters and commercial vehicles, resulting in increased productivity and economic activity.

The smart grid would also generate cost savings and new revenue streams by integrating renewable energy sources and reducing energy waste. This would result in reduced energy bills for consumers, as well as the potential to sell excess energy back to the grid.

Conclusion: *The financial benefits of these smart infrastructure initiatives would not only offset the initial investment costs but also provide long-term financial gains for the city and its residents.*

Smart Tourism: Igoumenitsa is a popular tourist destination, so we would leverage technology to enhance the tourist experience. We would develop a smart tourism platform that provides tourists with information about local attractions, events, and accommodations. The platform would be accessible through a mobile app and would use augmented reality to provide immersive experiences. We would also install smart kiosks and digital signage throughout the city that provide real-time information about local attractions, weather, and transportation.

Return on Investment (ROI): ROI for the City Council in implementing these smart tourism initiatives would include increased revenue from tourism, improved operational efficiency, and enhanced branding and marketing.

The smart tourism platform and mobile app would attract more tourists to the area, increase the duration of their stay, and encourage them to visit local attractions and participate in events. This would result in increased revenue for local businesses, such as hotels, restaurants, and tour operators.

The smart kiosks and digital signage would improve operational efficiency by reducing the need for printed materials and manual updates. This would result in cost savings and a more sustainable approach to tourism.

The smart tourism platform and digital signage would enhance the city's branding and marketing efforts, by providing a modern and innovative approach to showcasing local attractions and events. This would help to position Igoumenitsa as a desirable tourist destination and attract new visitors to the area.

Conclusion: *The financial benefits of these smart tourism initiatives would not only offset the initial investment costs but also provide long-term financial gains for the city and its tourism industry.*

Smart Sea Ports: Igoumenitsa has two commercial sea ports and several marinas, which are critical to the local economy. We would invest in developing a smart sea port infrastructure that improves efficiency and reduces congestion. This includes installing smart cranes that use sensors to optimize cargo handling, implementing an automated container tracking system, and deploying autonomous drones for surveillance and inspection. We would also develop a smart marina management system that provides boaters with real-time information about weather, tides, and docking availability.

Return on Investment (ROI): ROI for the City Council in implementing these smart sea port initiatives would include increased revenue from port operations, improved operational efficiency, and reduced maintenance costs.

The smart sea port infrastructure would improve efficiency by reducing cargo handling times, increasing port capacity, and improving safety and security. This would result in increased revenue from port operations, as well as attracting new businesses and industries to the area.

The automated container tracking system and autonomous drones for surveillance and inspection would improve operational efficiency by reducing labor costs, minimizing errors, and improving the accuracy of cargo tracking. This would result in cost savings and increased productivity.

The smart marina management system would enhance the boating experience for tourists and locals, resulting in increased revenue from marina operations and related businesses, such as restaurants and shops.

Conclusion: *The financial benefits of these smart sea port initiatives would not only offset the initial investment costs but also provide long-term financial gains for the city and its port and marina industries.*

Smart Agriculture: Igoumenitsa has a strong agricultural industry, so we would invest in developing a smart agriculture platform that uses IoT sensors to optimize farming practices. This includes monitoring soil moisture, temperature, and nutrient levels to improve crop yields and reduce water usage. We would also develop a smart irrigation system that uses real-time data to optimize water usage and reduce waste.

Return on Investment (ROI): ROI for the City Council in implementing these smart agriculture initiatives would include increased revenue from agriculture, improved operational efficiency, and reduced costs associated with resource usage.

The smart agriculture platform and IoT sensors would improve crop yields by providing farmers with real-time data on soil moisture, temperature, and nutrient levels. This would result in increased revenue from agriculture, as well as attracting new businesses and industries to the area.

The smart irrigation system would improve operational efficiency by reducing water waste, minimizing labor costs associated with manual irrigation, and minimizing crop damage from overwatering. This would result in cost savings and increased productivity.

The smart agriculture platform and irrigation system would reduce resource usage, such as water and fertilizer, resulting in cost savings and a more sustainable approach to agriculture. This would also help to position Igoumenitsa as a leader in sustainable agriculture practices, which could attract new markets and industries.

Conclusion: *The financial benefits of these smart agriculture initiatives would not only offset the initial investment costs but also provide long-term financial gains for the city and its agriculture industry.*

Smart Retail: Igoumenitsa has a thriving retail sector, so we would invest in developing a smart retail platform that provides personalized shopping experiences. This includes implementing a smart inventory management system that uses AI to optimize product placement and pricing, and deploying smart vending machines that use facial recognition technology to provide personalized recommendations. We would also develop a mobile app that provides real-time information about store promotions and deals.

Return on Investment (ROI): ROI for the City Council in implementing these smart retail initiatives would include increased revenue from retail, improved operational efficiency, and increased customer satisfaction.

The smart inventory management system and AI-based optimization of product placement and pricing would improve revenue by maximizing sales and minimizing costs associated with overstocking or understocking. Additionally, the smart vending machines with facial recognition technology would provide personalized recommendations, enhancing the customer experience and driving sales.

The mobile app would improve customer satisfaction by providing real-time information about store promotions and deals, helping customers make informed purchasing decisions. This would result in increased customer loyalty, repeat business, and positive word-of-mouth advertising for the city's retail sector.

The smart retail platform would improve operational efficiency by reducing labor costs associated with manual inventory management and increasing the accuracy of product placement and pricing. This would result in cost savings and increased productivity for retailers.

Conclusion: *The financial benefits of these smart retail initiatives would not only offset the initial investment costs but also provide long-term financial gains for the city and its retail sector.*

Smart Government Services: Igoumenitsa has several government services, including police, fire department, schools, and doctors. We would invest in developing a smart government platform that provides citizens with real-time information about government services and facilities. This includes implementing a smart emergency response system that uses AI to optimize response times and deploying smart police vehicles that use real-time data to prevent crime. Additionally, we would develop a smart healthcare system that uses telemedicine to provide remote medical consultations and monitoring for patients.

Return on Investment (ROI): ROI for the City Council in implementing these smart government initiatives would include increased efficiency and effectiveness of government services, reduced costs associated with emergency response, and improved healthcare outcomes for the community.

The smart emergency response system and smart police vehicles would reduce costs associated with emergency response and crime prevention. The AI-based optimization of response times and real-time data analysis would enable emergency responders to respond more quickly and effectively to emergencies, reducing the potential for property damage and saving lives. The smart police vehicles would use real-time data to prevent crime, reducing the need for costly investigations and court proceedings.

The smart healthcare system that uses telemedicine would improve healthcare outcomes for the community by providing remote medical consultations and monitoring for patients, reducing the need for costly and time-consuming in-person visits. This would result in cost savings for both patients and healthcare providers, as well as improved health outcomes for the community.

The smart government platform that provides citizens with real-time information about government services and facilities would improve the efficiency and effectiveness of government services. This would result in cost savings associated with reduced paperwork and more efficient service delivery, as well as increased satisfaction among citizens.

Conclusion: *The financial benefits of these smart government initiatives would not only offset the initial investment costs but also provide long-term financial gains for the city and its government services.*

Smart Citizen Engagement: We would invest in developing a smart citizen engagement platform that encourages citizen participation and involvement. This includes developing a mobile app that allows citizens to report issues and provide feedback to the government, and implementing a smart citizen survey system that uses AI to analyze citizen sentiment and identify areas for improvement. We would also develop a smart education system that uses digital technologies to enhance learning and promote student engagement.

Return on Investment (ROI): ROI for the City Council regarding Smart Citizen Engagement can be significant, not only in terms of improved communication and citizen engagement but also in potential advertising revenues from the developed mobile apps. By investing in the development of a smart citizen engagement platform that includes a mobile app for citizens to report issues and provide feedback, and a smart citizen survey system that uses AI to analyze citizen sentiment, the government can gain valuable insights from citizens. This can help identify areas for improvement and optimize resource allocation, leading to increased economic growth and sustainability.

Moreover, by implementing a smart education system that uses digital technologies to enhance learning and promote student engagement, the government can attract advertising partners looking to reach a targeted audience. The development of a popular and widely used mobile app can also generate advertising revenue, which can be reinvested in further improving the platform and providing better services to citizens.

Conclusion: *The investment in a Smart Citizen Engagement platform that includes mobile apps can provide a positive ROI for the City Council, not only through improved communication and citizen engagement but also through potential advertising revenues. The development of a mobile app and a smart citizen survey system can lead to increased citizen participation and better decision-making by the government. The implementation of a smart education system can improve educational outcomes and enhance the quality of life for citizens, making it a worthwhile investment for the City Council.*

Smart Accommodation: As a popular tourist destination, Igoumenitsa has a high demand for holiday accommodations, hotels, and restaurants. We would invest in developing a smart accommodation platform that provides tourists with real-time information about available accommodations, pricing, and amenities. This includes implementing a smart booking system that uses AI to optimize room availability.

Return on Investment (ROI): ROI for the City Council regarding Smart Accommodation can be significant for Igoumenitsa as a popular tourist destination, not only in terms of increased tourism and revenue but also in potential advertising revenues from the developed platform. By investing in the development of a smart accommodation platform that provides tourists with real-time information about available accommodations, pricing, and amenities, the government can enhance the overall visitor experience and generate additional revenue through the platform.

Moreover, the development of a popular and widely used platform can attract advertising partners looking to reach a targeted audience of tourists. The government can leverage the platform's popularity to offer targeted advertising space to interested parties, generating additional advertising revenue that can be reinvested in further improving the platform and providing better services to tourists.

Additionally, by implementing a smart booking system that uses AI to optimize room availability, the government can improve the efficiency of the tourism industry and attract more visitors, leading to increased revenue for the city.

Conclusion: The investment in a Smart Accommodation platform can provide a positive ROI for the City Council in Igoumenitsa, not only through increased tourism and revenue but also through potential advertising revenues. The development of a platform that provides real-time information about available accommodations and optimizes room availability can attract more visitors and enhance the overall visitor experience. The advertising revenue generated from the platform can be reinvested in further developing the tourism industry and improving the city's services for tourists, making it a worthwhile investment for the City Council.

Smart Waste Management: To improve sustainability and reduce waste, we would invest in a smart waste management system that uses sensors to optimize garbage collection and recycling. This includes installing smart trash cans that use IoT sensors to alert collection teams when they need to be emptied, and implementing a smart recycling system that sorts and separates recyclable materials more efficiently.

Return on Investment (ROI): ROI for the City Council regarding Smart Waste Management can be significant for a city looking to improve sustainability and reduce waste. By investing in a smart waste management system that uses sensors to optimize garbage collection and recycling, the government can reduce operational costs while improving the efficiency of waste management.

The installation of smart trash cans that use IoT sensors to alert collection teams when they need to be emptied can improve the efficiency of garbage collection and reduce the number of collection trips required. This can lead to cost savings in labor and fuel, resulting in a positive ROI for the government. Additionally, the implementation of a smart recycling system that sorts and separates recyclable materials more efficiently can increase the amount of waste that is recycled, further reducing waste and potentially generating revenue from the sale of recyclable materials.

The development of a smart waste management system can attract potential investors looking to tap into the growing market for sustainable technologies. These investors can contribute to the development of the system and provide additional funding to the government, resulting in a positive ROI for the city.

Conclusion: *The investment in a Smart Waste Management system can provide a positive ROI for the City Council, not only through reduced operational costs but also through potential revenue generation and the attraction of potential investors. The development of a system that optimizes garbage collection and recycling can improve the efficiency of waste management and reduce waste, making it a worthwhile investment for the City Council.*

Smart Water Management: To address the region's water supply, we would invest in developing a smart water management system that optimizes water usage and reduces waste. This includes implementing smart water meters that monitor consumption patterns and alert users when they are using too much water, and developing a smart irrigation system that uses real-time data to optimize watering schedules and reduce water usage.

Return on Investment (ROI): ROI for the City Council regarding Smart Water Management can be significant for a city looking to address water supply and reduce water waste. By investing in a smart water management system that optimizes water usage, the government can reduce operational costs while improving the efficiency of water management.

The implementation of smart water meters that monitor consumption patterns and alert users when they are using too much water can encourage water conservation and reduce wastage. This can lead to cost savings in water treatment and supply, resulting in a positive ROI for the government. Additionally, the development of a smart irrigation system that uses real-time data to optimize watering schedules and reduce water usage can reduce the amount of water used for landscaping and agricultural purposes.

The development of a smart water management system can attract potential investors looking to tap into the growing market for sustainable technologies. These investors can contribute to the development of the system and provide additional funding to the government, resulting in a positive ROI for the city.

Conclusion: *The investment in a Smart Water Management system can provide a positive ROI for the City Council, not only through reduced operational costs but also through potential revenue generation and the attraction of potential investors. The development of a system that optimizes water usage and reduces waste can improve the efficiency of water management and reduce water supply issues, making it a worthwhile investment for the City Council.*

Smart Transportation: To reduce traffic congestion and improve mobility, we would invest in a smart transportation system that uses real-time data to optimize routes and schedules.

Return on Investment (ROI): ROI for the City Council regarding Smart Transportation can be significant for a city looking to reduce traffic congestion and improve mobility. By investing in a smart transportation system that uses real-time data to optimize routes and schedules, the government can reduce operational costs while improving the efficiency of transportation.

The implementation of a smart transportation system can improve the accuracy of public transportation schedules and reduce wait times for commuters, leading to increased rider satisfaction and potential revenue generation from increased ridership. Additionally, the optimization of routes can reduce traffic congestion and improve traffic flow, leading to reduced travel times and fuel consumption, resulting in a positive ROI for the government.

The development of a smart transportation system can attract potential investors looking to tap into the growing market for sustainable transportation technologies. These investors can contribute to the development of the system and provide additional funding to the government, resulting in a positive ROI for the city.

Conclusion: *The investment in a Smart Transportation system can provide a positive ROI for the City Council, not only through reduced operational costs but also through potential revenue generation and the attraction of potential investors. The development of a system that optimizes routes and schedules can improve the efficiency of transportation and reduce traffic congestion, making it a worthwhile investment for the City Council.*

Smart Public Safety: To improve public safety, we would invest in a smart public safety system that uses real-time data to prevent and respond to emergencies. This includes deploying smart surveillance cameras that use AI to detect and prevent crime, and developing a smart emergency response system that integrates data from various sources, such as traffic cameras and weather sensors, to optimize response times.

Return on Investment (ROI): ROI for the City Council regarding Smart Public Safety can be significant for a city looking to improve public safety and emergency response. By investing in a smart public safety system that uses real-time data to prevent and respond to emergencies, the government can reduce operational costs while improving the effectiveness of emergency response.

The deployment of smart surveillance cameras that use AI to detect and prevent crime can lead to reduced crime rates and increased public safety, potentially resulting in cost savings from reduced law enforcement and emergency response. Additionally, the development of a smart emergency response system that integrates data from various sources can optimize response times and improve the effectiveness of emergency response, leading to potential cost savings from reduced damage and loss of life.

The development of a smart public safety system can attract potential investors looking to tap into the growing market for sustainable and innovative public safety technologies. These investors can contribute to the development of the system and provide additional funding to the government, resulting in a positive ROI for the city.

Conclusion: *The investment in a Smart Public Safety system can provide a positive ROI for the City Council, not only through reduced operational costs but also through potential revenue generation and the attraction of potential investors. The development of a system that uses real-time data to prevent and respond to emergencies can improve public safety and emergency response, making it a worthwhile investment for the City Council.*

Smart Street Lighting: To improve energy efficiency and reduce costs, we would invest in a smart street lighting system that uses sensors to adjust lighting levels based on real-time data, such as traffic flow and weather conditions. This includes implementing smart LED lights that use less energy and last longer than traditional streetlights.

Return on Investment (ROI): ROI for the City Council regarding Smart Street Lighting can be significant in terms of reducing operational costs and improving energy efficiency. By investing in a smart street lighting system that uses sensors to adjust lighting levels based on real-time data, the government can reduce electricity costs while improving the reliability and sustainability of the lighting infrastructure.

The implementation of smart LED lights that use less energy and last longer than traditional streetlights can result in significant energy savings, potentially leading to reduced operational costs over the long term. Additionally, the use of sensors to adjust lighting levels based on real-time data can help to reduce energy waste, leading to further cost savings for the city.

The development of a smart street lighting system can improve the safety and quality of life for citizens. By providing better lighting levels in areas with higher traffic flow or during adverse weather conditions, the government can improve road safety and reduce accidents. Additionally, the use of smart street lighting can improve the aesthetics of the city and make it more attractive to residents and tourists alike.

The implementation of a smart street lighting system can help the city meet its sustainability goals and attract potential investors interested in the development of innovative and sustainable technologies. These investors can contribute to the development of the system and provide additional funding to the government, resulting in a positive ROI for the city.

Conclusion: *The investment in a Smart Street Lighting system can provide a positive ROI for the City Council, not only through reduced operational costs but also through improved safety and aesthetics, as well as potential revenue generation and the attraction of potential investors. The development of a system that uses sensors to adjust lighting levels based on real-time data can improve energy efficiency and sustainability, making it a worthwhile investment for the City Council.*

Smart Civic Engagement: To promote civic engagement and participation, we would invest in a smart civic engagement platform that uses digital technologies to connect citizens with local government and community organizations. This includes developing a mobile app that allows citizens to submit feedback and participate in community events, and implementing a smart community forum that uses AI to analyze community sentiment and identify areas for improvement.

Return on Investment (ROI): ROI for the City Council regarding Smart Civic Engagement can be significant in terms of improving citizen participation, promoting community involvement, and identifying areas for improvement within the community. By investing in a smart civic engagement platform, the government can leverage digital technologies to connect citizens with local government and community organizations, ultimately leading to increased civic engagement and participation.

The development of a mobile app that allows citizens to submit feedback and participate in community events can lead to increased participation in local government and community affairs. The app can also provide a convenient platform for citizens to access important information about local government initiatives and services, which can help to build trust and transparency between the government and its constituents.

The implementation of a smart community forum that uses AI to analyze community sentiment and identify areas for improvement can provide valuable insights for the government to improve public services, infrastructure, and community development. The platform can also help to identify areas of concern and prioritize government initiatives accordingly, leading to more effective and efficient resource allocation.

Additionally, the development of a smart civic engagement platform can help to attract investment and businesses to the community, as it demonstrates a commitment to innovation and technology, which is becoming increasingly important for economic development. Moreover, the platform can help to improve the overall quality of life in the community, leading to increased economic growth and sustainability.

Conclusion: *The investment in a Smart Civic Engagement platform can provide a positive ROI for the City Council, not only through improved citizen participation and community involvement, but also through potential revenue generation and the attraction of investment and businesses. By leveraging digital technologies, the government can promote civic engagement and participation, leading to a more transparent, effective, and efficient local government.*

Smart Culture and Events: Smart Culture and Events play an essential role in promoting the city's cultural heritage and traditions while showcasing its modern and innovative outlook. Smart Culture and Events can help establish Igoumenitsa as a vibrant and dynamic smart city with a strong and unique cultural identity. By leveraging technology and innovation, the city can create unique and immersive experiences for visitors and residents alike while promoting local talent and creativity. Smart Culture and Events can also help attract visitors to the city's unique cultural events, festivals, and experiences, promoting the city's tourism industry.

Return on Investment (ROI): ROI for the City Council regarding Smart Culture and Events can have a significant impact on a city's cultural identity and tourism industry. By promoting local talent and creativity and leveraging technology and innovation, the city can create unique and immersive experiences for visitors and residents alike. Smart Culture and Events also play a crucial role in showcasing a city's cultural heritage and traditions while demonstrating its modern and innovative outlook.

One way to measure the ROI of Smart Culture and Events is by assessing their economic impact. By attracting visitors and increasing tourism revenue, these events can generate more economic activity for the city. Additionally, inviting international performers to use the amphitheater facilities offered by Igoumenitsa can further enhance the city's cultural offerings and attract more visitors. These events can contribute to long-term economic growth and generate more revenue for the city. To measure the ROI, the revenue generated by these events can be compared to the investment made by the City Council.

Another way to measure the ROI of Smart Culture and Events is through brand building. These events can help establish the city's identity as a vibrant and dynamic smart city with a strong and unique cultural identity. By showcasing both local and international talent in the amphitheater, Smart Culture and Events can enhance its brand image and reputation. The city can also build relationships with international performers, leading to potential collaborations and future events. To measure the ROI, surveys can be conducted to understand the public perception of the city and compare it to the pre-event perception.

Finally, Smart Culture and Events can have a positive social impact by promoting social cohesion, fostering community engagement, and improving the quality of life for residents. Inviting international performers can also expose residents to different cultures and perspectives, contributing to a more diverse and inclusive community. To measure the ROI, surveys can be conducted to understand the public perception of the events' social impact and compare it to the pre-event perception.

Conclusion: Smart Culture and Events can have a significant impact on a city's cultural identity, tourism industry, and social well-being. Inviting international performers to use the amphitheater facilities can further enhance the city's cultural offerings and attract more visitors. Measuring the ROI can involve assessing their economic impact, brand building potential, and social impact.

SEEKING THE INVESTMENT

The investment for the Smart City projects in Igoumenitsa can come from a variety of sources, including:

Public-Private Partnerships (PPPs): This involves partnering with private companies to jointly fund and manage Smart City projects. In this model, the public sector provides access to land and other resources, while private companies provide funding, expertise, and technology.

Government Funding: The national or regional government may provide funding for Smart City projects as part of their economic development or sustainability initiatives. This can include grants, loans, or tax incentives.

Venture Capital (VC) Funding: Venture capital firms may invest in Smart City startups or companies that are developing innovative solutions for the urban environment. These investments typically involve high-risk, high-reward opportunities.

Infrastructure Funds: Infrastructure funds invest in projects that generate long-term, stable returns, such as toll roads, airports, and utility networks. Smart City projects that have a revenue-generating component may be attractive to infrastructure funds.

Impact Investors: These investors seek to generate a social or environmental impact alongside a financial return. Smart City projects that have a positive impact on the environment, public health, or social equity may be attractive to impact investors.

Corporate Funding: Large corporations may invest in Smart City projects as part of their corporate social responsibility or sustainability initiatives. These investments can help companies improve their brand image and demonstrate their commitment to sustainability.

Citizen Investment Pool: The Smart City program in Igoumenitsa could potentially include a citizen investment pool where residents are given the opportunity to invest in local Smart City projects. This can help to build community support and engagement for the program while also providing a new source of funding. Through crowdfunding platforms or community investment trusts, residents can invest small amounts of money into projects they believe in, such as community solar installations or local transportation infrastructure. This approach not only helps to democratize investment opportunities but also increases the potential for community ownership and participation in the Smart City program.

European Union (EU) Grants: The EU has a range of funding programs available to support Smart City development across member states. These programs include Horizon 2020, the European Structural and Investment Funds, and the Connecting Europe Facility. EU grants can provide a significant source of funding for Smart City projects in Igoumenitsa, especially those that align with EU sustainability and innovation priorities.

Bequests: Another potential source of investment for Smart City programs in Igoumenitsa could be bequests. This refers to the act of leaving a gift or donation in one's will or estate plan to support a cause or organization. By building strong relationships with donors and philanthropists in the community, the Smart City program can leverage bequests to fund new initiatives or expand existing ones. This can also provide an opportunity for community members to leave a legacy and make a lasting impact on the development of Igoumenitsa.

Expats Living Abroad: Many cities with large expat communities have seen significant investments from these individuals who maintain a connection to their hometown. As Igoumenitsa has a significant diaspora, this presents an opportunity to leverage these connections and encourage investment in Smart City projects. By engaging with expats and providing them with opportunities to invest or participate in Smart City initiatives, the program can attract new sources of funding and expertise. This can also help to create a sense of pride and engagement among the diaspora, further strengthening the connection between Igoumenitsa and its global community.

Foreign Retiree Investors: Many retirees choose to invest in real estate or other projects in retirement destinations around the world. As Igoumenitsa has a growing tourism industry and a favorable climate, it may be attractive to foreign retirees looking for a new home or investment opportunity. By targeting this group and showcasing the benefits of investing in the Smart City program, such as improved infrastructure, safety, and quality of life, the program can attract new sources of investment. This can also help to bring new ideas and perspectives to the community, which can be beneficial for the long-term growth and sustainability of the Smart City program.

Conclusion: *The investment for the Smart City projects in Igoumenitsa will likely come from a combination of these sources, depending on the specific project and the preferences of the investors involved. The key is to develop a compelling business case for each project and to target investors who are aligned with the vision and goals of the Smart City program. Leveraging Federal Government and EU grants can also be an important source of funding for Smart City projects in Igoumenitsa.*

THE BUY-IN

Dear citizens of Igoumenitsa,

We are excited to introduce Igoumenitsa Smart City program, a comprehensive plan that aims to transform our city into a dynamic, modern, and sustainable urban center. Our program is designed to leverage technology and innovation to improve the quality of life for our citizens, promote economic growth, and enhance our city's unique cultural identity.

The Smart City program is built around multiple pillars. These pillars encompass a range of programs and initiatives that will benefit our citizens, businesses, and visitors alike. We have identified 15 key programs that will be implemented over the next five years to achieve our vision for a smarter and more sustainable Igoumenitsa.

Our Smart Infrastructure programs will improve our city's transportation, energy, and public safety systems. These programs include smart traffic management, renewable energy projects, and a citywide CCTV system that will enhance public safety.

Our Smart Economy programs will promote economic growth by supporting local businesses, attracting investment, and creating new jobs. These programs include a digital marketplace for local products, a startup accelerator program, and smart tourism initiatives that will attract visitors to our city.

Our Smart Culture and Events programs will celebrate our city's unique cultural identity and promote community engagement. These programs include a cultural center, online platform for local talent and events, and community-driven innovation program to promote smart city solutions.

Developing a Smart City can be a game-changer in reversing the brain-drain exodus of our youth to other countries. The lack of opportunities and a high-quality standard of living often prompts young people to leave their hometowns in search of greener pastures. A Smart City program that offers access to modern technology, job opportunities, and a high quality of life can help retain young talent in the region. By promoting innovation and entrepreneurship, the Smart City program can foster new ideas, businesses, and opportunities, creating a sustainable ecosystem of talent and innovation.

Igoumenitsa has the potential to become the model small town success, setting an example for other small towns to emulate. The Smart City program can serve as a catalyst for the growth and prosperity of Igoumenitsa and surrounding regions, bringing the benefits of modern technology and innovation to smaller communities. This can help to reduce the burden on large cities and revitalize small towns like Igoumenitsa. By showcasing the success of the Smart City program, we can attract investors, businesses, and talent to the region, creating a bright future of growth and prosperity.

CITY COUNCIL REVENUE STREAMS

Property Taxes: City councils collect property taxes from homeowners and businesses within their jurisdiction.

Fees and Charges: City councils charge fees for various services, such as building permits, parking, and garbage collection.

Sales Taxes: City councils may collect sales taxes on goods and services sold within their jurisdiction.

Grants: City councils may receive grants from the national government, the European Union, or other sources.

Rental Income: City councils may own and rent out properties, such as buildings or land, to generate income.

Tourist Taxes: In some cities, city councils may charge a tourist tax on visitors staying in hotels or other accommodations.

Fines: City councils may collect fines for various violations of city ordinances or regulations.

Business Taxes: City councils may levy taxes on businesses operating within their jurisdiction.

State Transfers: The Greek government may transfer funds to city councils to help fund their operations.

Parking Fees: City councils may charge for parking in public parking lots or on-street parking spaces.

Advertising: City councils may sell advertising space on public property, such as billboards or bus shelters.

Licenses and Permits: City councils may charge fees for licenses and permits required for various activities, such as starting a business or holding a special event.

Utility Fees: City councils may charge fees for providing water, sewer, or other utilities to residents and businesses.

Investments: City councils may invest funds in stocks, bonds, or other financial instruments to generate income.

Renting out Equipment: City councils may rent out equipment, such as trucks or construction equipment, to generate revenue.

Parking Fines: In addition to charging for parking, city councils may also collect fines for parking violations.

Donations and Sponsorships: City councils may receive donations or sponsorships from individuals or organizations to support specific projects or programs.

Public-Private Partnerships (PPPs): City councils can form PPPs with private companies that specialize in smart city technology. In exchange for the use of the city's data and infrastructure, the private company can pay the city a fee or a share of the revenue generated by the project.

Advertising and Sponsorship: Smart city projects often involve the use of digital displays or other technologies that can be used for advertising. City councils can sell advertising space on these displays or seek sponsorships from companies that want to associate themselves with the project.

User Fees: City councils can charge fees for the use of smart city services, such as public Wi-Fi or smart parking systems.

Data Monetization: Smart city projects generate a lot of data that can be valuable to companies and researchers. City councils can monetize this data by selling access to it or by using it to develop new products and services.

Energy Savings: Smart city projects can help cities save money on energy costs by optimizing energy use in buildings and public spaces. City councils can use the money saved to generate revenue or to invest in other projects.

Data Analytics: City councils can use data analytics to extract insights and create value from the data generated by smart city projects. This can involve selling data analytics services to businesses or other organizations that are interested in using the data for their own purposes.

Improved City Services: Smart city projects can improve the efficiency and effectiveness of city services, such as waste management and public transportation. This can result in cost savings for the city, which can be reinvested in other projects or used to generate revenue in other ways.

Increased Tourism: Smart city projects can help to promote tourism by providing visitors with better information and services. This can lead to increased revenue from tourism-related activities, such as hotels and restaurants.

Research Partnerships: City councils can partner with universities or research institutions to develop and test new smart city technologies. This can lead to the development of new products and services that can be commercialized, generating revenue for the city.

Smart Contracts: Smart city projects can use blockchain technology to create smart contracts that automate transactions and reduce costs. This can result in cost savings for the city, which can be reinvested in other projects or used to generate revenue in other ways.

Smart Meters: Smart city projects can include the installation of smart meters for electricity, water, and gas usage. This can enable the city to accurately bill residents and businesses for their usage, generating revenue for the city.

E-commerce: Smart city projects can enable the creation of online marketplaces that connect residents with local businesses. City councils can generate revenue by taking a commission on transactions that occur through the marketplace.

Open Data: Smart city projects generate a lot of data that can be valuable to businesses, researchers, and developers. City councils can make this data available to the public through open data initiatives, and charge fees for premium access or customized data sets.

Smart Waste Management: Smart city projects can optimize waste management processes, reducing costs and generating revenue from recycling and waste-to-energy initiatives.

Smart Mobility: Smart city projects can improve mobility options such as bike-sharing or car-sharing programs, which can generate revenue for the city through rental fees or commissions.

Smart Grid: Smart city projects can incorporate smart grid technology to improve energy distribution and management. City councils can generate revenue by selling excess energy generated by renewable sources back to the grid.

Smart Irrigation: Smart city projects can incorporate smart irrigation systems to optimize water usage in parks and public spaces. City councils can save money on water costs and generate revenue by selling any excess water to nearby agricultural businesses.

Telecommunications: Smart city projects can include the deployment of telecommunications infrastructure such as fiber optic cables, cell towers, and Wi-Fi hotspots. City councils can generate revenue by leasing this infrastructure to telecommunications companies.

Smart Building Management: Smart city projects can include the deployment of smart building management systems that optimize energy usage and reduce maintenance costs. City councils can generate revenue by offering these systems as a service to local businesses and property owners.

Environmental Monitoring: Smart city projects can incorporate sensors and other technologies to monitor air and water quality, noise levels, and other environmental factors. City councils can generate revenue by selling this data to environmental researchers and businesses.

HELLENIC CITY COUNCIL NATIONAL LAWS

Under Greek National Law (GNL), Greek City Councils are responsible for a range of local government functions, including:

Urban Planning and Zoning: GNL 2508/1997 regulates the spatial planning and urban development process in Greece, including the creation and implementation of master plans, the regulation of land use, and the control of building activity.

Public Works: GNL 3852/2010 establishes the framework for the organization and operation of local government in Greece, including the responsibilities of city councils for public works and infrastructure.

Waste Management: GNL 2939/2001 regulates the management of municipal solid waste in Greece, including the collection, transportation, and disposal of waste, as well as the establishment of recycling programs.

Public Transportation: GNL 3852/2010 establishes the framework for the organization and operation of local government in Greece, including the responsibilities of city councils for public transportation.

Public Safety: GNL 3852/2010 establishes the framework for the organization and operation of local government in Greece, including the responsibilities of city councils for public safety and security.

Education: GNL 1566/1985 establishes the framework for education in Greece, including the responsibilities of city councils for the operation and maintenance of local schools and educational facilities.

Cultural and Recreational Facilities: GNL 3852/2010 establishes the framework for the organization and operation of local government in

Greece, including the responsibilities of city councils for the provision and maintenance of cultural and recreational facilities.

Social Welfare: GNL 3852/2010 establishes the framework for the organization and operation of local government in Greece, including the responsibilities of city councils for social welfare services.

Business Development: GNL 3852/2010 establishes the framework for the organization and operation of local government in Greece, including the responsibilities of city councils for promoting local economic development.

Tourism: GNL 4177/2013 regulates tourism in Greece, including the responsibilities of city councils for promoting and managing tourism within their jurisdiction.

Environmental Protection: GNL 4014/2011 regulates environmental protection in Greece, including the responsibilities of city councils for managing waste disposal, regulating air and water quality, and preserving natural areas.

Cultural Heritage: GNL 3028/2002 regulates the protection and preservation of cultural heritage in Greece, including the responsibilities of city councils for the preservation and promotion of local cultural heritage.

Health: GNL 4268/2014 regulates public health in Greece, including the responsibilities of city councils for managing public health facilities and programs and regulating food and beverage service establishments.

Financial Management: GNL 3852/2010 establishes the framework for the organization and operation of local government in Greece, including the responsibilities of city councils for financial management.

Citizen Participation: GNL 3852/2010 establishes the framework for the organization and operation of local government in Greece, including the responsibilities of city councils for promoting citizen participation in local government decision-making processes.

These are some of the Greek National Laws that relate to the responsibilities of Greek City Councils. It's important to note that there may be other laws and regulations that also impact the responsibilities and activities of city councils in Greece.

CONCLUSION

The hypothetical transformation of Igoumenitsa into a Smart City provides an excellent model for other cities considering a similar transformation. This case study emphasizes the importance of a comprehensive and collaborative approach involving both public and private sectors to create a sustainable and technologically advanced urban environment. The success of the Smart City transformation in Igoumenitsa is based on a clear vision and roadmap that considers the city's unique needs and characteristics. By prioritizing smart infrastructure projects, the city significantly reduces its environmental footprint and improves residents' quality of life.

The Igoumenitsa case study offers two critical lessons for other cities looking to develop their Smart City program. The first lesson highlights the potential for Smart Cities to reverse the brain-drain by offering access to modern technology, job opportunities, and a high quality of life. This promotes entrepreneurship, retains young talent in the region, and leads to economic growth, improving overall community well-being. Igoumenitsa is a compelling case study for cities seeking to address brain-drain challenges and promote sustainable economic growth. The second lesson emphasizes the importance of citizen engagement and participation in the Smart City process. By involving residents in decision-making and promoting community-led initiatives, Igoumenitsa builds a sense of ownership and pride in the transformation, contributing to its success.

Additionally, the Igoumenitsa case study highlights the potential for Smart Cities to drive economic growth and job creation. By fostering innovation and entrepreneurship, the city attracts new businesses and investments while providing opportunities for its citizens to acquire new skills and participate in the digital economy.

The Igoumenitsa example demonstrates that Smart City transformation is not just about technology but creating a sustainable and inclusive urban environment that prioritizes citizens' well-being. Although each city will have its unique Smart City vision and requirements, the key principles and strategies outlined in the Igoumenitsa case study can serve as a starting point for other cities embarking on a similar journey.

Finally, the Greek National Laws that outline the responsibilities of City Councils can have a significant impact on Smart City development. As City Councils manage essential services such as transportation, waste management, and public safety, their decisions and actions play a crucial role in creating sustainable and innovative urban environments. The integration of technology and data-driven solutions is crucial, and City Councils must adapt to these changes and implement policies that encourage the use of new technologies. Therefore, careful consideration and implementation of Greek National Laws can help City Councils foster Smart City growth, ultimately leading to improved quality of life and economic prosperity for Greek citizens.

ONE SMART WORLD
ARE YOU READY FOR IT?

| xxi |
KEY TAKEAWAYS

Smart cities are cities that integrate technology and digital infrastructure into their functioning, with the aim of improving the quality of life for citizens, making the city more efficient, and reducing its environmental impact. This concept encompasses various aspects of city life, including transportation, energy, buildings, governance, and healthcare. The market for smart city solutions has grown rapidly in recent years, driven by factors such as increasing urbanization, the need for more sustainable and efficient cities, and advancements in technology.

KEY TAKEAWAYS

Following are 12 key takeaways regarding smart cities.

1. **Definition:** A smart city is a city that uses technology and digital infrastructure to improve the quality of life for its citizens, make the city more efficient, and reduce its environmental impact.

2. **Market Growth:** The market for smart city solutions has grown rapidly in recent years and is expected to reach a size of over $2.3 trillion by 2025.

3. **Key Buying Criteria:** When it comes to buying smart city solutions, cities and organizations consider factors such as scalability, interoperability, sustainability, cost-effectiveness, security, user-

centered design, data privacy and ownership, and innovation and advancement.

4. **Technologies Used:** Smart city solutions use various technologies, such as the Internet of Things (IoT), artificial intelligence (AI), big data analytics, and cloud computing, among others.

5. **Solutions Offered:** Smart city solutions aim to address various aspects of city life, such as smart transportation, smart energy, smart buildings, smart governance, and smart healthcare, among others.

6. **Advantages:** Smart cities offer several advantages, including increased efficiency and cost savings, improved quality of life for citizens, and reduced environmental impact.

7. **Challenges:** Implementing smart city solutions can be challenging and requires a significant investment in technology and infrastructure. Additionally, there are concerns about privacy, security, and data ownership.

8. **Government Role:** Governments play a crucial role in the development and implementation of smart city solutions and often provide funding and support for these initiatives.

9. **Private Sector Involvement:** The private sector also plays a significant role in the development and implementation of smart city solutions and offers a range of products and services.

10. **Citizen Engagement:** Engaging citizens in the development and implementation of smart city solutions is important for ensuring their adoption and success.

11. **Global Trends:** The trend towards smart cities is a global one, with cities in different regions of the world implementing smart city solutions to varying degrees.

12. **Future Outlook:** The future of smart cities is expected to be shaped by advancements in technology, increased citizen engagement, and a focus on sustainability and efficiency.

Summary: Smart cities use technology and digital infrastructure to improve the quality of life for citizens, make cities more efficient, and reduce their environmental impact. The smart city solutions market is growing rapidly, driven by urbanization and advancements in technology. When purchasing smart city solutions, cities and organizations must consider factors such as scalability, interoperability, sustainability, cost-effectiveness, security, user-centered design, data privacy, ownership, innovation and advancement. Smart city solutions use technologies like IoT, AI, big data analytics, and cloud computing to address various aspects of city life, including transportation, energy, buildings, governance, and healthcare. Governments and the private sector play crucial roles in the development and implementation of smart city solutions, and citizen engagement is essential for their success.

Conclusion: Smart cities offer numerous advantages, such as increased efficiency and cost savings, improved quality of life for citizens, and reduced environmental impact. However, implementing smart city solutions can be challenging and requires significant investment in technology and infrastructure. Privacy, security, and data ownership are also important concerns. Governments and the private sector should work together to address these issues and create more sustainable and efficient cities. The future of smart cities looks promising, with a focus on sustainability and efficiency, technological advancements, and increased citizen engagement.

Quote: "Smart cities require a visionary approach, integrating technology and digital infrastructure to enhance citizens' well-being, while ensuring sustainability and efficiency remain paramount in our quest for progress." Emin Hasic

EPILOGUE

As we come to the end of '**ONE SMART WORLD**', the author Emin Hasic makes it is clear that smart cities are no longer a futuristic concept, but rather a tangible reality. The world has embraced the idea of smarter, more efficient and sustainable cities, and this book is a testament to that.

From the historical timeline of ancient Greek philosophers to the modern-day projects in countries around the world, we have seen that the idea of smart cities is not new. It has evolved with the times, driven by technology and the ever-increasing demands of urbanization.

We have explored the benefits that smart cities offer, including increased efficiency, improved sustainability, and enhanced quality of life for citizens. However, we have also learned that there are challenges and pitfalls to be aware of, and that starting small and addressing barriers is essential for success.

This book has also highlighted the importance of embracing smart cities, and the trend towards embracing technology in all aspects of our lives. As we move towards a more connected world, the desire for smart cities is the new trend, and we must ensure that we are not left behind.

The future is bright for smart cities, with projects and programs by industry emerging every day. From the Smart City 5G Network to Smart City Agriculture, the possibilities are endless. However, we must not forget the importance of privacy, security, and ethical considerations as we continue to embrace technology.

As we look towards the future, it is clear that smart cities will play a crucial role in shaping our world. They will help us to address some of the most pressing issues of our time, from climate change to inequality. The world is changing, and the smart city revolution is leading the way. We must embrace this change NOW and WORK TOGETHER to build a better, smarter, and more sustainable world for all.

BONUS OFFER

Dear Reader,

I express my heartfelt gratitude for perusing the words inscribed in "ONE SMART WORLD". May its contents enlighten your understanding of the significance of a singular intelligent world, one composed of SMART CITIES. Let us seize the presented opportunities to pave a brilliant path towards a promising and bright future for all.

It is with great pleasure that I bring to your attention an offer from **GDPA**, the world's foremost platform for data privacy compliance in the domain of SMART CITIES. Their bonus package encompasses a personalized one-on-one consultation, designed to provide bespoke guidance on matters pertaining to data protection and privacy practices for your esteemed smart city and organizations, as well as an exclusive discount.

GDPA is offering a 40% discount on their Trust Platform, a comprehensive solution that includes cutting-edge software, education, and support services to assist you in ensuring Smart City compliance with data protection and privacy regulations. This package includes provisions for security, data governance, and a privacy management system that incorporates third-party risk monitoring. I urge you to seize this opportunity, which holds tremendous value for your smart city and organization's continued success.

To claim your bonus, visit:

Membership: www.trustgdpa.com/signup • promo code '**OSWSC**' and
Free Consultation: www.trustgdpa.com/letstalk • promo code '**OSWSC**'.

You can also purchase the **eBook** directly from: www.eminhasic.com/ebook

If you have any questions, please don't hesitate to contact me.

Thank you for your trust and support.

Emin Hasic (Data Privacy Executive & Author).
www.eminhasic.com/contact-us

VALUABLE EXTERNAL LINKS

The following links lead to smart city related sites

GDPA: www.trustgdpa.com
The Global Data Protection Agency. The GDPA platform offers a comprehensive and reliable solution to assist you in fulfilling your Smart City compliance requirements under the GDPR and other significant regulations such as the APP, CCPA etc. With 24/7 real-time monitoring and support from privacy experts and instructional designers with extensive expertise, GDPA's managed platform is designed to meet the strictest privacy regulations, standards, and audits. At GDPA, it is our core mission to help you effectively manage your privacy compliance.

DSAR Live: www.dsarlive.com
A complimentary free service created for individuals to assert their Privacy Rights with entities, organizations, public offices, or individuals that collect and utilize their data for commercial purposes.

Venetis: www.veneti.it
The architecture, design, and planning firm founded by Athanasios Venetis is a valuable contributor to the Smart City framework. The firm is committed to creating dynamic, socially responsible, and environmentally sustainable designs for projects of all scales, from urban design to construction details. They advocate for new approaches to architecture and design that are interdisciplinary, and their work reflects a deep study of the mental, intellectual, and spiritual production of humanity over time. By exploring traditional, modern, and futuristic technologies and materials, the firm's projects focus on the beauty of nature and the creative ability to deal with an invisible reality. As a licensed architect and engineer, Venetis firm is a valuable partner in the development of innovative, sustainable, and functional projects in the Smart City framework.

SIMI Connect Limited: www.simiconnect.com
SIMI Connect Limited (SCL) is an Australian company in the IOT, smart home, energy management and digital platform / digital economy sector. 12 Granted Patents across 3 continents (including Australia, New Zealand, USA, Canada, UK, Germany, France, Spain, Italy & Greece). Simi's platform is an eco-system of digital assets providing users with the ability to access thousands of digital assets and services from a single platform. At the heart is its open and ubiquitous IOT Energy Management application that allows users to manage thousands of devices and appliances from anywhere in the world. From monitoring and managing power consumption, integrating renewable energy generation, storage solutions, security surveillance to devices and appliances in homes, businesses, buildings, offices and across multiple sites. SIMI enhances the way people interact with their world helping them take control of their biggest assets, reduce their carbon footprint and assist them to make savings and informed choices along the way. Simi has recognized the significance of privacy and data protection for both its business and customers. As a result, Simi has become one of the few companies that prioritize their customers' privacy and data protection needs by partnering with GDPA (Global Data Protection Agency).

The Open University: www.open.edu/openlearn/science-maths-technology/smart-cities/
Free '24-hour online study course' on Smart Cities Level 1: Introductory

Smart Cities Dive: www.smartcitiesdive.com
A daily newsletter covering news and trends related to smart cities and urban technology.

Smart Cities Council: www.smartcitiescouncil.com
A global coalition of companies, cities, and organizations dedicated to promoting smart city development and best practices.

The Smart City Journal: www.thesmartcityjournal.com
An online magazine featuring articles, interviews, and case studies related to smart city development and urban innovation.

Urban Land Institute: www.uli.org
A non-profit organization that provides research, education, and networking opportunities for professionals involved in real estate and urban development.

Smart Cities Connect: www.smartcitiesconnect.org
A community of smart city stakeholders, including city officials, industry leaders, and researchers, providing resources, events, and news related to smart city development.

IEEE Smart Cities: www.smartcities.ieee.org
A global initiative by the Institute of Electrical and Electronics Engineers (IEEE) to promote the development of smart city technologies and standards.

World Smart Sustainable Cities Organization: www.wssco.org
A non-profit organization dedicated to promoting sustainable urban development and smart city initiatives worldwide.

Smart Cities Information System: www.smartcities-infosystem.eu
An online platform funded by the European Union that provides information and tools for smart city planning and implementation.

wannalOOk: www.wannalook.com
The leading data privacy compliant short URL service, where your data is fully under your control and not shared with any third parties without your explicit consent. Take advantage of this genuine offer and receive a **50% discount** on your membership to **wannalOOK** when you use promo code "**OSWSC**".

Emin Hasic: www.eminhasic.com
The author of 'One Smart World 'and 'We Are Data Subjects.'

RESOURCES

CITIES, DISTRICTS, MUNICIPALITIES, STATES, GOVERNMENTS

Aalborg, Denmark; Aarhus, Denmark; Abu Dhabi, United Arab Emirates; Adana, Turkey; Adelaide, Australia; Agadir, Morocco; Aguascalientes, Mexico; Al Daayen, Qatar; Al Hail, Oman; Al Khuwair, Oman; Al Seeb, Oman; Al-Dasma, Kuwait; Al-Faisaliah, Saudi Arabia; Al-Farwaniya, Kuwait; Almere, Netherlands; Al-Rai, Kuwait; Al-Sulaymaniya, Kuwait; Al-Zour, Kuwait; Amsterdam, Netherlands; Andhra Pradesh, India; Antalya, Turkey; Antwerp, Belgium; Arnhem, Netherlands; Athens, Greece; Auckland, New Zealand; Austin, United States of America; Bærum, Norway; Bandung, Indonesia; Bangkok, Thailand; Banská Bystrica, Slovakia; Barcelona, Spain; Barka, Oman; Basel, Switzerland; Beijing, China; Belo Horizonte, Brazil; Berlin, Germany; Bhopal, India; Bilbao, Spain; Birmingham, United Kingdom; Bologna, Italy; Bordeaux, France; Boston, United States of America; Brampton, Canada; Brasília, Brazil; Bratislava, Slovakia; Bristol, United Kingdom; Brussels, Belgium; Bucharest, Romania; Budapest, Hungary; Bursa, Turkey; Busan, South Korea; Cagayan De Oro, Philippines; Cairo, Egypt; Calgary, Canada; Cambridge, United Kingdom; Can Tho, Vietnam; Cape Town, South Africa; Casablanca, Morocco; Central Java, Indonesia; Chandigarh, India; Charleroi, Belgium; Chiang Mai, Thailand; Chicago, United States of America; Christchurch, New Zealand; Cluj-Napoca, Romania; Columbus, United States of America; Copenhagen, Denmark; Corinth, Greece; Cork, Ireland; Coruña, Spain; Curitiba, Brazil; Cyberjaya, Malaysia; Da Nang, Vietnam; Daegu, South Korea; Darwin, Australia; Davao, Philippines; Debrecen, Hungary; Denizli, Turkey; Doha, Qatar; Drammen, Norway; Dubai, United Arab Emirates; Dublin, Ireland; Dubrovnik, Croatia; Dunedin, New Zealand; Durban, South Africa; Edmonton, Canada; Eindhoven, Netherlands; Eskisehir, Turkey; Espoo, Finland; Fez, Morocco; Flanders, Belgium; Florence, Italy; Frankfurt, Germany; Freiburg, Germany; Gangwon, South Korea; Geneva, Switzerland; Ghent, Belgium; Glasgow, Scotland; Gold Coast, Australia; Gothenburg, Sweden; Government, Austria; Government, Belarus; Government, Bulgaria; Government, Canada; Government, Costa Rica; Government, Denmark; Government, Egypt; Government, Finland; Government, Iceland; Government, Indonesia; Government, Ireland; Government, Israel; Government, Lithuania; Government, Luxembourg; Government, Malaysia; Government, Mexico; Government, Norway; Government, Philippines; Government, Poland; Government, Qatar; Government, Romania; Government, Serbia; Government, Singapore; Government, Slovenia; Government, South Africa; Government, Sweden; Government, Switzerland; Government, Thailand; Government, United Arab Emirates; Government, United Kingdom; Government, United States of America; Government, Vietnam; Graz, Austria; Grenoble, France; Groningen, Netherlands; Guadalajara, Mexico; Guangzhou, China; Gujarat, India; Ha Noi, Vietnam; Halandri, Greece; Halden, Norway; Halifax,

Canada; Hamburg, Germany; Hamilton, New Zealand; Hasselt, Belgium; Hawke's Bay, New Zealand; Helsinki, Finland; Heraklion, Greece; Hermosillo, Mexico; Ho Chi Minh City, Vietnam; Hobart, Australia; Hong Kong, China; Iasi, Romania; Ibra, Oman; Incheon, South Korea; Indore, India; Istanbul, Turkey; Izmir, Turkey; Jahra, Kuwait; Jaipur, India; Jakarta, Indonesia; Jeddah, Saudi Arabia; Jeju, South Korea; Johannesburg, South Africa; Kansas City, Usa; Kazan Republic Of Tatarstan, Russia; Kenitra, Morocco; Kerala, India; King Abdullah, Saudi Arabia; Kochi, India; Konya, Turkey; Košice, Slovakia; Krakow, Poland; Krasnodar, Russia; Kristiansand, Sweden; Kuala Lumpur, Malaysia; Lancaster, United States of America; Larissa, Greece; Las Vegas, United States of America; Lausanne, Switzerland; Leeds, United Kingdom; León, Mexico; Leuven, Belgium; Liège, Belgium; Limerick, Ireland; Linz, Austria; Ljubljana, Slovenia; London, United Kingdom; Los Angeles, United States of America; Lund, Sweden; Lusail, Qatar; Lyon, France; Macao, China; Madhya Pradesh, India; Madrid, Spain; Maharashtra, India; Makassar, Indonesia; Makkah, Saudi Arabia; Malaga, Spain; Malang, Indonesia; Malmö, Sweden; Manchester, United Kingdom; Marrakech, Morocco; Marseille, France; Martin, Slovakia; Mechelen, Belgium; Medina, Saudi Arabia; Melbourne, Australia; Miami, United States of America; Milan, Italy; Milton Keynes, United Kingdom; Minsk, Belarus; Miskolc, Hungary; Mohenjo-Daro, Pakistan; Monaco, Monaco; Montreal, Canada; Morelia, Mexico; Moscow, Russia; Munich, Germany; Muscat, Oman; Nador, Morocco; Nagpur, India; Nantes, France; Naples, Italy; Neom, Saudi Arabia; New Capital City, Egypt; New Delhi, India; New York City, United States of America; Newcastle, Australia; Newcastle, United Kingdom; Niagara Falls, Canada; Nice, France; Nitra, Slovakia; Nizhny Novgorod, Russia; Nordhavn, Denmark; Ohio, United States of America; Osijek, Croatia; Oslo, Norway; Ostend, Belgium; Otago, New Zealand; Ottawa, Canada; Oujda, Morocco; Oulu, Finland; Paris, France; Parma, Italy; Patras, Greece; Pécs, Hungary; Pennsylvania, United States of America; Perth, Australia; Philadelphia, United States of America; Pisa, Italy; Plovdiv, Bulgaria; Poprad, Slovakia; Port Said, Egypt; Poznan, Poland; Prešov, Slovakia; Puebla, Mexico; Pula, Croatia; Pune, India; Queenstown, New Zealand; Queretaro, Mexico; Rabat, Morocco; Reykjavik, Iceland; Rhodes, Greece; Rijeka, Croatia; Rio De Janeiro, Brazil; Riyadh, Saudi Arabia; Rome, Italy; Rotterdam, Netherlands; Rouen, France; Ruter, Norway; Sabah Al-Salem, Kuwait; Sakarya, Turkey; Salalah, Oman; Salzburg, Austria; San Diego, United States of America; San Francisco, United States of America; San Jose, Costa Rica; Santa Cruz, United States of America; Santander, Spain; Santos, Brazil; São Gonçalo Do Amarante, Brazil; São Paulo, Brazil; Seattle, United States of America; Sejong, South Korea; Semarang, Indonesia; Seoul, South Korea; Seville, Spain; Shanghai, China; Shenzhen, China; Silkeborg, Denmark; Singapore, Singapore; Sixth Of October, Egypt; Ski, Norway; Småland, Sweden; Sochi, Russia; Södermalm, Sweden; Sofia, Bulgaria; Sohar, Oman; Songdo, South Korea; South Sabah Al-Ahmad, Kuwait; South Sulawesi, Indonesia; South Surra, Kuwait; Split, Croatia; St. Petersburg, Russia; Stockholm, Sweden; Strasbourg, France; Stuttgart, Germany; Sunshine Coast, Australia; Sur, Oman; Surabaya, Indonesia; Surat, India; Sydney, Australia; Szeged, Hungary; Tampere, Finland; Tangier, Morocco; Taupo, New Zealand; Tauranga, New Zealand; Tel Aviv, Israel; Tétouan, Morocco; The Hague, Netherlands; Thessaloniki, Greece; Tijuana, Mexico;

Timisoara, Romania; Tokyo, Japan ; Tomsk, Russia; Toronto, Canada; Toulouse, France; Trabzaon, Turkey; Trenčín, Slovakia; Trnava, Slovakia; Trondheim, Norway; Turin, Italy; Ulyanovsk, Russia; Utrecht, Netherlands; Uttar Pradesh, India; Valencia, Spain; Vancouver, Canada; Vantaa, Finland; Varanasi, India; Varaždin, Croatia; Venice, Italy; Victoria, Australia; Vienna, Austria; Visakhapatnam, India; Vitória, Brazil; Vitoria-Gasteiz, Spain; Warsaw, Poland; Wellington, New Zealand; Western Sydney, Australia; Wroclaw, Poland; Yogyakarta, Indonesia; Zagreb, Croatia; Zaragoza, Spain; Žilina, Slovakia; Zug, Switzerland; Zurich, Switzerland.

BODIES & ORGANISATIONS

IBM; The European Commission; Indian Government; The European Union; The United Nations; The Smart Cities Council; The World Economic Forum; U.S. Department of Transportation; GDPA; Venetis; Smart Cities Dive; Smart City Expo World Congress; Smart Cities Council; Urban Land Institute; The Smart City Journal; Smart Cities Connect; CityLab; IEEE Smart Cities; World Smart Sustainable Cities Organization; Smart Cities Information System.

INDIVIDUALS

ABU YUSUF: Islamic Jurist and Scholar; ADAM GREENFIELD: American Writer and Urbanist; AL-FARABI: Islamic Philosopher and Political Theorist; AL-KINDI: Islamic Philosopher, Mathematician, and Scientist; ANTHONY TOWNSEND: American Urbanist and Futurist; ARISTOTLE: Greek Philosopher; AUGUSTINE OF HIPPO: Christian Theologian and Philosopher; CHARLEMAGNE: Frankish King; CHRISTOPHER ALEXANDER: British Architect and Urban Designer; CONFUCIUS: Chinese Philosopher; CYRUS THE GREAT: Persian Empire Founder; DAVID HARVEY: British Geographer; DIOCLETIAN: Roman Emperor; DON IHDE: American Philosopher of Science and Technology; DONELLA MEADOWS: American Environmental Scientist; EBENEZER HOWARD: Author and Publisher; FRIEDRICH ENGELS: German Philosopher and Social Theorist; GEORGE PULLMAN: Developer; HENRI LEFEBVRE: French Philosopher and Sociologist; HENRY DAVID THOREAU: American Author, Philosopher, and Naturalist; HIPPODAMUS OF MILETUS: Greek Architect; IVAN ILLICH: Austrian Philosopher; JANE BENNETT: American Political Theorist; JANE JACOBS: American-Canadian Journalist, Author, and Activist; KAUTILYA: Indian Philosopher; LEWIS MUMFORD: American Historian, Sociologist, and Philosopher; MANUEL CASTELLS: Spanish Sociologist; MANUEL DE LANDA: Mexican-American Philosopher and Urbanist; MARTIN HEIDEGGER: German Philosopher; MICHEL FOUCAULT: French Philosopher and Social Theorist; PAOLO SOLERI: Italian Architect and Urban Designer; PLATO: Greek Philosopher; PROCOPIUS: Byzantine Historian; REM KOOLHAAS: Dutch Architect and Urbanist; RICHARD SENNETT: American Sociologist and Urbanist; SASKIA SASSEN: Dutch Sociologist; SIDDHARTHA GAUTAMA: BUDDHA Spiritual Leader; SUN TZU: Chinese General and Military Strategist; VITRUVIUS: Roman Architect and Engineer; WILLIAM J. MITCHELL: MIT Professor of Architecture, Media Arts and Sciences.

ACRONYMS

To view associated acronyms, scan the following QR code

or go to

wannalook.com/oswr

IMPORTANT

Kindly be informed that the domain name and services offered on wannalook.com are owned and operated by the author of this book, Emin Hasic. It is important to note that wannalook.com is a legitimate website and not a fraudulent spamming third-party site.

The purpose of Emin's creation of wannalook.com was to provide user-friendly short links and QR codes that are easy for users to interact with.

SMART CITY ALGORITHMS

The author Emin Hasic is an accomplished software developer for over 30 years. By scanning the code below, you will find a series of algorithms relating to Smart City Programs which you will find very useful.

An algorithm is a set of instructions or a step-by-step procedure for solving a problem or performing a task. Algorithms can be thought of as recipes for performing a specific task, where each step is well-defined and unambiguous.

The purpose of an algorithm is to solve a problem efficiently and accurately, often in a way that can be automated or programmed. Algorithms are used in a wide range of applications, from data processing and analysis to machine learning and artificial intelligence. They are also used in a variety of industries, including finance, healthcare, and transportation, to improve efficiency and productivity.

or go to

wannalook.com/oswa

SELF ASSESSMENT

There are many factors that contribute to success in a career, including education, experience, skills, and personal attributes, and no self-assessment can fully capture all of these factors. However, a self-assessment can be a useful tool for individuals to evaluate their strengths and weaknesses and make informed decisions about their career paths.

Here is a self-assessment with 50 questions related to the field of Smart Cities. Each question has four options to choose from, plus a further 12 Assessments and Techniques for Personal and Career Development within Smart Cities.

You are welcome to take these self-assessments as many times as you like via the use of a digital device by scanning the following QR code.

or go to

wannalook.com/osws

SMART CITY CONTACTS

If you wish to contact any of the Smart Cities, Districts, Municipalities, States, or Governments listed in the RESOURCES section, scan the following QR code:

or go to

wannalook.com/oswc

Disclaimer: The information contained in this book has been obtained from resources believed to be reliable. While every effort has been made to ensure the accuracy of the information contained within this book, the author and publisher cannot guarantee its completeness or accuracy and assume no liability for any errors or omissions. The views and opinions expressed in this book are those of the author and do not necessarily reflect the views or opinions of the resources used. The author has made reasonable efforts to ensure the information contained in this book is up-to-date and applicable at the time of publication, but is not responsible for any errors or inaccuracies. The author does not endorse or make any warranties or representations about the accuracy, reliability, completeness, or timeliness of the information contained in this book, or about the results to be obtained from using it. This book is intended for informational purposes only and is not a substitute for professional advice. The reader should always seek the advice of a qualified professional.

INDEX

To streamline this publication, a print index has not been included. Instead, use the QR code below to access the digital index. We believe this will serve you well.

or go to

wannalook.com/oswi

Ο δὲ ἀνεξέταστος βίος οὐ βιωτὸς ἀνθρώπῳ

(μια ανεξέταστη ζωή δεν αξίζει να τη ζεις)

(an unexamined life is not worth living)

SOCRATES

Greek Philosopher 5th Century BC